Boundary-Layer Separation

International Union of Theoretical and Applied Mechanics

F. T. Smith, S. N. Brown (Eds.)

Boundary-Layer Separation

Proceedings of the IUTAM Symposium
London, August 26-28, 1986

Springer-Verlag
Berlin Heidelberg New York
London Paris Tokyo

Prof. Frank T. Smith

Prof. Susan N. Brown

Mathematics Department
University College
Gower St.
London WC1E 6BT
United Kingdom

ISBN-13:978-3-642-83002-0 e-ISBN-13:978-3-642-83000-6
DOI: 10.1007/978-3-642-83000-6

Library of Congress Cataloging in Publication Data

2161/3020 5 4 3 2 1 0

Scientific Committee

F.T.Smith, UK (Chairman)
B.Fornberg, USA
K.Gersten, FRG
Sir James Lighthill, UK
R.E.Melnik, USA
M.Nishioka, JAP
O.S.Ryzhov, USSR
V.V.Sychev, USSR
A.I.Van de Vooren, NETH

Local Organizing Committee

D.C.Belinfante
S.N.Brown
B.R.Lankester
T.W.Ng
F.T.Smith

Symposium Sponsors

International Union of Theoretical and Applied Mechanics (IUTAM)
University College London

Introduction

The IUTAM Symposium on Boundary-Layer Separation, suggested by the UK National Committee of Theoretical and Applied Mechanics and supported by the International Union of Theoretical and Applied Mechanics, was held at University College London on August 26-28, 1986.

The proposed theme and scope of the Symposium were designed to help to bring about the necessary interaction between experimentalists, computationalists and theoreticians for the furthering of understanding in this challenging subject. The talks and discussions were aimed at representing the very wide range and application of separating-flow phenomena, which often substantially affect the whole of fluid dynamics at medium to large Reynolds numbers, covering in particular both laminar and turbulent flow, steady or unsteady, two- or three-dimensional, small- or large-scale, incompressible or compressible, external or internal, from the experimental, computational and theoretical standpoints.

It was intended that about 80 scientists would participate in the Symposium, with about 25 talks being delivered, to which poster sessions with 8 contributions were added subsequently. All the speakers and poster presenters were selected by the scientific committee, although two late replacements of speakers were required. Fruitful discussions, well led by the session chairmen, took place formally after each talk and after the poster sessions and informally on other occasions including the social events.

The present proceedings of the Symposium appear to reflect much of the current state of experimental, computational and theoretical work and progress in boundary-layer separation. We hope that they provide also ideas, questions and stimulation, in addition to major recent developments.

F.T. Smith & S.N. Brown

University College London

Participants

Acrivos, A.	USA	Department of Chemical Engineering, Stanford University, Stanford, California, 94305.
Adamson, Jr., T.C.	USA	Department of Aerospace Engineering, The University of Michigan, Ann Arbor, Michigan 48109.
Ashill, P.R	UK	Royal Aircraft Establishment, Bedford.
Avis, R.	UK	Department of Mathematics, University College London, Gower Street, London WC1E 6BT.
Badr, H.M.	SAUD	Department of Mechanical Engineering University of Petroleum & Minerals, UPM Box 322, Dhahran-31261.
Belinfante, D.C.	UK	Department of Mathematics, University College London, Gower Street, London WC1E 6BT.
Bippes, H.	FRG	DFVLR Institut fur Experimentalle Stromungsmechanik, Bunsenstrasse 10, D-3400 Gottingen.
Blennerhassett, P.J.	AUS	School of Mathematics, University of New South Wales, P.O. Box 1, Kensington, N.S.W. 2033.
Bodonyi, R.J.	USA	Department of Aeronautical & Astronautical Engineering, Ohio State University, 2036 Neil Ave.Mall, Columbus Ohio 3210.
Bogdonoff, S.	USA	Princeton University, Princeton, New Jersey 08540.
Brotherton-Ratcliffe, R.V.	UK	Department of Mathematics, University College London, Gower Street, London WC1E 6BT.
Brown, E.F.	UK	Office of Naval Research, 223 Old Marylebone Rd., London NW1 5TH.
Brown, S.N.	UK	Department of Mathematics, University College London, Gower Street, London WC1E 6BT.

Burggraf, O.R.	USA	Department of Aeronautical & Astronautical Engineering, Ohio State University, 2036 Neil Ave.Mall, Columbus, Ohio 43210.
Cebeci, T.	USA	Douglas Aircraft Co., CR-E83 36-81, 3855 Lakewood Boulevard, Long Beach, California 90846.
Chanetz, B.	F	Office National d'etudes et de Recherch Aerospatiales, ONERA - 72, 92322 Chatillon Cedex.
Cheng, H.K.	USA	Department of Aerospace Engineering, University of Southern California, University Park, Los Angeles, California 90089-0192.
Cowley, S.J.	UK	Department of Mathematics, Imperial College of Science and Technology, London SW7 2BZ.
Dallmann, U.	FRG	Institut fur Theoretische Stromungsmechanik, DFVLR, Bunsenstr. 10, D-3400 Gottingen.
Dennis, S.C.R.	CAN	Department of Applied Mathematics, University of Western Ontario, London, Ontario N6A 5B9.
Deriat, M.	F	ONERA - BP 72, 92322 Chatillon Cedex, France.
Doorly, D.J.	UK	Department of Mathematics, University College London, Gower Street, London WC1E 6BT.
Dovgal, A.V.	USSR	Institute of Theoretical and Applied Mechanics, Siberian Branch of U.S.S.R. Academy of Sciences, 6300090 Novosibirsk.
Drazin, P.G.	UK	School of Mathematics, The University, University Walk, Bristol BS8 1TW.
Duck, P.W.	UK	Department of Mathematics, University of Manchester, Oxford Road, Manchester, M13 9PL.
Edwards, D.E.	USA	United Technologies Research Center, Hartford, Conn. 06108.
Elliott, J.W.	UK	Department of Applied Mathematics, The University, Cottingham Rd., Hull HU6 7RX.

Fornberg, B.	USA	Exxon Research and Engineering Co., Clinton Township, Route 22 East, Annandale, New Jersey 08801.
Gajjar, J.	UK	Department of Mathematics, University of Exeter, North Park Road, Exeter EX4 4QE.
Gersten, K.	FRG	Institut fur Thermo-und Fluiddynamik, Ruhr-Univ. Bochum, Postfach 102148, D-4630 Bochum.
Ghia, K.N	USA	Department of Aerospace Engineering and Applied Mechanics, University of Cincinnati, Cincinnati, Ohio 45221.
Gittler, P.	AUT	Institute fur Stromungslehre und Wamebertagung, Technische Universitat Wien, Wiedner Haupstrasse A-1040 Vienna.
Girodroux-Lavigne, P.	F	ONERA, B.P.72, 92322 Chatillon Cedex.
Henkes, M.	NETH	Delft University of Technology, Delft.
Hornung, H.	FRG	Institut fur Theoretische Stromungsmechanik, DFVLR, Bunsenstr. D-3400, Gottingen.
Humphreys, D.	SWED	FFA, Box 11021, S-16111, Bromma.
Ingham, D.B.	UK	Department of Applied Mathematics, University of Leeds, Leeds LS2 9JT.
Justesen, P.	DEN	Inst. Hydrodynamics and Hydraulic Engineering, DK-2800 Lyngby.
Khorrami, A.F.	UK	Department of Mathematics, University College London, Gower Street, London WC1E 6BT.
Kiya, M.	JAP	Faculty of Engineering, Hokkaido University, Sapparo 060.
Kluwick, A.	AUT	Institut fur Stromungslehre Warmiebertagrung, Technische Universitat Wien, Wiedner Haupstrasse 7, A-1040.
Kozlov, V.V.	USSR	Institute of Theoretical and Applied Mechs., Siberian Branch of USSR Academy of Science, 630090, Novosibirsk.

Kuwahara, K.	JAP	Institute of Space and Aeronautical Research, Komaba, Meguro-ku, Tokyo 153.
Lam, S.T.	UK	Department of Mathematics, Imperial College of Science and Technology, London SW7 2BZ.
Lighthill, Sir James.	UK	University College London, Gower Street, London WC1E 6BT.
Lock, R.C.	UK	74 Ormond Ave., Hampton, Middx. TW12 2RX.
Maslowe, S.A.	CAN	Department of Math. Stat., 805 Sherbrooke St. West, McGill University, Montreal, Quebec H3A 2K6.
Melnik, R.E.	USA	Grumman Aerospace Corporation, Corporate Mailstop A08-35, Bethpage, New York 11714.
Merkin, J.H.	UK	Department of Applied Mathematics, University of Leeds, Leeds LS2 9JT.
Michael, D.H.	UK	Department of Mathematics, University College London, Gower Street, London WC1E 6BT.
Mikkelsen, H.E.	DEN	Department of Geology, Lange-landsgade Building 520, DK-8000 Aarhus C.
Minten, I.A.	FRG	Institut für Dampf- und Gasturbinen der RWTH Aachen, Templergraben 55, 5100 Aachen.
Mitra, N-K.	FRG	Bochum AR am Institut fur Thermo und Fluiddynamik der Ruhr-Universitat Bochum.
Neiland, V. Ja	USSR	The Central Institute of Aero- and Hydrodynamics, 140160 Zhukosky-town 3.
Neish, A.J.	UK	Department of Mathematics, University College London, Gower Street, London WC1E 6BT.
Ng, T.W.	UK	Department of Mathematics, University College London, Gower Street, London WC1E 6BT.
Page, M.A.	AUS	Department of Mathematics, Monash University, Clayton, Victoria 3168.

Papageorgiou, D.T.	USA	Courant Institute of Mathematical Sciences, 251 Mercer Street, New York, New York 10012.
Patrick, T.J.	UK	Department of Physics, University College London, Gower Street, London WC1E 6BT.
Pedley, T.J.	UK	D.A.M.T.P., University of Cambridge, Silver Street, Cambridge CB3 9EW.
Reed, H.L.	USA	Department of Mechanical and Aerospace Engineering, Arizona State University, Tempe, AZ 85287.
Riley, N.	UK	School of Mathematics and Physics, University of East Anglia, University Plain, Norwich NR4 7TJ.
Rothmayer, A.P.	USA	Computational Fluid Dynamics Centre, Iowa State University, College of Engineering, 403 Town Engineering Building, Ames, Iowa 50011.
Ryzhov, O.S.	USSR	Computing Centre of the Academy of Sciences of the USSR, 44 Vavilov Street, 11733 Moscow.
Schmitt, R.	F	ONERA, 29 Avenue de la Division, Cedex 92320 Chatillon -sous Bagneux.
Schultz, H.D.	FRG	Institut für Strahlantriebe und Turboarbeitsmaschinen, Rhein.-Westf. Techn. Hochschule, D-5100 Aachen.
Simpson, R.L.	USA	Department of Engineering Science & Mechanics, V.P.I., Blacksburg, 24061, Virginia.
Smith, F.T.	UK	Department of Mathematics, University College London, Gower Street, London WC1E 6BT.
Smith, J.H.B.	UK	Royal Aircraft Establishment, Farnborough, Hants. GU14 6TD.
Smith, P.D.	UK	Aero Department, R141 Building, Royal Aircraft Establishment Farnborough, Hants. GU14 6TD.
Stewart, P.A.	UK	Department of Mathematics, University College London, Gower Street, London WC1E 6BT.
Sutton, P.	UK	Pembroke College, Cambridge, CB2 1RF.

Teipel, I.	FRG	Institut fur Mechanik der Universitat Hannover, Appelstr. 11, D-3000 Hannover 1.
Thiele, F.	FRG	TU Berlin Hochschulassistent am Hermann - Fottinger - Institut, Sekr. HF1, D-1000 Berlin 12.
Tutty, O.R.	UK	D.A.M.T.P., University of Cambridge, Silver Street, Cambridge CB3 9EW.
Van de Vooren, A.	NETH	Department of Mathematics, Rijksuniv Groningen, P.O. Box 800, 9700 AV Groningen.
Van Dommelen, L.L.	USA	FAMN/FSC Inst. of Engineering, Florida State University, Tallahassee, Florida 32306.
Veldman, A.E.P.	NETH	National Aeronautical Institute, Anthony Fokkerweg 2, 1059 CM Amsterdam.
Wheeler, A.	UK	Mathematical Institute, St. Giles, Oxford.
Williams, B.R.	UK	Royal Aircraft Establishment, Farnborough, Hants. GU14 6TD.
Williams, P.G.	UK	Department of Mathematics, University College London, Gower Street, London WC1E 6BT.

Table of Contents

---

* Paper not available at time of printing

Session 1

Steady Viscous Flow Past a Cylinder and a Sphere at High Reynolds Numbers

BENGT FORNBERG

Exxon Research and Engineering Company
Annandale, NJ 08801, USA.

ABSTRACT:

Numerical solutions are presented for steady flow past a circular cylinder and past a sphere for Reynolds numbers up to 600 and 2000 resp. These steady (but unstable) flow fields can not at present be obtained by any other means. We observe distinctive trends for increasing Reynolds numbers which may be significant in devising asymptotic wake models.

INTRODUCTION:

Viscous flow past a circular cylinder becomes unstable around Reynolds number Re = 40 and past a sphere around Re = 130. In this work, we are interested in the structure of the steady (but unstable) solutions above these Reynolds numbers. Apart from studies by the present author, the upper limit for (reasonably accurate) solutions for the cylinder in the literature appears to be Re = 120 [1] and for the sphere Re = 400 ([2]; unpublished results [3] are quoted in [4])

In this paper we summarize the calculation [5] for the cylinder up to Re = 600 and also give preliminary results from a similar calculation for the sphere, presently carried out up to Re = 2000. In both cases, the main reason we have been able to obtain converged solutions well past the 'barrier' where instabilities start to occur is our use of Newton's method to solve the discretized equations. Conventional iterative techniques tend, in the artificial time introduced by the iterations, to pick up instabilities reminiscent of temporal ones. The quadratic convergence of Newton's method precludes this from happening. With the present development of large computers, the computational cost of this approach is no longer prohibitive.

F. T. Smith, S. N. Brown, (Eds.)
Boundary-Layer Separation
IUTAM Symposium London 1986

In the cylinder case, the wake bubble (region with re-circulation flow) is found to grow linearly with Re in both length and width above Re ≈ 300. In the sphere case, both growth rates are much lower, possibly like log Re. In both cases, it appears likely that very high Reynolds number wake bubbles (regions with recirculating fluid) will be large, wide and similar in structure to Euler solutions which can translate through the fluid without any bodies present.

MATHEMATICAL FORMULATION:

The governing equations and the ideas for the code implementation are nearly identical in 2-D and 3-D (axial symmetry). Since we already have described the 2-D case in [5], we give a brief account of the 3-D case here. With a sphere of radius 1 and the Reynolds number based on the diameter, the steady state Navier Stokes equations in cylindrical coordinates take the form

$$\frac{\partial^2\Psi}{\partial x^2} + \frac{\partial^2\Psi}{\partial y^2} - \frac{1}{y}\cdot\frac{\partial\Psi}{\partial y} + y\,\omega = 0 \tag{1}$$

$$\frac{\partial^2\omega}{\partial x^2} + \frac{\partial^2\omega}{\partial y^2} + \frac{1}{y}\cdot\frac{\partial\omega}{\partial y} - \frac{\omega}{y^2} - \frac{Re}{2y}\left\{\frac{\partial\Psi}{\partial y}\cdot\frac{\partial\omega}{\partial x} - \frac{\partial\Psi}{\partial x}\cdot\frac{\partial\omega}{\partial y} + \frac{\omega}{y}\cdot\frac{\partial\Psi}{\partial x}\right\} = 0\,. \tag{2}$$

Here, Ψ denotes streamfunction and ω vorticity. The x-axis is the downwind symmetry axis and y the distance from this axis.

Near the sphere, all the vorticity is concentrated within a region like the one sketched in Figure 1 a. A conformal mapping $Z = X^{1/2} + X^{-1/2}$ maps such regions in the physical $X = x + iy$ plane to rectangles in the $Z = \xi + i\eta$ plane (Figure 1 b).

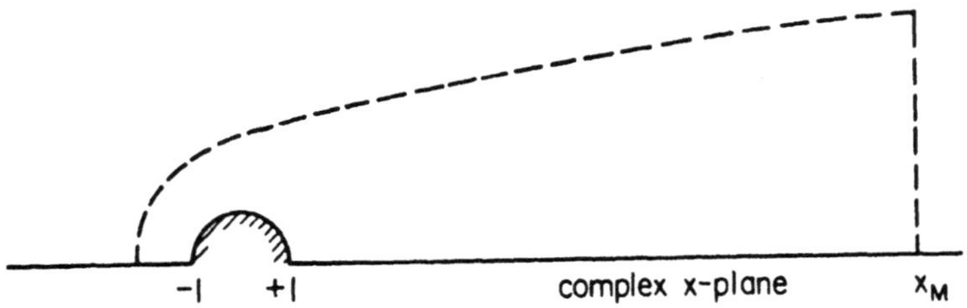

Figure 1 a. Computational domain in physical X-plane.

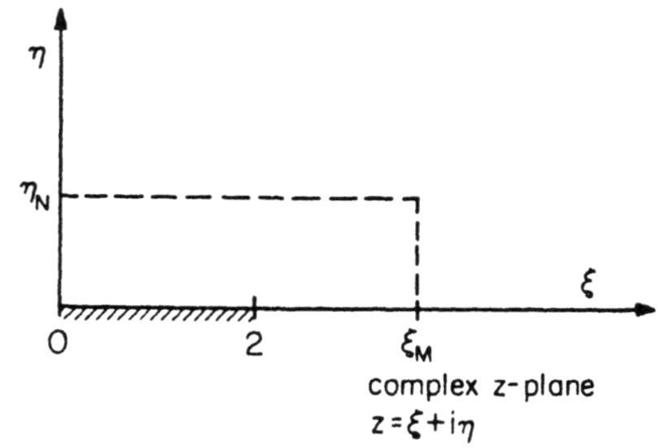

Figure 1 b. Domain above conformally mapped to a rectangle.

In the new coordinates, the governing equations (1), (2) take the form

$$\left\{ \frac{\partial^2\Psi}{\partial\xi^2} + \frac{\partial^2\Psi}{\partial\eta^2} \right\}\cdot J(\xi,\eta) + \frac{1}{y}\left\{ \frac{\partial\eta}{\partial x}\cdot\frac{\partial\Psi}{\partial\xi} - \frac{\partial\xi}{\partial x}\cdot\frac{\partial\Psi}{\partial\eta} \right\} + y\cdot\omega = 0 \quad (3)$$

$$\left\{ \frac{\partial^2\omega}{\partial\xi^2} + \frac{\partial^2\omega}{\partial\eta^2} \right\}\cdot J(\xi,\eta) - \frac{1}{y}\left\{ \frac{\partial\eta}{\partial x}\cdot\frac{\partial\omega}{\partial\xi} - \frac{\partial\xi}{\partial x}\cdot\frac{\partial\omega}{\partial\eta} \right\} - \frac{\omega}{y^2} + \quad (4)$$
$$+ \frac{Re}{2y}\left\{ \left(\frac{\partial\Psi}{\partial\xi}\cdot\frac{\partial\omega}{\partial\eta} - \frac{\partial\Psi}{\partial\eta}\cdot\frac{\partial\omega}{\partial\xi}\right)\cdot J(\xi,\eta) - \frac{\omega}{y}\left(\frac{\partial\xi}{\partial x}\cdot\frac{\partial\Psi}{\partial\xi} + \frac{\partial\eta}{\partial x}\cdot\frac{\partial\Psi}{\partial\eta}\right) \right\} = 0$$

where $J(\xi,\eta) = (\partial\xi/\partial x)^2 + (\partial\eta/\partial x)^2$. These equations were further modified by separate changes of variables in the ξ and η-directions. This allows us to increase the resolution along the body surface and, in particular, near the rear of the sphere (where the conformal mapping has a singularity). Figure 2 shows the image in the X-plane near the sphere of a grid like the ones we used (but, for clarity, with only 55×19 points).

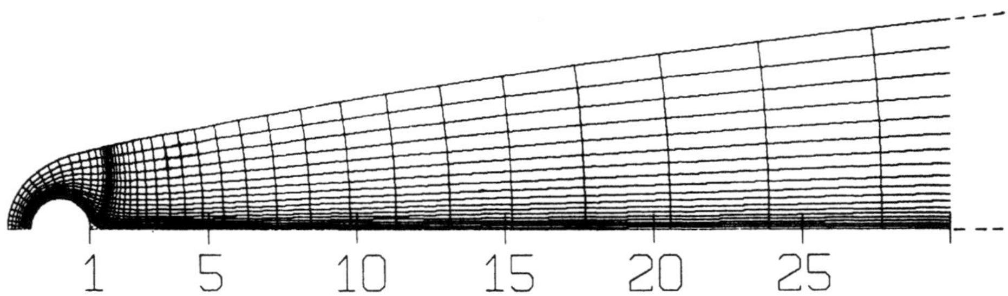

Figure 2. Part near sphere of 'typical' grid in X-plane.

NUMERICAL APPROXIMATIONS:

The governing equations are approximated by centered second order finite differences at all interior points. At each boundary, two conditions are supplied. At the body surface, they are $\Psi=0$ and $\partial\Psi/\partial\eta=0$. Along the axis of symmetry, $\Psi=0$ and $\omega=0$. At the outflow end, simple extrapolations proved satisfactory (cf. [5]).

For increasing values of η, the vorticity decays exponentially to zero. The 'top' boundary is located sufficiently far out that the vorticity on or above it can be ignored. Figure 2 showed the location used at all Reynolds numbers in this study (of the sphere). On this boundary, ω is set to zero. The crucial issue is which second condition to apply. At least in 2-D, use of 'free stream' or leading terms in far field asymptotic expansions is unsatisfactory even at large distances [5]. This becomes even more the case if the boundary is not uniformly far out but comes close to the sphere, as in our case. The following approach applies equally well to 2-D as to 3-D. It is also very easy to implement, even where (possibly non-conformal) mappings have given the top boundary an arbitrary shape.

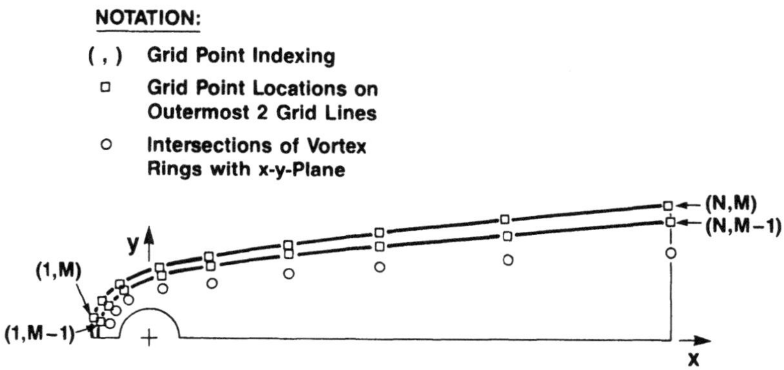

Figure 3. Implementation of boundary condition for Ψ on the top boundary.

Figure 3 illustrates our computational domain with the top two grid lines marked. If we imagined marching equation (3) (with $\omega=0$) upwards from these two lines, the initial conditions for Ψ on them should be such that we do not pick up any exponentially growing modes. That is the case only if the values on these two lines obey a certain set of N linear constraints. With the notation in Figure 3, it should be possible to write such constraints as

$$\begin{bmatrix} \Psi_{1,M-1} \\ \Psi_{2,M-1} \\ \vdots \\ \cdot \\ \Psi_{N,M-1} \end{bmatrix} = \begin{bmatrix} \\ \\ A \\ \\ \\ \end{bmatrix} \cdot \begin{bmatrix} \Psi_{1,M} \\ \Psi_{2,M} \\ \vdots \\ \cdot \\ \Psi_{N,M} \end{bmatrix} \tag{5}$$

It remains to determine the matrix A. For this, we note that we easily can find N linearly independent sets of vector pairs which, when placed side by side, give N×N matrices B and C such that B = A×C. This relation can then be solved for A. Any solution to (3), with $\omega = 0$ and decaying in Ψ for increasing η on and above the top two grid lines, gives such a vector pair when we inspect its values for Ψ at the grid points along these lines. One possibility (which gives a well-conditioned matrix C) is to consider circular line vortices. The Ψ-fields around such vortex rings are explicitly known [6]. Placing a sequence of vortex rings as indicated in Figure 3 provides all the data needed to solve for A. In 2-D, pairs of point vortices can be similarly used. This matrix A constitutes the bottom right square block in the matrix in Figure 4 b (described below).

SOLUTION OF ALGEBRAIC SYSTEM:

The discretized equations are nonlinear. One convenient way to order the equations and unknowns is described in [5]. It gives a Jacobian matrix with the structure shown in Figure 4 a. This coefficient matrix is first reduced to roughly half its size by eliminating all entries below the single diagonal (located in the top right corner block). The new structure is shown in Figure 4 b. The sizes of the blocks are given both for a general M×N grid and for our 325×109 grid. Row pivoted Gaussian elimination was employed (first for the banded part). Each linear system of this size requires 2.1×10^{10} arithmetic operations. On a 2-pipe CDC Cyber 205, this takes 234 seconds, i.e. the sustained average computational speed is 90 Mflops (million floating point operations per second, 64 bit precision). Times for other parts of the calculation are negligible. For example the generation of each reduced Jacobian requires only 0.02 seconds.

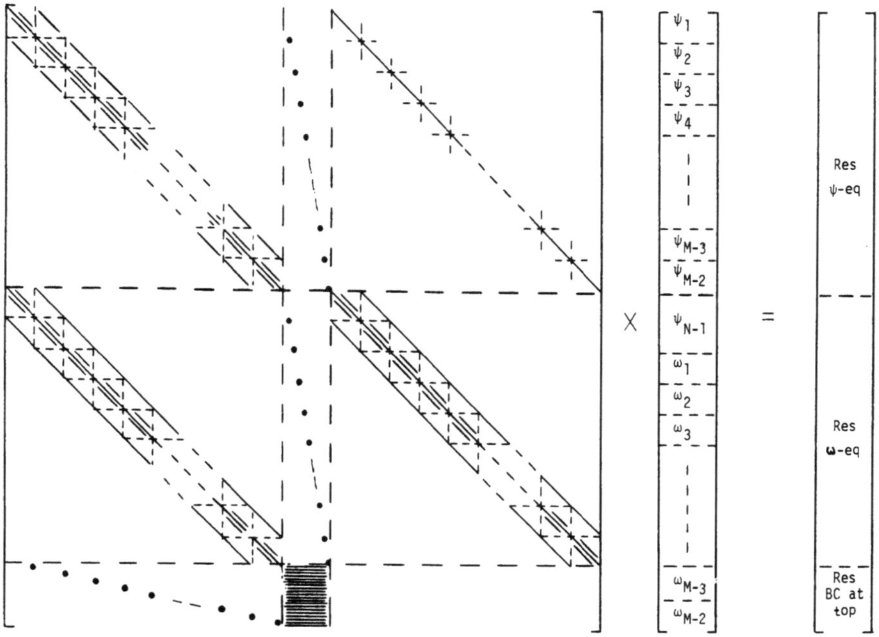

Figure 4 a. Structure of the linear system in Newton's method.

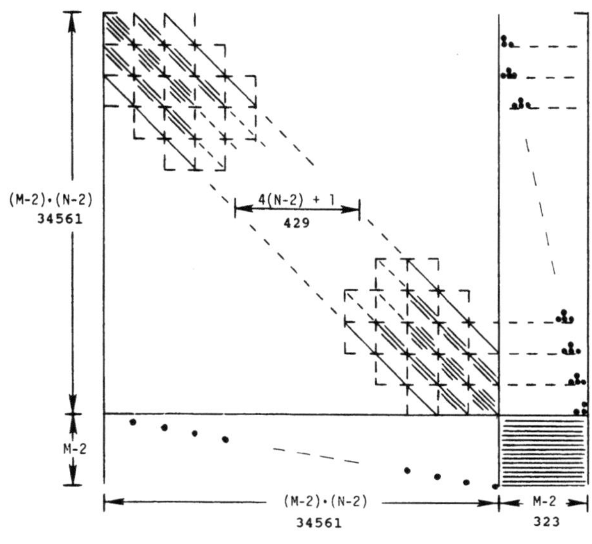

Figure 4 b. Structure of the reduced Jacobian.

RESULTS IN 2-D:

Figures 5 a and b show contour lines for the streamfunction and the vorticity at Reynolds numbers 100 to 600. In Figure 6, the vorticity fields are illustrated as surface elevations, in order to highlight the emergence of the large area with nearly constant vorticity inside the wake bubble. The lengths and widths of the wake bubbles are shown in Figures 7 a and b (measured from the center of the cylinder and to the center line resp.) In order to give an impression of the truncation error level, results from three different grids are shown (241×49, 361×73 and 541×109 points resp.). As we noted in the introduction, the length of the wake bubble is found to grow as O(Re) throughout the complete range considered. The width starts out as $O(Re^{1/2})$ but also becomes O(Re) following a transition around Re=300. Other results, (on pressure, drag etc.), as well as further error tests (effects of boundary conditions etc.), are reported in [5]. It is interesting to note that the linear trends above Re = 300, as seen in Figures 7 a and b, point to wakes with length/width tending to 1.67 (based on Richardson extrapolation over the three grids; the width is now measured across the full wake bubble, not just the top half). For translating Euler solutions with piecewise constant vorticity (no vortex sheet at their edge; vorticity discontinuity across the symmetry line), this ratio is a constant, 1.66915 ± .00005 according to [7]. For translating Euler solutions with a vortex sheet at the edge [8], this ratio is higher. Recent theoretical studies attempting to describe high Re wakes as perturbations of Euler solutions of these kinds include [9] and [10].

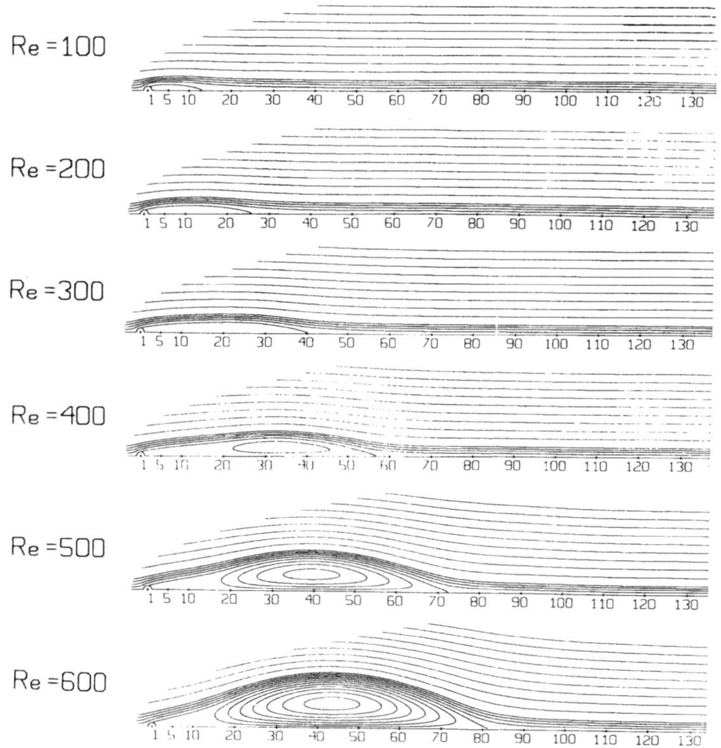

Figure 5 a.
Streamlines past the cylinder.

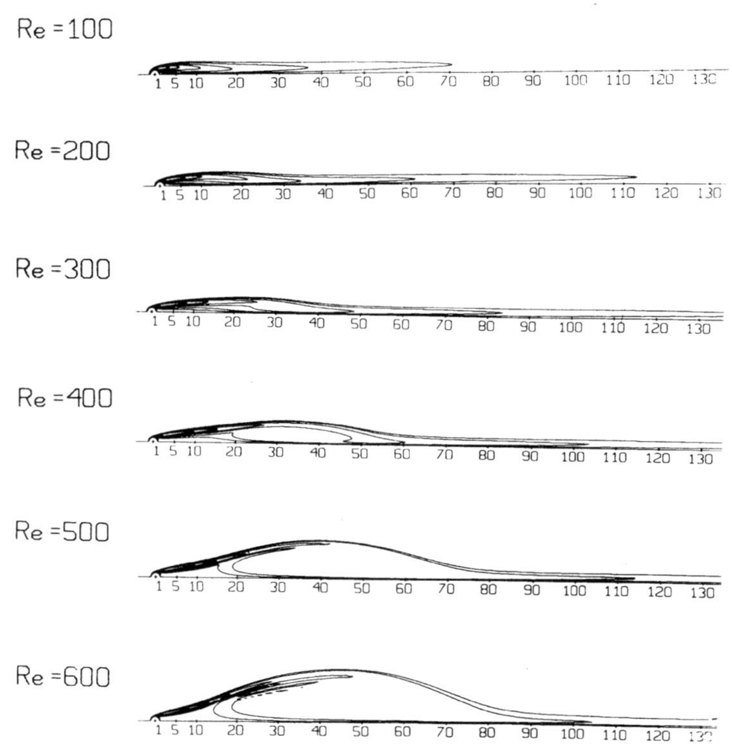

Figure 5 b.
Contours of constant vorticity (cylinder).

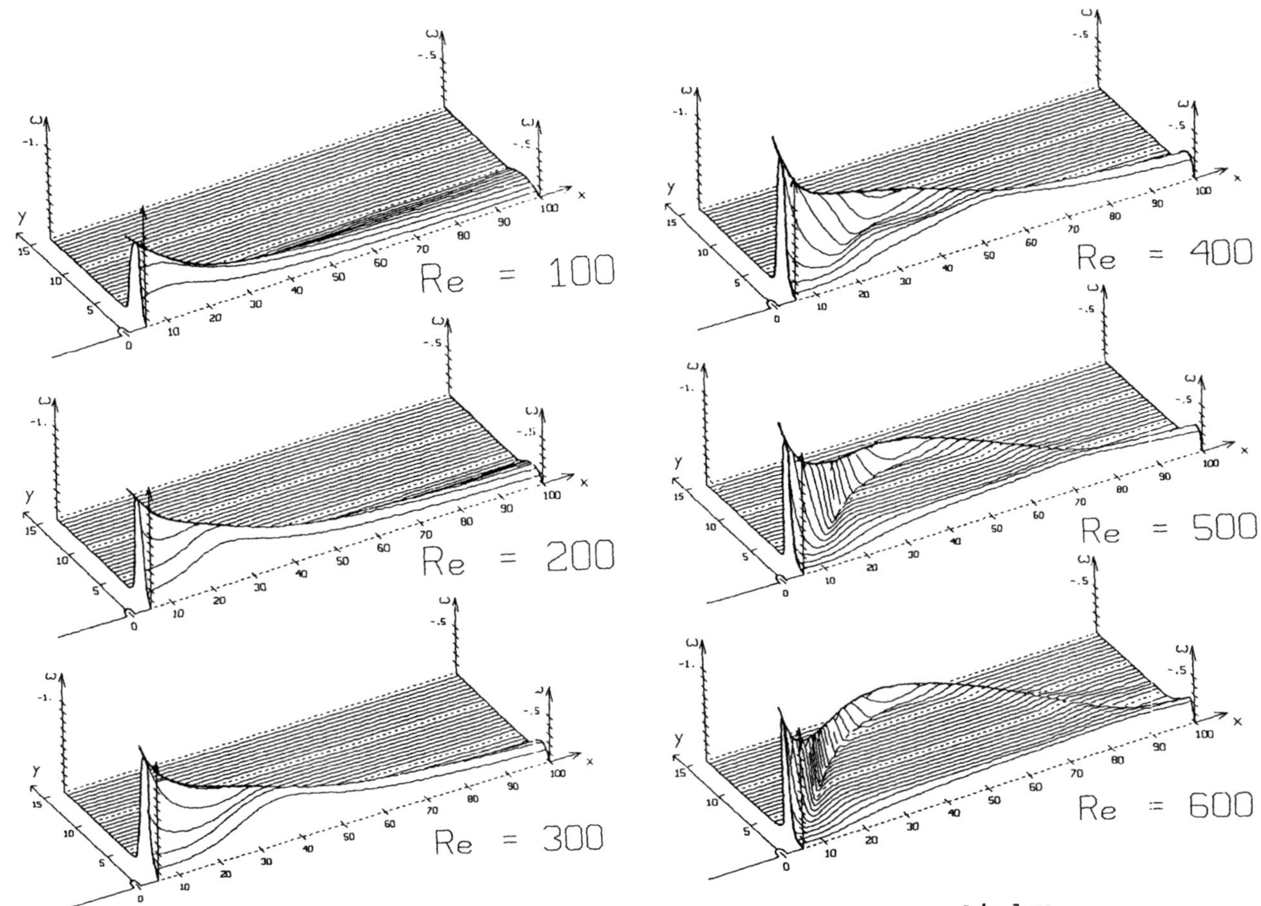

Figure 6. Distribution of vorticity in the wake behind the cylinder.

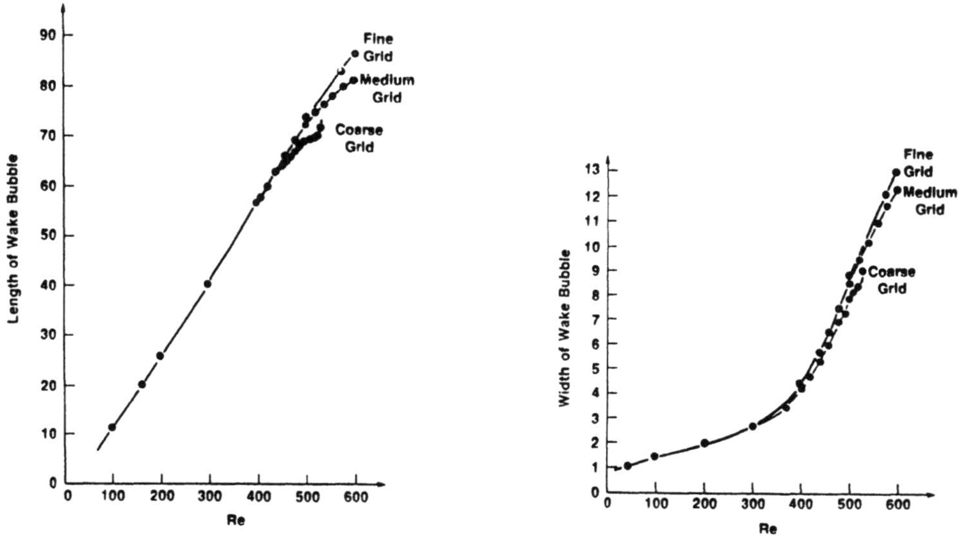

Figure 7. Length and width of the wake bubble behind the cylinder.

RESULTS IN 3-D:

Figures 8 a and b show the streamfunction and vorticity fields at Reynolds numbers 50, 100, 200, 500, 1000 and 2000. Figure 9 illustrates the vorticity fields as surface projections. At the higher Reynolds numbers, the vorticity distribution in the wake bubble resembles that of a Hill's spherical vortex [6] (which has ω/y = const. inside a spherical flow domain, $\omega = 0$ outside it). Wake bubble lengths and widths as functions of log Re are shown in Figures 10 a and b. (These dimensions should be quite accurate; for ex. changing from the 325×109 grid to a coarser 217×73 grid at Re=2000 affected both length and width less than 1 %). Although less clear than in 2-D, the trends seem to point towards large and wide wakes at still higher Reynolds numbers.

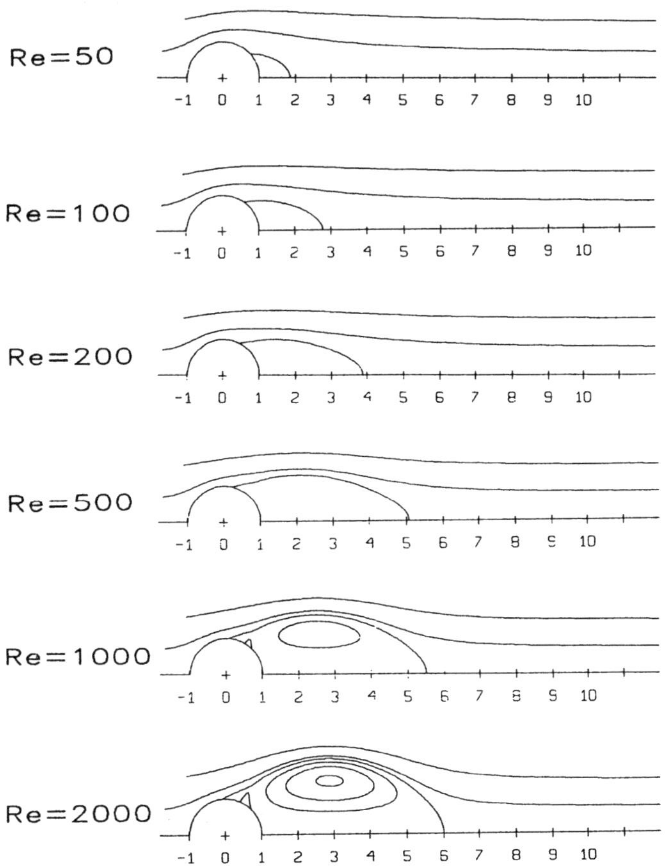

Figure 8 a. Streamlines past a sphere.

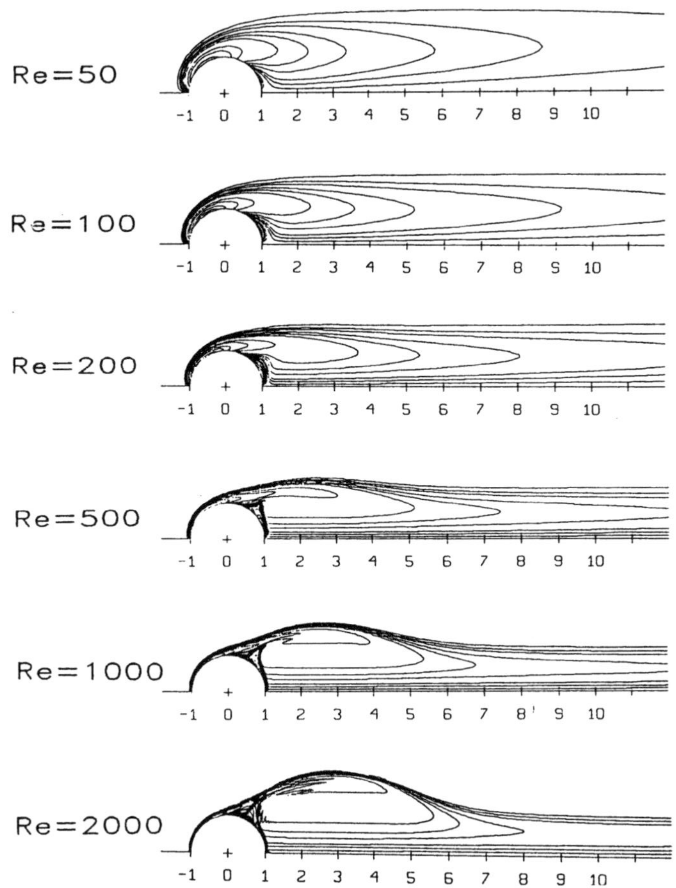

Figure 8 b. Contours of constant vorticity (sphere).

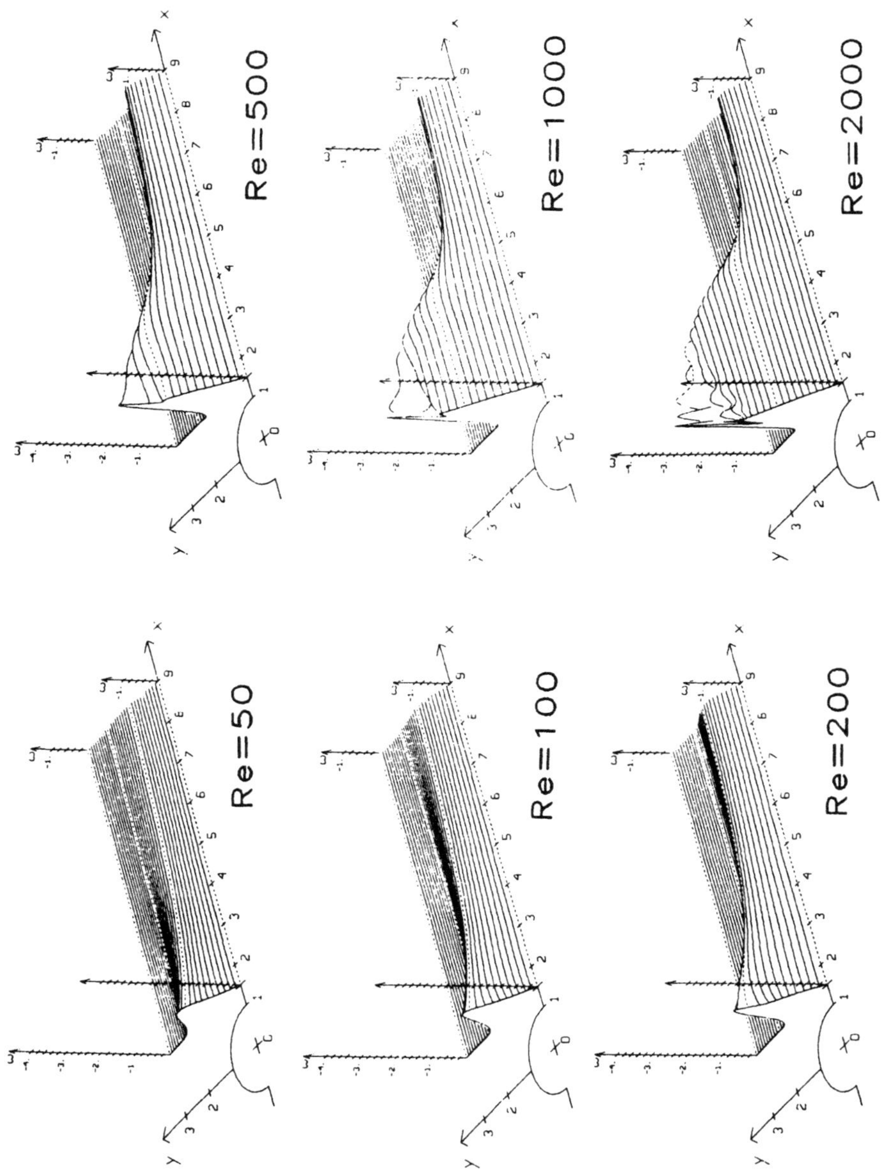

Figure 9. Distribution of vorticity in the wake behind the sphere.

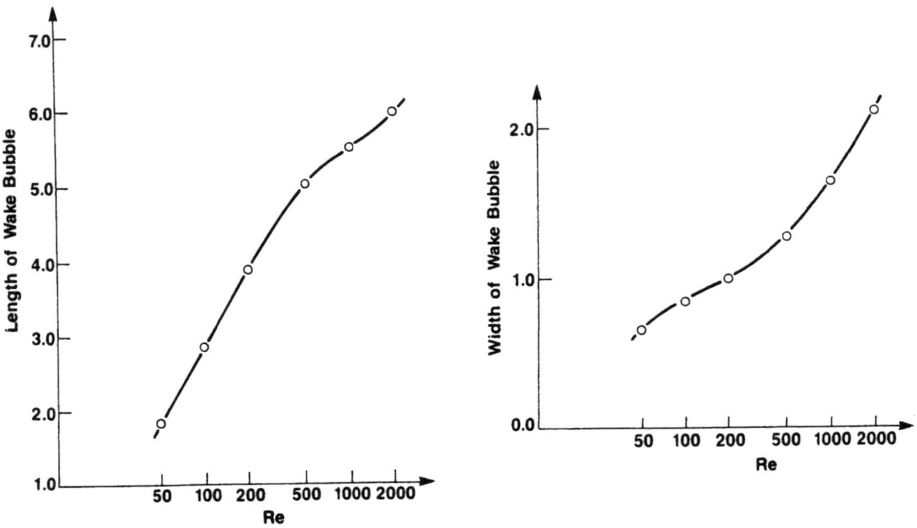

Figure 10. Length and width of the wake bubble behind the sphere.

ACKNOWLEDGEMENTS:

The study of the cylinder case was initialized while the author was working at the California Institute of Technology. It was then supported by Control Data Corporation, Department of Energy (Office of Basic Energy Sciences) and the John Simon Guggenheim Memorial Foundation. Figures 1 and 4-7 are reproduced from [5] with permission from Journal of Computational Physics (Academic Press Inc.).

REFERENCES:

1. Ta, P.L.: Étude numérique de l'écoulement d'un fluide visqueux incompressible autour d'un cylindre fixe ou en rotation. Effet Magnus. J. Méc. 14 (1975), 109-134.

2. Le Clair, B.P., Hamielec, A.E. & Pruppacher, H.R.: A numerical study of the drag on a sphere at low and intermediate Reynolds numbers. J. Atmos. Sci. 27 (1970), 308-315.

3. Woo,S.-W.: Ph.D. thesis, McMaster Univ., Hamilton, Ontario (1971).

4. Clift, R., Grace,J.R. & Weber, M.E.: Bubbles, Drops and particles. Academic Press (1978).

5. Fornberg, B.: Steady viscous flow past a circular cylinder up to Reynolds number 600. J. Comp. Phys. 61 (1985), 297-320.

6. Batchelor, G.K.: An introduction to fluid mechanics. Cambridge University Press (1967).

7. Wu, H.M., Overman II, E.A. & Zabusky, N.J.: Steady state solutions of the Euler equations in two dimensions: Rotating and translating V-states with limiting cases. 1. Numerical algorithms and results. J. Comput. Phys. 53 (1984), 42-71.

8. Sadovskii, V.S.: Vortex regions in a potential stream with a jump of Bernoulli's constant at the boundary. Appl. Math. Mech. 35 (1971), 729-735.

9. Peregrine, D.H.: A note on the steady high-Reynolds-number flow about a circular cylinder. J. Fluid Mech. 157 (1985), 493-500.

10. Smith, F.T.: A structure for laminar flow past a bluff body at high Reynolds number. J. Fluid Mech. 155 (1985), 175-191.

Prediction of Buffeting and Calculation of Unsteady Boundary Layer Separation over Airfoils

J.C. LE BALLEUR and P. GIRODROUX-LAVIGNE

ONERA BP 72, 92322 Chatillon Cedex, France.

Abstract.

The unsteady boundary layer separation is computed with a fully time-consistent viscous-inviscid interaction method. The new "Semi-Implicit" method is used to get time-consistently the strong coupling. The Defect Formulation theory is approximated with an unsteady defect integral method for attached or separated turbulent layers, interacted with a potential small perturbations solver.

The time-dependent method discriminates between the steady solutions and the unsteady "Buffet" generated by transonic shock-induced separation over a steady lifting airfoil, depending on Mach number and incidence. From the first results, the unsteady pressure distributions display time-histories and spectral properties which compare rather well with buffet experiments, in two-dimension, providing some hope in further buffeting prediction.

1 - Introduction

Much progress has been obtained in calculating steady boundary layer separation by development of viscous-inviscid interaction methods. Examples are given by Melnik [1], Carter, Edwards [2], F. Smith [3], Lock, Firmin [4], Veldman [5], Yoshihara [6], Cebeci [7], Le Balleur [8,9], among others. If the computation of steady separation in three-dimension still requires much research, the calculation of unsteadiness in boundary layer separation over steady geometries seems also to be an important area for theory and applications, if we attempt for example to understand how far the separation is connected with the buffet and buffeting phenomena.

Recently, we have suggested a new coupling method, the "Semi-Implicit" method [10,11], which is capable to solve fully time-consistently the viscous-inviscid interaction. This coupling method was developed together with an unsteady integral viscous calculation, direct or inverse, in Defect Formulation, using an unsteady TSP-potential inviscid solver for interaction. With this method, it is attempted here to predict the self-induced buffeting separation over steady airfoils, at lifting conditions, with unsteady transonic shock wave-boundary layer interaction. The answer of the calculations is analyzed with the help of the transonic buffet experiments of Benoit [12] and Schmitt, and of Dor et al. [13].

2 - Governing Equations.

The method is based on the Defect Formulation of Le Balleur [14,15], where

(*) *Study financially supported by French STPA*

F. T. Smith, S. N. Brown, (Eds.)
Boundary-Layer Separation
IUTAM Symposium London 1986

the Navier-Stokes equations are split into a strictly equivalent system involving an Euler system, a viscous Defect system, and strong coupling relations, with a full overlap of computational domains for the viscous and inviscid systems.

In the case of unsteady viscous flows, the inviscid solution is presently approximated by solving a TSP equation, see [16], and the viscous flow field is computed by solving integral defect equations, based on a first order expansion of the viscous defect system with respect to the boundary layer thickness δ, see [15].

In spite of these approximations, the method retains a Navier-Stokes-like capability to solve the strong viscous interactions, such as separation and shock-boundary layer interactions, so long as a strong coupling is fully achieved with a time-consistent numerical technique.

Denoting by x and y the coordinates along and normal to the wall or wake, t the time, $\bar{u}$, $\bar{v}$ the velocity components in the x and y directions, $\bar{p}$ the pressure, $\bar{\rho}$ the density, $\bar{\tau}$ the shear stress, and by u,v,p,ρ the corresponding inviscid quantities, the integral defect equations of continuity, momentum, plus the local x-momentum defect equation (entrainment) at $y = \delta$ are :

$$\left\{\frac{\partial}{\partial t}[\rho\delta_o] + \frac{\partial}{\partial x}[\rho q\delta_1] - [\rho v - \bar{\rho}\bar{v}]\right\}_{(x,0,t)} = 0 \qquad (1)$$

$$\left\{\frac{\partial}{\partial t}[\rho q\delta_1] + \frac{\partial}{\partial x}[\rho q^2(\delta_1+\theta_{11})] - [\rho uv - \bar{\rho}\bar{u}\bar{v}] = \rho q^2\frac{Cf}{2}\right\}_{(x,0,t)} \qquad (2)$$

$$\bar{p}(x,y,t) \simeq p(x,y,t) \qquad (3)$$

$$\left\{\frac{\partial}{\partial t}[\rho(\delta-\delta_o)] + \frac{\partial}{\partial x}[\rho q(\delta-\delta_1)] = \rho q[E+\frac{\bar{\rho}\bar{v}}{\rho q}]\right\}_{(x,0,t)} \qquad (4)$$

$$\delta_{o(x,t)}[\rho]_{(x,0,t)} = \int_0^\infty [\rho - \bar{\rho}]_{(x,y,t)}\, dy$$

$$\delta_{1(x,t)}[\rho q]_{(x,0,t)} = \int_0^\infty [\rho u - \bar{\rho}\bar{u}]_{(x,y,t)}\, dy \qquad (5)$$

$$[\delta_1 + \theta_{11}]_{(x,t)}[\rho q^2]_{(x,0,t)} = \int_0^\infty [\rho u^2 - \bar{\rho}\bar{u}^2]_{(x,y,t)}\, dy$$

$$E_{(x,t)} = \left\{\frac{\dfrac{\partial\bar{\tau}}{\partial y}}{\rho q\dfrac{\partial}{\partial y}[u-\bar{u}]}\right\}_{(x,\delta,t)}, \quad \left[\frac{Cf}{2}\right]_{(x,t)} = \left[\frac{\bar{\tau}}{\rho q^2}\right]_{(x,0,t)}, \quad q^2 = u^2 + v^2.$$

Within the small disturbances approximations, the modified boundary conditions of the pseudo-inviscid problem can be written :

$$\text{wall} : \frac{\partial \phi}{\partial y} = \frac{\partial H}{\partial x} + k\frac{\partial H}{\partial t} + \left\{\frac{v}{q}\right\}_{(x,0,t)}$$

$$\text{wake} : <\frac{\partial \phi}{\partial y}>_{(x,0,t)} = \left\{\frac{v}{q}\right\}_{(x,0^+,t)} + \left\{\frac{v}{q}\right\}_{(x,0^-,t)} \qquad (6)$$

where ϕ is the disturbance potential, k the reduced frequency based on the chord, H(x,t) the airfoil contour, and <f> the discontinuity of f along the wake-cut .

3 - Calculation of unsteady separation.

Attention is given here on the three important points of the method, which allow to compute strong viscous interactions such as unsteady separations.

3.1 - Turbulent closure

The viscous defect system (1),(2),(3),(4),(5) is closed by the velocity profile modelling suggested by Le Balleur for attached or separated 2D or 3D turbulent layers, [15,17]. This modelling, designed for steady flows, is assumed to be valid in the case of unsteady flows, for the instantaneous velocity profile in the frame of reference linked to the body.

The instantaneous and relative mean velocity profiles assume a two-parameter description $[\delta, C_2]_{(x,t)}$, where δ_{1i} is the incompressible displacement and $y = \delta\eta$:

$$\bar{q} = q\left[1 - C_2\tilde{F}(\eta) + C_1 Log\eta\right]$$

$$\tilde{F}(\eta) = F\left[\frac{\eta - \eta^*}{1 - \eta^*}\right] \ , \ F(\eta) = \left[1 - \eta^{3/2}\right]^2 \ , \ \tilde{F} = 1 \quad \text{if} \quad \eta < \eta^* \qquad (7)$$

$$1 - C_2 = C_1\left[Log(R_\delta \sqrt{|Cf|/2} - 5.25 \times 0.41\right] \ , \ C_1 = \frac{1}{0.41}\frac{Cf}{\sqrt{2}\,|Cf|} \qquad (8)$$

$$\begin{aligned} \delta\eta^* &= 0 && ,\text{if} \quad 0 < \delta_{1i} < 0.44\delta \\ \delta\eta^* &= 4.598\,(\delta_{1i} - 0.44\delta)^2 && ,\text{if} \quad 0.44\delta < \delta_{1i} < 0.69\delta \qquad (9) \\ \delta\eta^* &= 2.229\,(\delta_{1i} - 0.565\delta) && ,\text{if} \quad 0.69\delta < \delta_{1i} < \delta \ . \end{aligned}$$

For adiabatic walls, the density profiles are deduced from the velocity profiles with the isoenergetic approximation. From the profiles, a mixing length model has provided an algebraic entrainment closure for equilibrium turbulence [15] :

$$E_{eq} = \left[C_4\,(1 - \tilde{u}_w) - C_5\frac{Cf}{\sqrt{|Cf|}}\right]\lambda_1\lambda_2\lambda_3 \qquad \tilde{u}_w = 1 - \frac{2.22}{1 + 1.22\eta^*}\frac{\delta_{1i}}{\delta} \qquad (10)$$

with $\lambda_1 = \lambda_2 = \lambda_3 = 1$ for usual boundary layers, $\lambda_1 = 2$, $Cf = 0$ for usual wakes, λ_2 and λ_3 corresponding to curvature and external turbulence effects. As initially suggested, the present computations have been performed with $C_4 = 0.053$, $C_5 = 0.182$, although a better tuning will be probably used in the near future. Two additional transport integral equations [15,17] can be used to compute the departure from equilibrium $\lambda_{(x,t)}$ of the turbulence, such that $E_{(x,t)} = \lambda_{(x,t)} \cdot E_{eq(x,t)}$, but they were switched off in the present calculations.

The velocity profile (7),(8),(9) generalizes the Coles profiles by an $\eta^*(\delta_1/\delta)$ law which modifies the usual wake function. The η^* law eliminates the non-physical infinite reverse velocity which would appear with $\eta^* = 0$ for infinitely separated boundary layers, and allows the velocity profiles to describe accurately the two limiting test-cases for equilibrium turbulence : the flat plate boundary layer in the attached flow range, and the isobaric mixing layers in the highly separated flow range, see [8].

As shown previously by Le Balleur [18], if $\delta - \delta_1 = \tilde{H} \cdot \theta_{11}$ and $\delta_1 = H \cdot \theta_{11}$, the characteristic cone of dependence of the unsteady integral system, when solving the problem in Direct mode, is such that no upstream influence is present if $\partial\tilde{H}/\partial H < 0$, and that the upstream influence is possible if $\partial\tilde{H}/\partial H > 0$. The limiting case $\partial\tilde{H}/\partial H = 0$ does not degenerate, in the case of unsteady flows, into a Goldstein-like singularity as for steady flows, but indicates the beginning of the upstream influence, and is considered as the separation point.

The present closure provides in the diagram $\tilde{H}(H)$ a minimum, and only one, for any Mach and Reynolds numbers, always about $H_i = 2.7$, insuring the existence of the upstream influence in the whole separated flow range, see [8].

3.2 - Numerical solution of the viscous integral equations.

The integral Defect system (1),(2),(3),(4) can be written in the following form

$$\left\{\frac{1}{q} C^i_{j\,(a,m,\delta)} \frac{\partial f^j}{\partial t} + A^i_{j\,(a,m,\delta)} \frac{\partial f^j}{\partial x} = b^i\right\}_{(x,0,t)} \qquad (11)$$

$$f^j = \begin{bmatrix} \hat{m} \\ \hat{h}_t \\ \delta \\ a \end{bmatrix} \quad , \quad b^i = \begin{bmatrix} \dfrac{\hat{v}}{q} \\ \dfrac{Cf}{2} \\ E \end{bmatrix} \quad , \quad \begin{matrix} i = 1,3 \\ j = 1,4 \end{matrix}$$

The viscous unknowns are the reduced Mach number $\hat{m} = .5(\gamma-1)\hat{M}^2$, the transpiration velocity $\hat{v}$, the total enthalpy $\hat{h}_t$, the boundary layer thickness δ and a shape parameter $a = \delta_{1i}/\delta$. The three unknows $\hat{m},\hat{v},\hat{h}_t$ have to be coupled with the corresponding variables in the pseudo-inviscid problem, m,v,h_t :

$$\hat{m} \rightarrow m \quad , \quad \hat{h}_t \rightarrow h_t \quad , \quad \hat{v} \rightarrow v \quad . \qquad (12)$$

The rank-3 system (11), involving 5 unknowns,is solved with the 2 following coupling conditions :

$$\left\{\hat{h}_t = h_t\right\}_{(x,0,t)} \quad , \quad \left\{\epsilon\frac{\hat{v}}{q} + (1-\epsilon)\frac{1}{\hat{m}}\frac{\partial \hat{m}}{\partial x} = \epsilon\frac{v}{q} + (1-\epsilon)\frac{1}{m}\frac{\partial m}{\partial x}\right\}_{(x,0,t)} \quad . \qquad (13)$$

The resolution of the viscous system in Direct mode ($\epsilon = 0$) with a space-marching technique is possible only for attached flows. In order to maintain a space-marching solution in the reverse flow range, the system is solved in Inverse mode ($\epsilon = 1$). The upstream influence, which is eliminated in the viscous step of the calculation by the space-marching technique, is in fact recovered in the coupling step.

The marching integration scheme is implicit in time and space, and uses an hermitian technique to link the time and space derivatives, see Le Balleur, Girodroux-Lavigne [10,11]. The non-linearities due to C_j^i, A_j^i, b^i in system (11) are solved by a Newton method.

3.3 - Numerical coupling : Semi-Implicit method.

The coupling step is built on the "viscous influence function" which connects the three unknowns $[\hat{m},\hat{v},\hat{h}_t]$, and which is deduced from system (11) by eliminating the two purely viscous unknowns δ, a, see [10]. This function can be written at time level n, and node i,1 :

$$R_{i,1}^{n}\left(\frac{\hat{v}}{q},f^{j}\right) = \left[\frac{\hat{v}}{q} + \alpha_j \frac{\partial \hat{f}^j}{\partial t} + \beta_j \frac{\partial \hat{f}^j}{\partial x}\right]_{i,1}^{n} = 0 \qquad \left[\hat{f}^j = \hat{m}\ ,\ \hat{h}_t\right] \quad j=1,2\ . \quad (14)$$

Before convergence of the strong coupling, relation (14) is not verified by the inviscid quantities $[m,v,h_t]$, and non-zero residuals $R_{i,1}^{n}(v/q,f^j)$ are obtained at time level n and coupling iteration ν.

The "Semi-Implicit" method [10]is a relaxation technique for these coupling residuals, where the non-linear terms α_j,β_j and the residuals $R_{i,1}^{n}$ are provided by the previous viscous uncoupled solution, in Direct or Inverse mode, and the previous inviscid solution :

$$\left[\left(\frac{v^{\nu+1}}{q} - \frac{v^{\nu}}{q}\right) + \alpha_j^{\nu}\frac{\partial}{\partial t}\left(f^{j,\nu+1} - f^{j,\nu}\right) + \beta_j^{\nu}\frac{\partial}{\partial x}\left(f^{j,\nu+1} - f^{j,\nu}\right)\right]_{i,1}^{n} = -\overline{\omega}\left[R^{\nu}\right]_{i,1}^{n}$$

$$\left\{f^j = m\ ,\ h_t\right\} \qquad j=1,2 \qquad \overline{\omega}=1 \qquad (15)$$

$$\left[R^{\nu}\right]_{i,1}^{n} = (1-\epsilon)\left[\frac{\hat{v}}{q} - \frac{v^{\nu}}{q}\right]_{i,1}^{n} + \epsilon\left[\alpha_j^{\nu}\frac{\partial}{\partial t}\left(\hat{f}^j - f^{j,\nu}\right) + \beta_j^{\nu}\frac{\partial}{\partial x}\left(\hat{f}^j - f^{j,\nu}\right)\right]_{i,1}^{n} .$$

The conditioning terms on the left hand side are discretized with the potential variable of the inviscid solver, using upwind and centered (for $\partial m/\partial x$) schemes, but in the centered case, a Gauss-Seidel discretization based on iterations ν and ν+1 (at the same time-level n) is involved. The relaxation equation (15) is then solved space-marching, simultaneously with the y-sweep of the ADI inviscid solver, and the coupling procedure is iterated, at a given time step, until convergence of the strong coupling.

Although involving a step where the inviscid field is semi-simultaneously relaxed with a viscous influence equation, the "Semi-Implicit" method is different from the "Quasi-Simultaneous" method of Veldman [5], and maintains a step where an uncoupled Direct or Inverse viscous problem is solved. Roughly, Veldman's method solves a viscous problem plus an inviscid influence function. On the contrary, the "Semi-Implicit" method solves an inviscid problem plus a viscous influence function, deduced from direct or inverse viscous solutions.

In the case of unsteady flows, the viscous upstream influence is here fully recovered from the discretization of the residuals $R_{i,1}^{n}$, and from the time-consistent iterative coupling which is converged at each time step. If the coupling was not converged within a time step, the upstream influence would be only recovered in the case of steady flows, from the residuals, when the time-derivatives vanish, as in the method of Houwink, Veldman [19].

4 - Numerical results : prediction of buffeting flows.

4.1 - Symmetrical airfoil.

Very encouraging results were previously found with the present method to discriminate between steady and unsteady separated flows on the steady symmetric 18% circular-arc airfoil at zero angle of attack, [11]. Figure 1 shows that a steady symmetric flow is obtained for $M_\infty = 0.788$, and that an unsteady configuration is found for $M_\infty = 0.760$.

4.2 Lifting airfoil.

The method is used here to predict the onset of buffeting flow in the case of the RA16SC1 airfoil with positive incidences, in the transonic regime. Steady flows as well as unsteady shock-induced separations were found experimentally on this airfoil, depending on the Mach number and angle of attack, by Benoit [12] and Dor et al. [13], in two different wind tunnels. The calculations are here compared with the experiments of Benoit.

The computations were carried out time-consistently on a 160x42 cartesian grid, with 100 nodes on each side of the airfoil and 30 nodes in the wake. The top and bottom boundaries are located at $\pm$ 40 chord lengths. In the x-direction, the boundaries are located at -20 and +10 chord lengths. All the calculations were made over a great number of time-steps (3400 time-steps), in order to have a good description of the eventual unsteady phenomena. The initial state is a uniform flow over an airfoil of zero-thickness at zero incidence. The thickness and incidence are continuously increased from zero to the required values within 32 time-steps.

Several calculations were performed for a Mach number $M_\infty = 0.723$, a Reynolds number of 4.2 millions, and incidences between $\alpha = 0^\circ$ and $\alpha = 4^\circ$. The time histories of the lift coefficient, figure 2, show that the steady solutions which are predicted for the lowest incidences, are progressively transformed into unsteady solutions when the incidence is increased. A good agreement is found, figure 3, between the calculated lift coefficients and the experimental values, corrected for wall effects.

The hatched regions which visualize the extremal values of the lift coefficient, figure 3, and the envelope of the pressure distributions, figure 4, point out the importance of the unsteady fluctuations for high angles of attack. The unsteadiness of the flow is principally due to the pressure fluctuations in the separated zone, and to the moving of the shock position. For the highest incidence, $\alpha = 4^\circ$, the extent of the domain of variation of the shock location is about 20% of the chord length, and the relative value of the lift variation is of the same order.

Details of the time fluctuations are shown in figure 5, which displays the non-linear histories of the distributions of pressure coefficient, incompressible shape parameter, displacement thickness and skin-friction, for $\alpha = 4^\circ$.

The unsteady character of the flow field is visualized, figure 6, by the instantaneous iso-Mach lines over one pseudo-cycle of oscillation, showing the important variations of the extent of the supersonic pocket, and of the reverse flow region.

4.3 - Time-histories of the pressure.

The highly non-harmonic time-histories of the unsteady pressure at different locations (figure 7) show that a good agreement is found between the calculation and the experiment, not only on the qualitative aspect, but also on the quantitative levels. The RMS values of the unsteady pressure, P_{RMS} :

$$P_{RMS} = \left(\frac{1}{T} \int_0^T P'^2 dt \right)^{\frac{1}{2}}$$

where P' is the unsteady part of the pressure P, which give a good estimate of the intensity of fluctuations, are displayed on figure 8 for different angles of attack. They are similar to the experimental RMS distributions, except in the separated region where the experimental minimum located at $x/c \simeq 0.75$ does not appear in the calculations.

When studying the unsteady process, an important point is the distribution of power of the signal versus frequency, which is measured by the power spectral density function (DSP) G_{xx}, defined by :

$$G_{xx}^2(f) = \frac{2}{T^2} X(f) \, . \, X^*(f)$$

where $X(f)$ is the Fourier transform of a function $x(t)$ with $0<t<T$.

Figure 9 shows the DSP distributions of the pressure coefficient on the upper surface of the airfoil for the calculations at $\alpha = 3°$ and $4°$, and the experimental results at $\alpha = 4°$. For $\alpha = 3°$, the calculation predicts non-zero amplitudes only in the shock region and at low frequencies. For $\alpha = 4°$, substantial amplitudes appear in the whole frequency range, in the region corresponding to the displacement of the shock. Moreover, important peaks of the DSP function (with harmonics) are present, not only in the shock region, but also in the whole shock-induced separated zone.

The computation and the experimental results are found qualitatively similar. The method predicts a fundamental frequency of 90 Hz, very close to the experimental value of 100 Hz. The levels of the harmonics in the calculation are however higher than in the experiment. This fact can perhaps be explained if we consider that a period is described in the experiment with about 10 points, whereas the calculation, using about 200 points per cycle, can give a better resolution of the high frequencies than the experiment.

5 - Conclusions

A time consistent viscous-inviscid interaction method has been developed. The convergence of the strong coupling at each time step allows the method to take into account the viscous upstream influence in unsteady flows, and gives a Navier-Stokes-like capability to predict steady and unsteady separations.

The prediction of self-induced unsteady separated flows was previously obtained at non-lifting conditions over a symmetrical circular-arc airfoil.

The time-dependent method is found here to discriminate also between steady flows and unsteady buffeting flows over the RA16SC1 airfoil, depending on the incidence and Mach number, at transonic speed and lifting conditions. An overall good agreement for the pressure is found between calculations and buffet experiments.

Although it is obviously difficult to know how far the quantitative prediction of buffet is possible based on a computing capability in unsteady separation, and although more investigation is necessary on the calculations, especially on the influence of the initial conditions, the good agreement with experiment obtained here on the time-histories and spectral properties of the pressure distributions is believed very encouraging for further buffeting calculation.

References

[1] Melnik R.E., Brook J.W. - The computation of Viscid/Inviscid Interaction on airfoils with separated flow. - Proceed. 3rd Symp. Numerical and Physical Aspects of Aerodynamic Flows. Long Beach (jan. 1985), T. Cebeci ed., Springer Verlag (1986).

[2] Carter J.E., Edwards D.E., Davis R.L., Hafez M.M. - Analysis of strongly interacting inviscid-viscid flows including separation. - Proceed 9th ICNMFD, Saclay (1984), Lecture Notes in Physics, 218, Springer Verlag (1985).

[3] Smith F.T., Papageorgiou D., Elliott J.W. - An alternative approach to linear and nonlinear stability calculations at finite Reynolds numbers. - J. Fluid Mech. (1984), vol.146, pp.313-330.

[4] Lock R.C., Firmin M.C.P. - Survey of techniques for estimating viscous effects in external aerodynamics. - RAE Tech. Memo. Aero 1900 (1981).

[5] Veldman A.E.P. - New, quasi-simultaneous method to calculate interacting boundary layers. - AIAA J. Vol. 19, no 1 (1981), p.79-85.

[6] Yoshihara H., Wai J. - Transonic turbulent separation on swept wings - A return to the Direct formulation... AIAA Paper no 84-0265, Reno, (1984).

[7] Cebeci T., Clark R.W., Chang K.C., Halsey N.D., Lee K. - Airfoils with separation and the resulting wake. Procced. 3rd Symp. Numerical and Physical Aspects of Aerodynamics Flows, Long Beach (Jan.1985), T. Cebeci ed., Springer-Verlag 1986.

[8] Le Balleur J.C. - Viscous-Inviscid interaction solvers and computation of highly separated flows. - Proceed. of ICASE workshop on Vortex Dominated Flows, NASA Langley Field, USA, (july 9-10, 1985), Springer-Verlag (1986).

[9] Le Balleur J.C. - Numerical flow calculation and viscous-inviscid interaction techniques. - Recent Advances in Numerical Methods in Fluids, Vol. 3 : Computational methods in viscous flows, p.419-450, W. Habashi ed., Pineridge Press (1984).

[10] Le Balleur J.C., Girodroux-Lavigne P. - A semi-implicit and unsteady numerical method of viscous-inviscid interaction for transonic separated flows. - La Recherche Aerospatiale 1984-1, p.15-37, English and French editions, (1984).

[11] Le Balleur J.C., Girodroux-Lavigne P. - A viscous-inviscid interaction method for computing unsteady transonic separation. - Proceed. 3rd Symp. Numerical and Physical Aspects of Aerodynamics Flows, Long-Beach (jan.1985), T. Cebeci ed., Springer-Verlag (1986).

[12] Benoit B. - Etude du champ de pression instationnaire sur le profil RA16SC1 en regime de tremblement a S3MA. - Rapport ONERA RTS no 17/3423A (juin 1986).

[13] Dor J.B., Gobert J.L., Plazanet M., Mignosi A. - Etude experimentale des instabilites auto-induites sur un profil RA16 de corde 180 mm en ecoulement transsonique, a la soufflerie T2. - Rapport Technique ONERA-CERT-10/5017D. (juin 1986).

[14] Le Balleur J.C. - Computation of flows including strong viscous interactions with coupling methods. - AGARD-CP-291, General Introduction, Lecture 1, Colorado-Springs (1981), or ONERA TP 1980-121.

[15] Le Balleur J.C. - Strong matching method for computing transonic viscous flows including wakes and separations. Lifting airfoils. - La Recherche Aerospatiale 1981-3, p.21-45, English and French editions, (1981).

[16] Couston M., Angelini J.J. - Solution of nonsteady two-dimensional transonic small disturbances potential flow equation. - ASME Conf., San Francisco (1978) or J. of Fluids Engin., vol.101, no 3, (1979).

[17] Le Balleur J.C. - Numerical viscid-inviscid interaction in steady and unsteady flows. - Proceed. 2nd Symp. Numerical and Physical Aspects of Aerodynamic flows, Long-Beach, (1983), Springer-verlag, T. Cebeci ed., chapt.13, p.259-284 (1984), or ONERA-TP 1983-8.

[18] Le Balleur J.C. - Viscous-inviscid flow matching : Analysis of the problem including separation and shock waves. - La Recherche Aerospatiale 1977-6, p.349-358 (Nov.1977). English transl. ESA-TT-476.

[19] Houwink R., Veldman A.E.P. - Steady and unsteady separated flow computations for transonic airfoils. - NLR MP 84028U, or AIAA 17th Fluid-Plasma Conf., Snowmass, Colorado (june 1984).

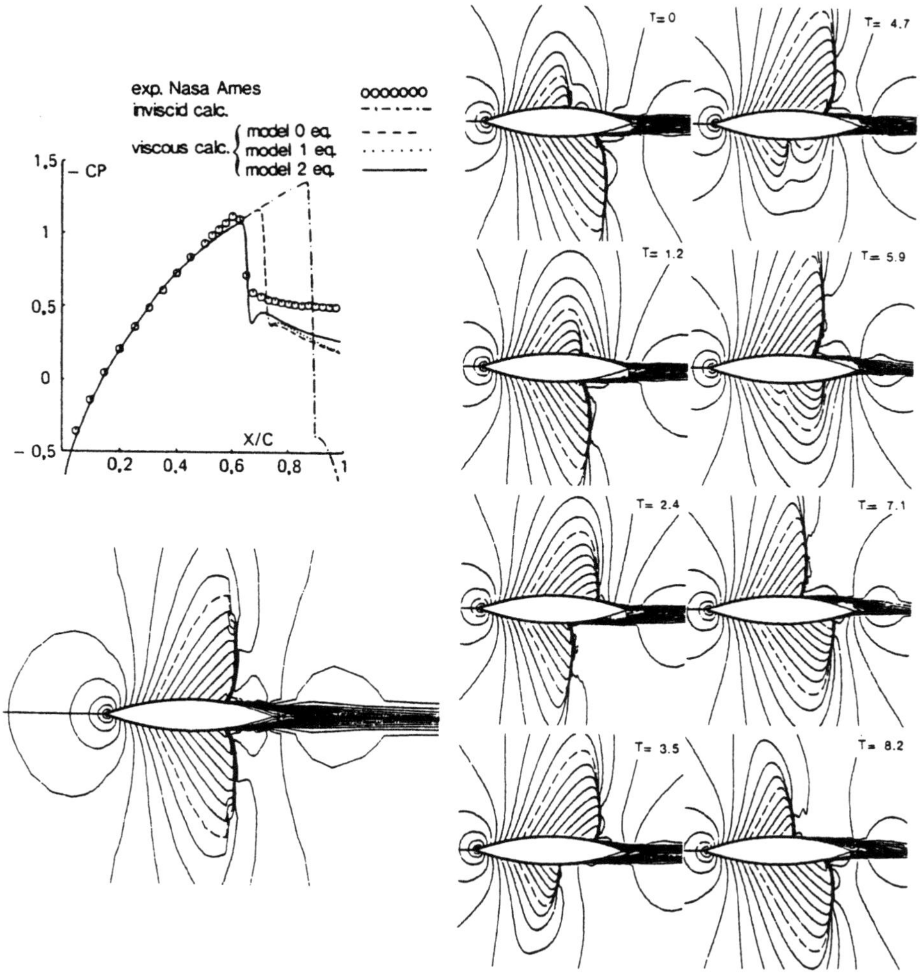

steady flow
$M_\infty = 0.788$

self-induced
unsteady flow
$M_\infty = 0.760$

Fig.1 - 18% circular-arc airfoil. Pressure distributions. Iso-Mach lines. ($\alpha = 0°$, $\mathrm{Re} = 11 \times 10^6$).

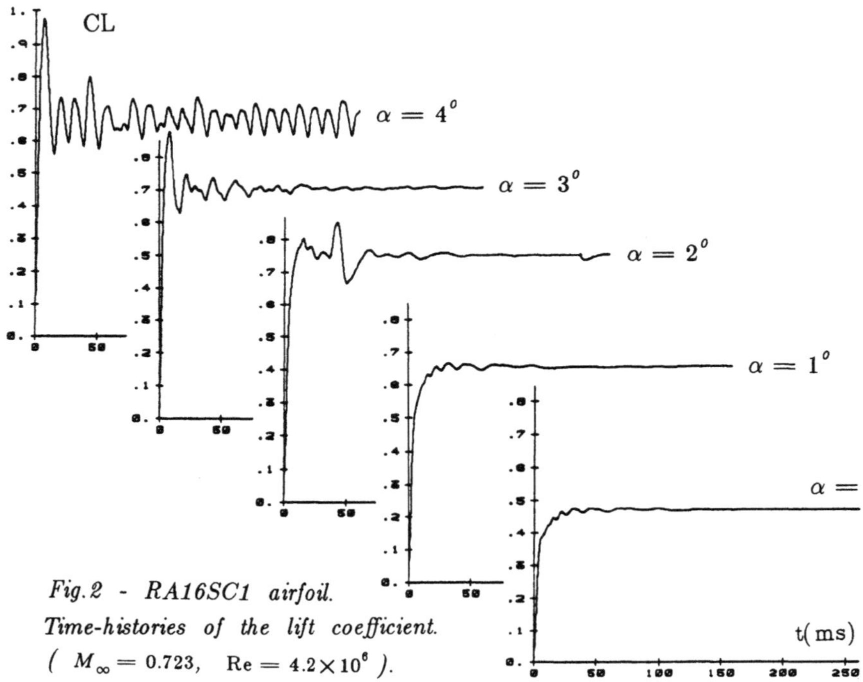

Fig. 2 - RA16SC1 airfoil. Time-histories of the lift coefficient. ($M_\infty = 0.723$, $\mathrm{Re} = 4.2 \times 10^6$).

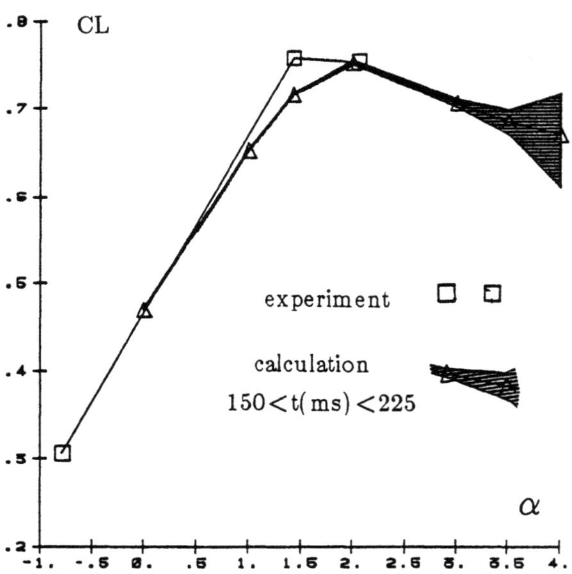

Fig. 3 - RA16SC1 airfoil. Evolution of the extremal values of the lift coefficient with the incidence. ($M_\infty = 0.723$, $\mathrm{Re} = 4.2 \times 10^6$).

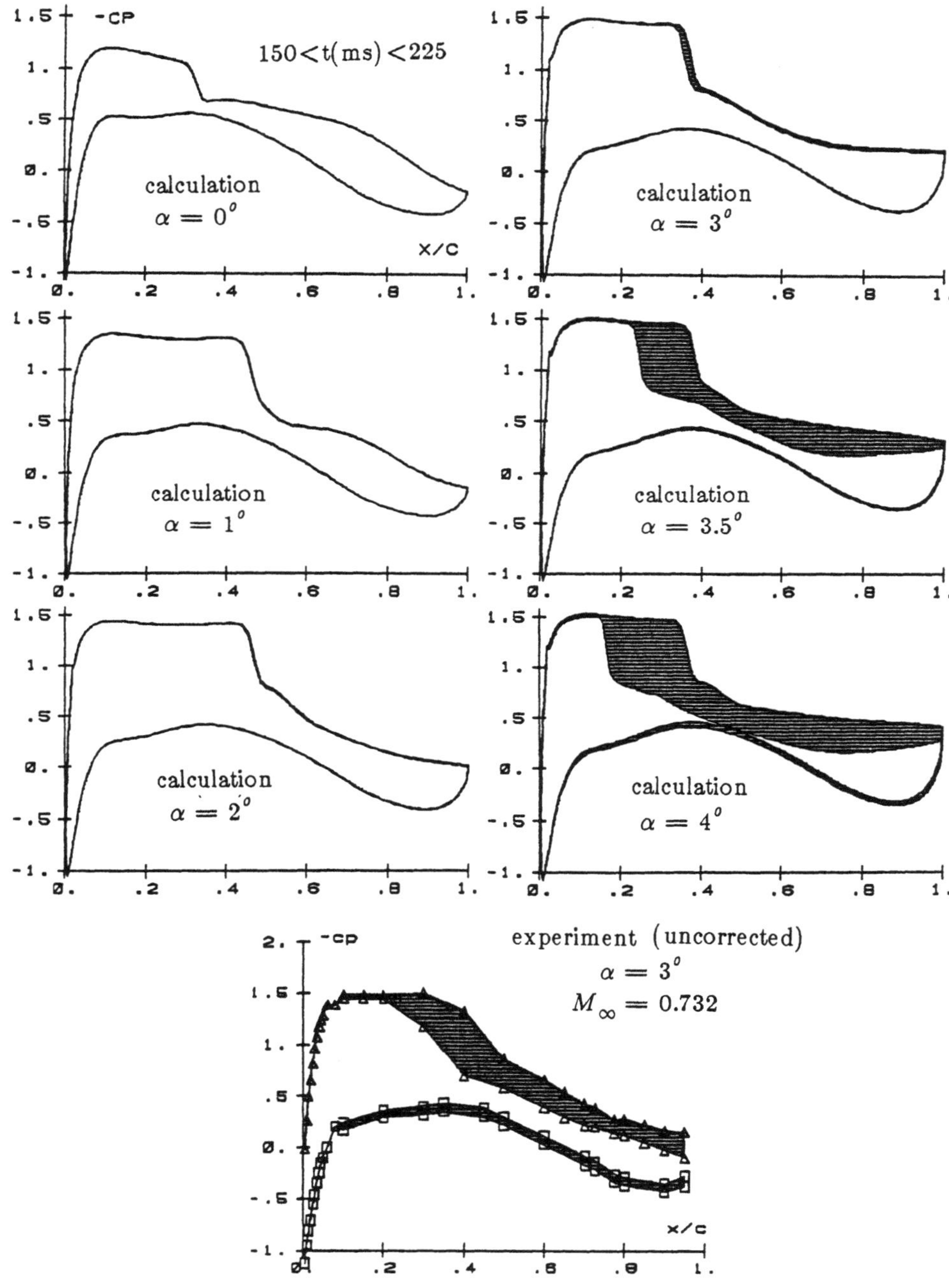

Fig.4 - RA16SC1 airfoil. Evolution of the extremal values of the pressure distributions with the incidence. ($M_\infty = 0.723$, $\mathrm{Re} = 4.2\times10^6$).

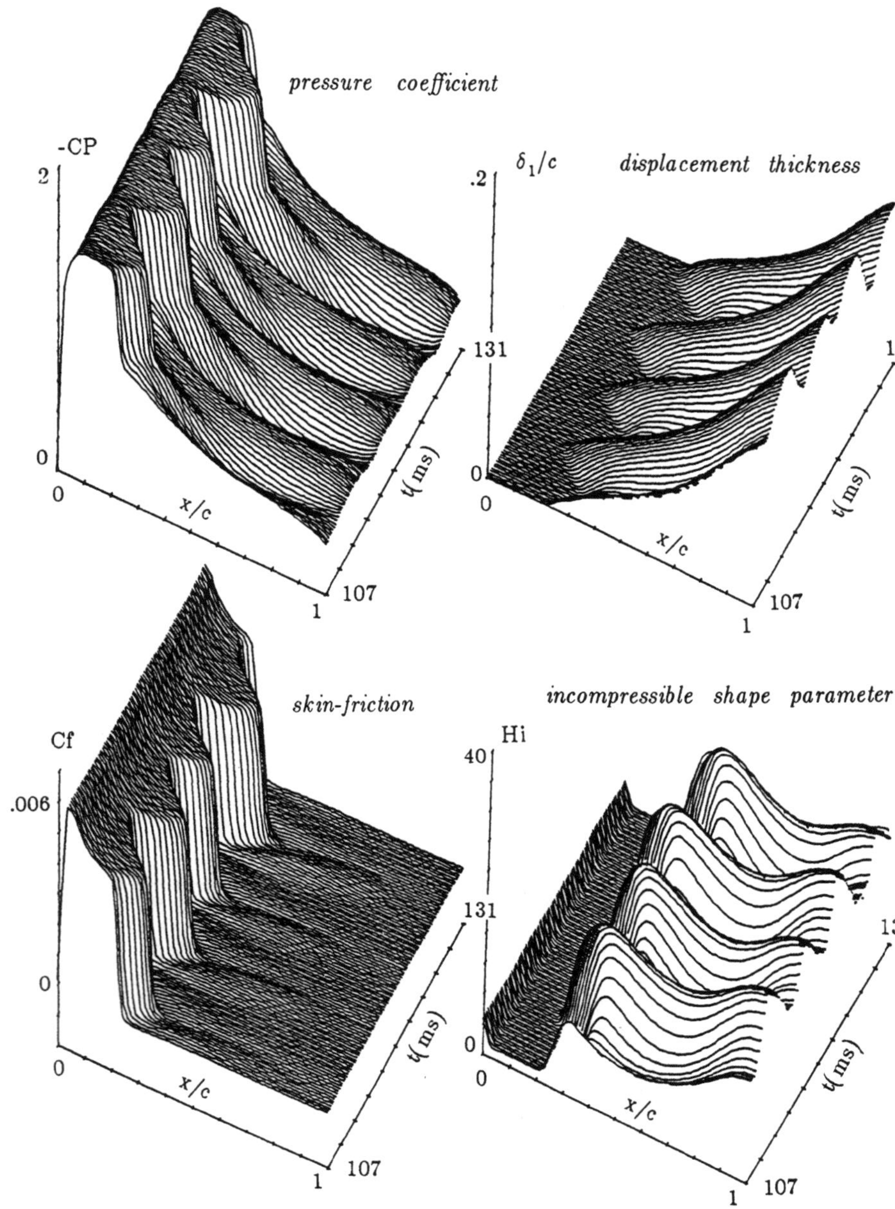

Fig. 5 - RA16SC1 airfoil. Evolution in time of the pressure coefficient, incompressible shape parameter, displacement thickness and skin-friction distributions on the upper surface. ($M_\infty = 0.723$, $\alpha = 4°$, $\mathrm{Re} = 4.2 \times 10^6$).

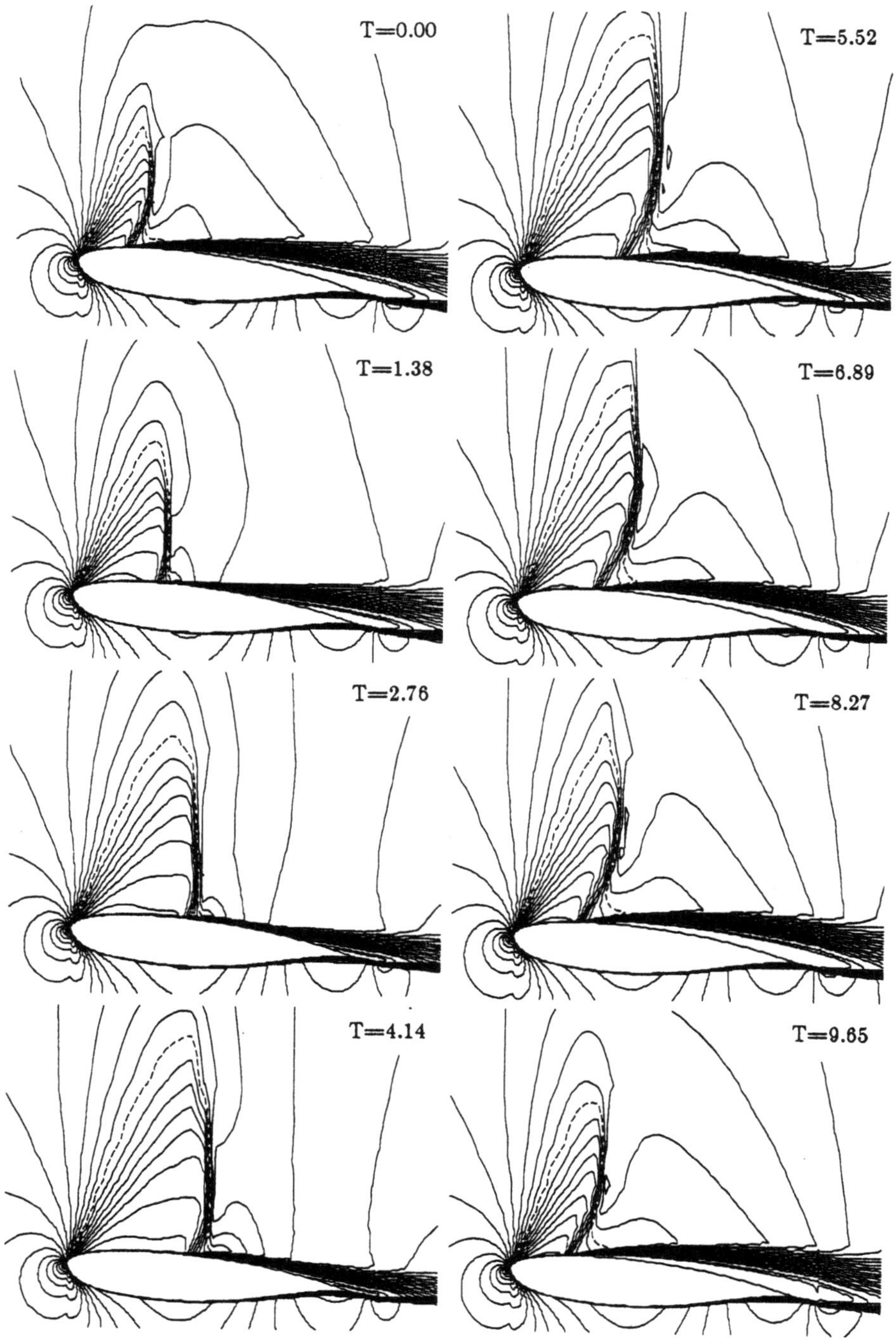

Fig. 6 - RA16SC1 airfoil. Instantaneous iso-Mach lines over one pseudo-cycle.
($M_\infty = 0.723$, $\alpha = 4°$, $\mathrm{Re} = 4.2 \times 10^6$).

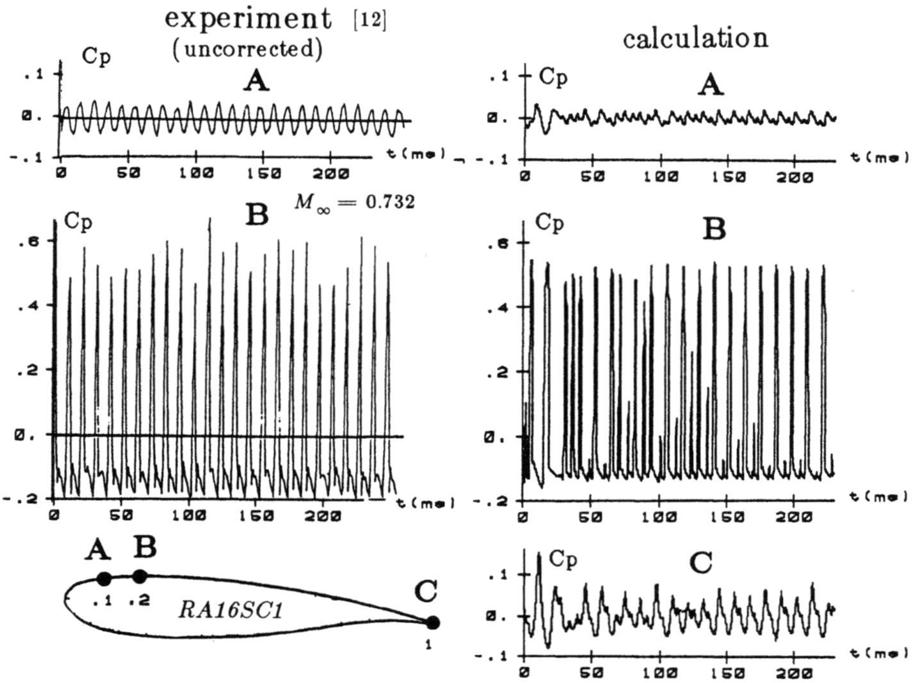

Fig.7 - RA16SC1 airfoil. Time-evolution of the unsteady pressure coefficient. Comparaison with experiment. ($M_\infty = 0.723$, $\alpha = 4°$, Re $= 4.2 \times 10^6$).

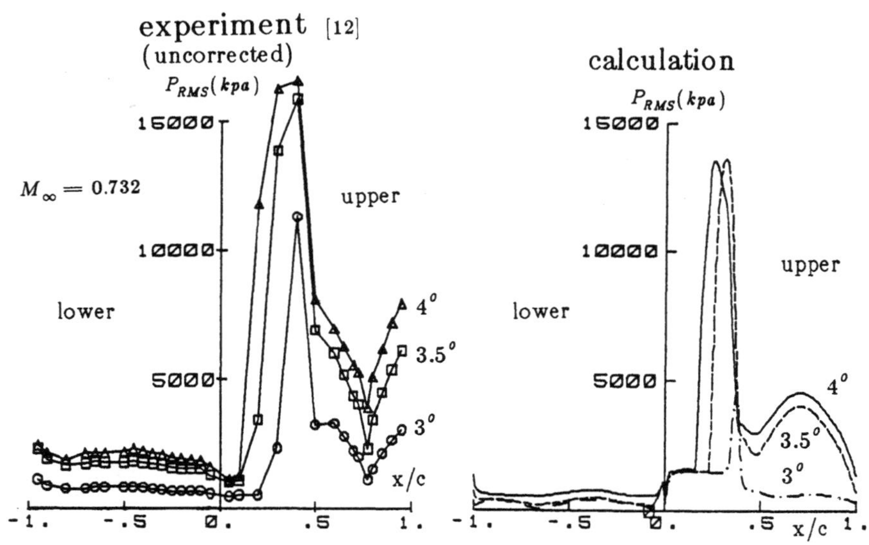

Fig.8 - RA16SC1 airfoil. Distributions of the RMS values of the pressure coefficient. ($M_\infty = 0.723$, Re $= 4.2 \times 10^6$).

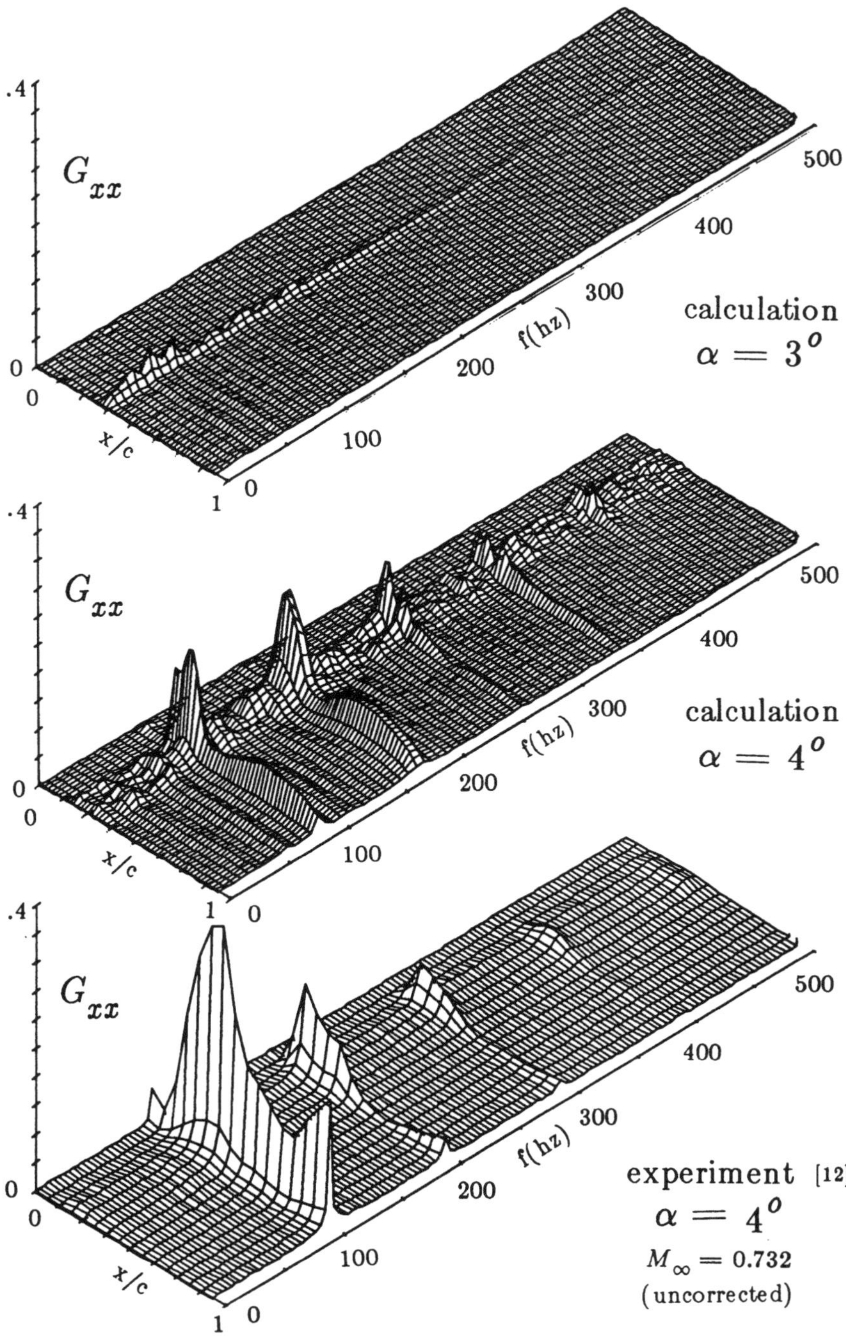

Fig. 9 - RA16SC1 airfoil. Evolution with frequency of the power spectral density of the pressure coefficient on the upper surface. ($M_\infty = 0.723$, $Re = 4.2\times10^6$).

Observation of Three-dimensional "Separation" in Shock Wave Turbulent Boundary Layer Interactions

S. M. BOGDONOFF

Department of Mechanical and Aerospace Engineering
Princeton University
Princeton, NJ 08544

Abstract

On the basis of the present analysis of specific two and highly swept three-dimensional shock wave and turbulent boundary layer interactions, it has been observed that: 1) three-dimensional flows are radically different than the "classical" two-dimensional flows, 2) the scale, pressure gradients, unsteadiness, and computability are quite different, 3) the designation of "separation" in three dimensions is not realistic, and 4) a concept of vorticity rearrangement is proposed to describe the physics of the interactions in three dimensions.

I. Introduction

Separation, in fluid mechanical terms, is generally used to describe the detachment of a flow from a surface. It is most often induced by an abrupt change in the wall shape or the imposition of a strong adverse pressure gradient. Once a flow detaches from a surface, its subsequent behavior is quite unpredictable. There are losses due to mixing, there is the question of where the flow reattaches to the surface (if it does), and then there is the problem of the downstream flows and how they develop. Separation and subsequent development of the flowfield is dependent, to a large extent, on the characteristics of the boundary layer and laminar, transitional, and turbulent flows can all occur. In the present paper, our discussion is limited to boundary layers which are fully turbulent, somewhat simplifying the complexity of the overall problem. Most of the studies on separation, both experimental and theoretical, have been concerned with two-dimensional flows, and much of the flow physics comes from the modelling of these flows. Many of the concepts developed from these flows have been carried over to three-dimensional flows.

Our consideration in the present paper will be limited further to supersonic flows with turbulent boundary layers. In this area, boundary

F. T. Smith, S. N. Brown, (Eds.)
Boundary-Layer Separation
IUTAM Symposium London 1986

layer theory and computation have not been adequate to predict, in any general sense, either the conditions at which the boundary layer will separate nor the details of the complex flows which ensue. Detailed fundamental experimental data is quite sparse. Most experiments are engineering tests primarily concerned with mean surface data, with very little in the way of high frequency investigations. Surface flow visualization by oil films is a simple, widely used technique which has very low frequency response. Flowfield visualization is very limited, and the interpretation of the results is extraordinarily difficult because of the three-dimensionality of the flow (the visualization techniques usually provide two-dimensional cuts) or the low frequency of observation. As a result, three-dimensional flows, with possible separation, have generated a great deal of conjecture and innovative modelling, but a rather limited number of direct measurements to truly establish the flow details.

The Gas Dynamics Laboratory at Princeton University has, for many years, been carrying out experimental studies, in two and three dimensions, on high Reynolds number supersonic turbulent boundary layers subjected to adverse gradients caused by shock waves. The Laboratory has used many of the classical experimental tools and has, in recent years, added significant high frequency capability in both flowfield diagnostics and surface static pressures. The experimental program has been closely related to large computational programs solving the Navier-Stokes equations, NASA-Ames with Dr. Horstman, and Rutgers University with Prof. Knight.

The present paper grows out of studies of the past several years re-examining many of the older data sets, carrying out new experiments with high frequency instrumentation, and close interactions with the computational programs. These studies suggest a premise quite different than that usually accepted by the community working in this area. Three-dimensional "separation" is found to be quite different from that which is called "separated" flows in two dimensions. A careful examination of two-dimensional geometries shows that these separated flowfields are not two-dimensional. Flows are highly unsteady, and most of the modelling based on mean flows neglects important physical parameters. In the three-dimensional flows which have been examined in detail, it is found that the flows are <u>not</u> separated in any two-dimensional concept. There are serious questions of the interpretation of surface flow and flowfield visualization as they are usually carried out. Implying flowfield details from surface flow visualization is not supported by the flowfield experiments or computations, and leads to concepts which are not supported by the data.

II. Data Base

1. Introduction

There have been many studies of two- and three-dimensional separated flows and no attempt will be made to include the results of all of these studies in the present paper. Excellent survey papers have been presented in Refs. 1-3. The present paper will limit the discussions to a series of studies which have been carried out at the Gas Dynamics Laboratory at a Mach number of about 3, with high Reynolds number turbulent boundary layers and pressure gradients caused by shock waves of various strengths. We are using these limited sets of experiments because they are probably the most highly instrumented and closely coordinated with the extensive computational studies. The results and conclusions which are drawn are compatible with the data from other experimental studies, although the interpretation, in many cases, is different. Two-dimensional interactions are examined as a base for comparison with the extensive three-dimensional studies, since many of the concepts of separated flow have been developed from the two-dimensional studies.

2. Two-Dimensional Studies

a) Geometry

The discussion of two-dimensional geometries will be limited to the wedge or ramp flow noted in Fig. 1. This geometry has consistently given better "two-dimensional" results than the incident shock counterpart, and extensive experimental studies are available, Refs. 4 & 5. Wedge angles from 8° to 24° cover the regime from attached to significantly separated flows. Early work, Ref. 6, was primarily concerned with the determination of where separation began as shock strength was increased. High resolution studies in the corner showed that separation was <u>always</u> present, but that the extent of separation grew exponentially with shock strength. The definition of "incipient separation" was determined by instrumentation resolution, rather than physics. The general flowfield sketched in Fig. 1a is an example of "separated" results obtained from mean surface and flowfield studies noted in the following section.

b) Mean measurements

The classical data set obtained to study such interactions is the surface static pressure distribution, surface flow visualization, and total head and pressure surveys of the flowfield. The sketch of Fig. 1a and 1b was constructed from such data. A set of wall static pressure distributions for varying wedge angles is shown in Fig. 2. Figure 3 presents one typical result of the surface flow visualization technique which is an improvement on the usual "oil flow" technique, Ref. 7. The surface flow visualization technique used an evaporating "carrier" which leaves fine graphite markings on the surface. It eliminates the small surface bumps associated with oil streak or oil dot techniques.

c) High frequency measurements

The primary high frequency data has been obtained from individual or arrays of small static pressure gauges (Kulites) in the surface, and hot-wire surveys in the flowfield. Although the mean wall static pressure data system was not designed with any specific high frequency characteristics, the ensemble averaged high frequency data falls quite close to the mean wall static pressure results. The rms fluctuations, however, show a major change through the interaction, Fig. 1c, Ref. 8. Very large peaks in the fluctuations were found upstream of the surface flow visualization accummulation line (called "separation") which appears to occur near the downstream boundary of a highly fluctuating flow. The data from the high frequency gauges arranged laterally shows that the flow is far from two-dimensional, Ref. 9. The classical two-dimensional schlieren photograph is usually considered part of high frequency data acquisition but it should be noted that it is a two-dimensional integrater, although the exposure is very short. The results of the high frequency cross flow and longitudinal instrumentation indicate that the instantaneous static pressure is far from the usual accepted mean distribution. It is very abrupt at any instant, and the mean smooth pressure distribution is the result of the statistics of the movement of a very abrupt disturbance. The lateral "waviness" is seen in the two-dimensional schlieren photograph as a compression fan. It appears, from the surface data, that a shock wave extends far into the boundary layer. What looks like a compression fan is probably a series of shock waves at different positions across the tunnel.

3. Three-Dimensional Studies

a) Geometry

The Gas Dynamics Laboratory has studied a wide range of three-dimensional geometries, Fig. 4. For the present discussion we will concentrate on the sharp unswept fin which has been tested at a range of angles of attack from 2 to 20°. Several conditions have been the subject of full flowfield surveys, and this configuration has had the most attention from the computational studies.

b) Mean measurements

Typical results of a series of fin angles are shown in the pressure distribution plots of Fig. 5. These wall pressure distributions were obtained along lines parallel to the initial flow using conventional static pressure orifices, transducers, and scanivalves to provide the mean results. A surface flow visualization is shown in Fig. 6. A sketch of the main surface features for the higher angle fins is presented in Fig. 7. Contour plots for the flowfield, along a plane parallel to the initial flow, are shown for two parameters in Fig. 8, Ref. 10.

c) High frequency measurements

The primary high frequency measurements were obtained using arrays of high frequency surface pressure gauges aligned both parallel to the initial flow direction and parallel to some of the surface features found from the flow visualization, Refs. 11-13. The rms fluctuating wall pressures are shown in Fig. 9. Other information from these references show that, in addition to the movement of the structure in this plane, data from the gauges oriented parallel to the surface features shows "ripples or waviness" somewhat similar to that found in the two-dimensional study.

III. Discussion

1. Two-Dimensional Flows

The mean flow, as shown in Fig. 1, characterizes the flow with a separation point, the generation of a free shear layer over a subsonic bubble, reverse flow, reattachment of the free shear layer, and then recovery of the boundary layer downstream of the interaction. Computation of this flow has not been satisfactory, with no consistency with regards to the determination of the separation point or the reattachment point. The flow downstream is quite different from that measured. In reality, the flow is highly unsteady and not two-dimensional. The smooth pressure rise in the initial part of the interaction is, as noted previously, simply caused by the statistics of what appears to be a very sharp rise in pressure in instantaneous coordinates. The point at which the flow is actually separated is not known, although Gramann and Dolling, Ref. 14 , have made the first attempts to quantify this important physical phenomena. The flow appears to be separated far upstream of the line indicated by surface flow visualization. The flow is unsteady, however, and the surface flow visualization seems to indicate approximately where the flow is probably separated all of the time. The mean static pressure curves, Fig. 2, show that the mean gradients in the initial part of the interaction become weaker as the shock strength increases, reaching a constant value as noted by Chapman et al. many years ago in his "free interaction concept", Ref. 15. The major gradients take place within a few boundary layer thicknesses. The unsteady data shows a major peak in the rms fluctuations in the initial part of the "separation" area, and unpublished data shows that the reattachment region is even more violently unsteady.

To summarize, the so-called two-dimensional flow is highly unsteady, three-dimensional, and probably strongly dependent on turbulence, mixing, and is definitely dissipative. The separation and reattachment points are very unsteady, and surface flow visualization does not clearly indicate these key physical phenomena. We do not know at the moment whether the separated region is "puffing" or "sloshing", since there has not yet been extensive studies of the correlation between high frequency measurements in the separation or reattachment regions. It seems clear that computations based on Reynolds averaged characteristics and current turbulence models are not designed to give detailed results for such interactions. The whole unsteady, turbulent, dissipative, interaction region will have a major effect on the flow development downstream, where the computations have been shown to be far from correct.

2. Three-Dimensional Flows

It is important to note that, in the three-dimensional studies, all of the data sets are obtained along arbitrary two-dimensional planes, none of which are directly connected to the complex flowfield. Whether the planes are parallel to the freestream direction or normal to the shock (the two most widely used coordinates), no streamline remains in that plane for any distance. Each cut is a "footprint," which makes interpretation and understanding of the flowfield quite difficult. Several critical observations can be made from an examination of the pressure distribution data of Fig. 5. The gradients experienced in three-dimensional interactions are very much smaller than those experienced in two-dimensional interactions for the same strength shock waves. As the shock wave strength increases, the gradients in the interaction increase, in direct opposition to the characteristics in two dimensions. The scale of the three-dimensional interactions, when considered in boundary layer thickness terms, is much larger. However, since the three-dimensional flowfield grows in a lateral dimension, comparison with two-dimensional studies requires an understanding of this three-dimensional pattern. There is no a priori "direct comparison" that can be made because the scale of the interaction depends on the position in the flowfield for three-dimensional flows.

One of the main connections to two-dimensional flows has been the reference to "separation." From surface flow visualization, a line of convergence of surface streaks is found upstream of the shock location for shock wave strengths (for the present conditions) above that generated by an 8° fin. This line of convergence has generally been called "separation" although there is, for some conditions, another line further downstream in the interaction which has been labeled "secondary separation," Ref. 16. Korkegi, Ref. 17, has noted that the convergence line occurs at a static pressure ratio of approximately 1.5.

The flowfield surveys were presented in the form of contour maps in Fig. 8. Along the horizontal scale, the characteristics noted from the surface flow visualization have been indicated. "Secondary separation" has been observed from surface flow visualization for the case of thin boundary layers, but was not found for the same shock wave conditions for thick boundary layers (presented data) and, thus far, has not been indicated by the computations.

The unsteady characteristics of the three-dimensional flows was indicated in Fig. 9. The level of unsteadiness, as measured by pressure fluctuations, is about half that found for the two-dimensional case for the same strength shock wave. The initial region is still "rippled" and the effect of this unsteadiness has yet to be evaluated in any critical way.

The computations of Knight and Horstman, Ref. 18, was the culmination of a whole series of studies and provided, for the first time, a direct assessment of assumed turbulence models. In summary, it was found that, in a large part of the outer region of the interaction, (not close to the wall), the results of the two computations, both carried out with refined grid systems but with different turbulence models, agreed reasonably well with each other and with the experiments. The agreement of the two sets of computations took place even though the local eddy viscosity varied by a factor of almost 20 in some of the calculations. This is a clear indication that this outer flow is primarily an inviscid rotational flow. Close to the surface, the two computations with different turbulence models no longer agreed, and there was some significant difference with the experimental results. The prediction of the line of convergence differed considerably between the two computations, and did not agree with the experiments. Outer flow streamlines determined from the computations agreed well with each other, even with the significant differences close to the surface. The computations were used to construct a flowfield model shown in Fig. 10. The primary feature is that a large part of the initial boundary layer wraps up to form a vortical structure along the direction of the generated shock wave. Above this surface, the initial flow flows over the vortical structure, comes to the wall, and forms a new boundary layer, downstream of the interaction, which has little knowledge of the history of the initial boundary layer which is swept off laterally. The computations also show a second surface, starting near the line of convergence, which wraps up in the vortical structure. It should be noted, however, that the details close to the wall do not have the same "validation" of check with experiments or independence from turbulence models.

3. "Separation"

The concept of "separation" in three-dimensional flows is primarily based on the observations of surface flow visualization, in particular, the line of convergence. The early studies of topology, Refs. 19 & 20, and many which followed, considered flows separated when such lines of convergence were observed or when surface streaks approached the shock wave from the "downstream" direction. In more recent work, Refs. 21 & 22, it has been shown that this surface feature is a necessary but not sufficient condition to support the contention that there is a "separation sheet." The computations of Ref. 18 indicated a sheet which leaves the surface at the line of convergence, Fig. 10. The computations show that the sheet noted is a surface developed by streamlines which come, in large part, from a region near the apex and which were above the surface in the initial flow. Surface streamlines enter the sheet (see dashed lines) and remain parallel to the surface. The computed contours of pitch angle close to the surface show no change in passing over the surface feature of the line of convergence. George Inger, who has studied these three-dimensional flows analytically for some time, has recently, Ref. 23, carried out a specific analysis at the line of convergence and finds no vertical velocity.

Tran, in a new study, Ref. 13, has shown that the combination of smaller gradients and lower level of flow unsteadiness suggests that the strongest three-dimensional interaction studied, that of the 20° fin, is the equivalent of only a 12° wedge.

Flowfield visualization, using a vapor screen technique, has also been quoted as supporting the "separation" concept, Ref. 24, Fig. 11. It must be noted that this technique illuminates a plane in which no particle remains. No particle follows the illuminated path. Vapor screen techniques for these three-dimensional flows give no indication of vertical velocity components, nor can they define the structure of the flow since the flowfield normal to the plane of observation is not observed.

If one considers the physics of the so-called "separated flow," there appears to be no adverse pressure gradient which would cause any surface streamline to leave the surface. In contrast to two-dimensional flows, the low energy air in the lower part of the entering boundary layer simply turns and flows along lines of constant pressure. It should also be noted that the three-dimensional flows show very little dissipation. The flow is almost entirely supersonic, and a major area of the flow has been shown to be primarily inviscid-rotational. This is in major contrast to two-dimensional flows where there is significant dissipation and an inability to compute because of the inadequacy of current dissipation models. Three-dimensional flows of the type examined seem to have much in common with the "secondary flows" so often observed in ducts.

4. Vorticity Model

Stimulated primarily by the streamline tracing studies of Horstman and Knight, Ref. 18, and the observation that the flows are primarily inviscid, the concept of vorticity rearrangement has been developed. The vorticity in the original two-dimensional stream must be re-organized in the imposed three-dimensional flowfield. The original two-dimensional vorticity is re-formed into a vortical structure, parallel to the constant pressure lines which are oriented approximately along the shock direction. The interaction grows laterally as more fluid is entrained (with its vorticity). As seen from the computations, the outer surface (which separates the flow which enters into the vortical structure and that which passes over it) is near the outer edge of the boundary layer and encompasses the vorticity of the original flow. The flow from the top of the boundary layer, which probably still includes some of the outer edge of the larger scale turbulent structure, passes over the laterally flowing vortical structure, flows downward to the wall, and continues downstream. The original heat transfer data of Oskam, Ref. 25, shows very high heat transfer rates in the downstream region which are probably due to this high energy air being brought close to the surface, probably still turbulent, and with only a thin viscous layer developed after the laterally moving vortical structure.

If the vortical structure concept is correct, then all three-dimensional flowfields of the type examined should be considered, not as viscous interaction regions, but vortical interaction regions which can be computed and perhaps even controlled. This concept suggests that the turbulent structure, so important to two-dimensional flows, would not be a primary consideration for three-dimensional flows except close to the surface. Here they would be important in the deterination of heat transfer and skin friction.

IV. Conclusions/Observations

On the basis of the present analysis of specific detailed two and highly swept three-dimensional shock wave turbulent boundary layer interactions, it has been observed:

1) Three-dimensional flows are radically different than the classical two-dimensional flows which have been the basis of interaction structures modeling in the past. There are major differences in scale, unsteadiness, dissipation, computability, and viscous or inviscid importance.

2) Three-dimensional flow interactions are of large scale compared to their two-dimensional counterparts. The gradients normal to the shock wave are much less for the same shock strength and the extent of the interaction is relatively independent of shock strength. The unsteadiness is only half that of the equivalent two-dimensional interaction. Computations show little effort of turbulence models, indicating these flows are primarily inviscid-rotational dominated in contrast to two-dimensional flows.

3) Three-dimensional "separation" is not a realistic designation of flows which have been so described on the basis of surface flow visualization. None of the direct measurements, calculations, or computations have established any vertical velocity components close to the surface, and computation without dissipation can describe the flow structure.

4) A vorticity rearrangement model is proposed as the basic physics of the interaction on the basis of calculations and the observed experimental data.

References

1. Green, J. E.: Interactions Between Shock Waves and Turbulent Boundary Layers. Progress in Aerospace Sciences, Vol. 11, 1970. D Kuchemann edt.
2. Peake, D. J.; Tobak, M.: Three-Dimensional Interactions and Vortical Flows with Emphasis on High Speeds. NASA TM 81169, 1980.
3. Settles, G. S.; Dolling, D. S.: Swept Shock Wave/Boundary-Layer Interactions. Tactical Missile Aerodynamics Volume, AIAA Progress in Astronautics and Aeronautics Series, M. Hemsch and J. Nielsen, eds. To be published 1986/87.
4. Settles, G. S.; Fitzpatrick, T. J.; Bogdonoff, S. M.: Detailed Study of Attached and Separated Compression Corner Flowfields in High Reynolds Number Supersonic Flow. AIAA Journal, Vol. 17, No. 6, June 1979.
5. Smits, A. J.; Hayakawa, K.; Bogdonoff, S. M.: Turbulence Measurements in Two Shock-Wave/Shear-Layer Interactions. Presented at the I.U.T. A.M. Symposium on the Structure of Complex Turbulent Shear Flow, Marseille, 1982. Published by Springer-Verlag.
6. Settles, G. S.; Bogdonoff, Seymour M.; Vas, Irwin E.: Incipient Separation of a Supersonic Turbulent Boundary Layer at High Reynolds Numbers. AIAA Journal, Vol. 14, No. 1, January 1976.
7. Settles, G. S: An Experimental Study of Compressible Turbulent Boundary Layer Separation at High Reynolds Numbers. Ph. D. Dissertation, Princeton University, September 1976.
8. Dolling, D. S.; Murphy, M. T.: Unsteadiness of the Separation Shock Wave Structure in a Supersonic Compression Ramp Flowfield. AIAA Journal, Vol. 21, No. 12, December 1983.
9. Muck, K. C.; Dussauge, Jean-Paul; Bogdonoff, S. M.: Structure of the Wall Pressure Fluctuations in a Shock-Induced Separated Turbulent Flow. AIAA Paper No. 85-0179, January 1985.
10. Shapey, B.: Three-Dimensional Shock Wave Turbulent Boundary Layer Interaction for a 20° Sharp Fin at Mach 3. MSE Thesis, Princeton University, Oct. 1986, in preparation.
11. Tan, D.K.M.; Tran, T. T.; Bogdonoff, S. M.: Surface Pressure Fluctuations in a Three-Dimensional Shock Wave Turbulent Boundary Layer Interaction. AIAA Paper 85-0125, January 1985.
12. Tran, T. T.; Tan, D.K.M.; Bogdonoff, S. M.: Surface Pressure Fluctuations in a Three-Dimensional Shock Wave/Turbulent Boundary Layer Interaction at Various Shock Strengths. AIAA Paper 85-1562, July 1985.
13. Tran, T. T.: Experimental Investigation of Unsteadiness in Swept Shock Wave/Turbulent Boundary Layer Interactions. Ph.D. Thesis, Princeton University, October 1986, in preparation.
14. Gramann, R. A.; Dolling D. S.: Unsteady Separation in Shock Wave Turbulent Boundary Layer Interaction. AIAA Paper No. 86-1033, May 1986.
15. Chapman, D. R.; Kuehn, D. M.; Larson, H. K.: Investigation of Separated Flows in Supersonic Stream with Emphasis on the Effect of Transition. NACA Rept. 1356, 1958.
16. Zheltovodov, A. A.: Regimes and Properties of Three-Dimensional Separation Flows Initiated by Skewed Compression Shocks. Zhurnal Prikladnoi Mekhaniki i Tekhnicheskoi Fiziki, No. 3, May-June 1982.
17. Korkegi, R. H.: A Lower Bound for Three-Dimensional Turbulent Separation in Supersonic Flow. AIAA Journal, Vol. 23, No. 3, March 1985.

18. Knight, D.; Horstman, C.; Shapey, B.; Bogdonoff, S. M.: The Flowfield Structure of the 3-D Shock Wave-Boundary Layer Interaction Generated by a 20 deg Sharp Fin at Mach 3. AIAA Paper No. 86-0343, January 1986.
19. Maskell, E. C.: Flow Separation in Three Dimensions. R.A.E. Rept. Aero. No. 2565, 1955.
20. Lighthill, M. J.: Attachment and Separation in Three-Dimensional Flow. In Laminar Boundary Layers, Oxford University Press, 1963.
21. Dallman, U.: Structural Stability of Three-Dimensional Vortex Flows. Nonlinear Dynamics of Transcritical Flows, edited by Jordaon, H. L., Oretel, H., and Robert, K., Springer-Verlag, Berlin, 1984.
22. Chapman, Gary T.: Topological Classification of Flow Separation on Three-Dimensional Bodies. AIAA Paper No. 86-0485, January 1986.
23. Inger, G.: Private communication, January 1986.
24. Lu, F. K.; Settles, G. S.: Conical Similarity of Shock/Boundary Layer Interactions Generated by Swept Fins. AIAA Journal, Vol. 83, No. 7, July 1985.
25. Oskam, B.: Three-Dimensional Flow Fields Generated by the Interaction of a Swept Shock Wave with a Turbulent Boundary Layer. Ph.D. Dissertation, Princeton University, September 1976.

Acknowledgements

This work was supported by the U. S. Air Force Office of Scientific Research, Contract AF49620-84-C-0086, monitored by Dr. James D. Wilson.

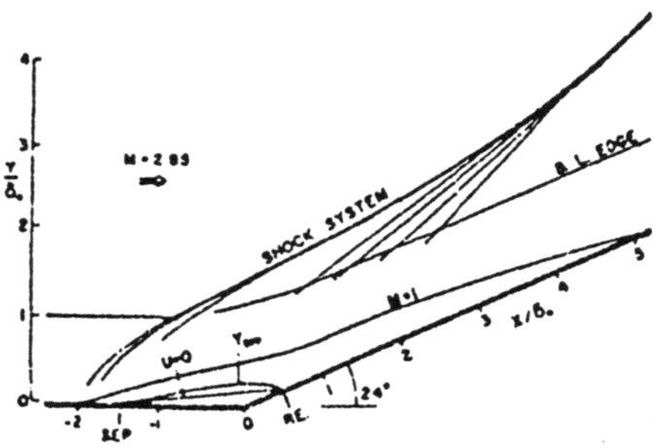

a) Flowfield model from surveys.

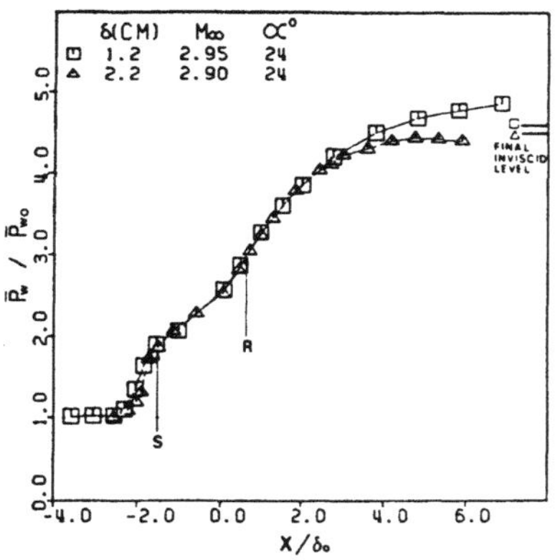

b) Mean wall pressure distribution.

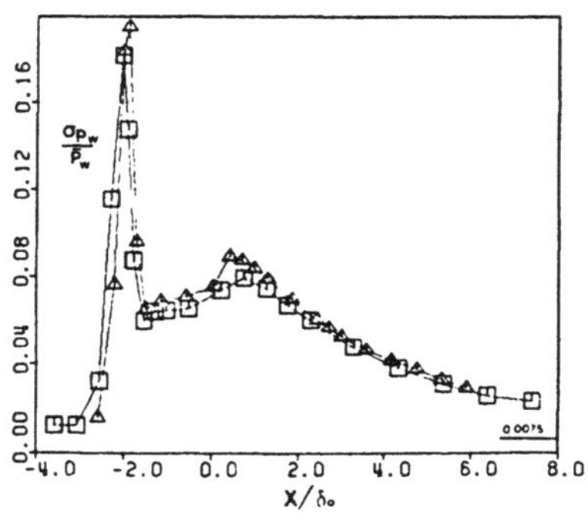

c) High frequency wall pressure standard deviation distribution.

Fig. 1. Characteristics of a 24° Compression Corner at M = 2.93.

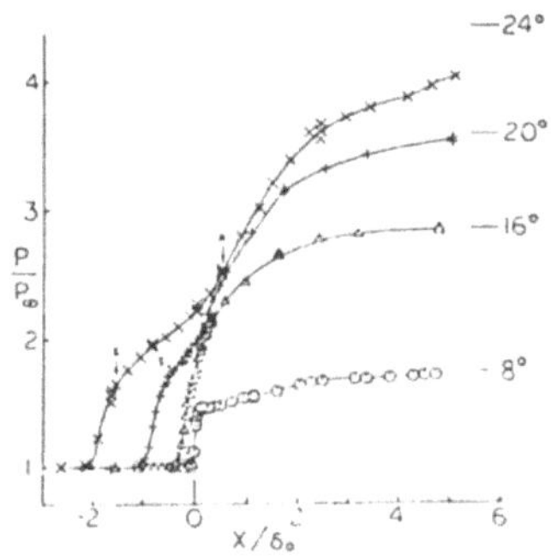

Fig. 2. Surface Pressure Distributions on the 8, 16, 20, and 24° compression corner models.

Fig. 3. Surface Flow Visualization for the 24° Compression Corner.

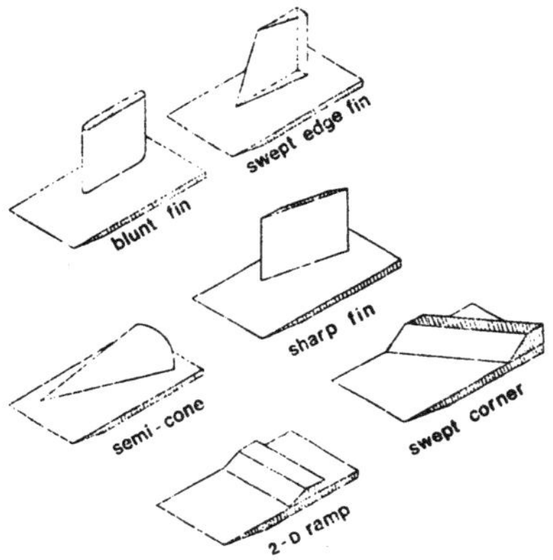

Fig. 4. Shock Generator Configurations Studied at the Gas Dynamics Lab.

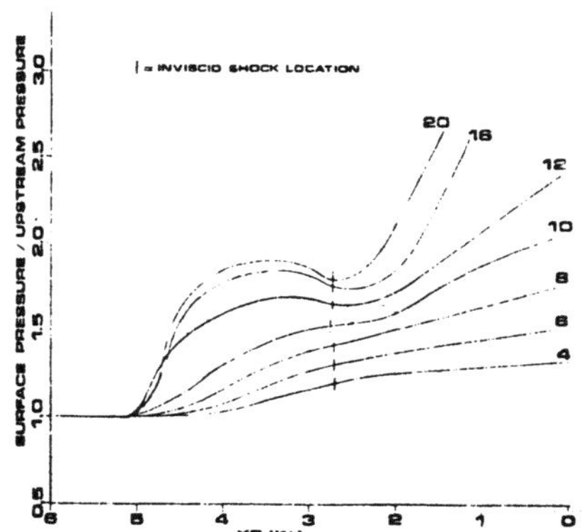

Fig. 5. Surface Pressure Distributions in Plane of Initial Flow for Sharp Fin Interactions of Different Strengths, α.

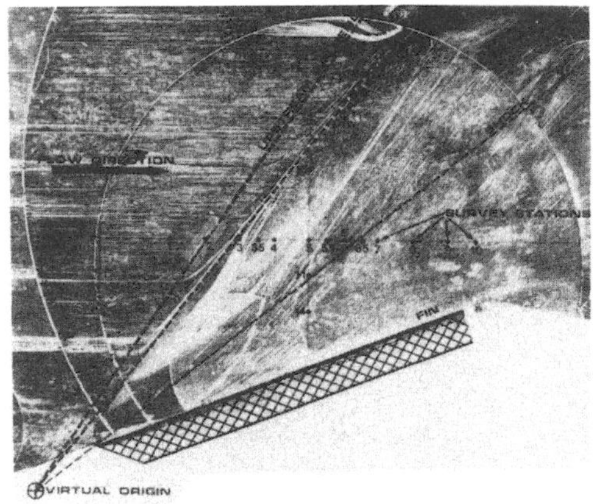

Fig. 6. Surface Flow Visualization for a 20° Fin.

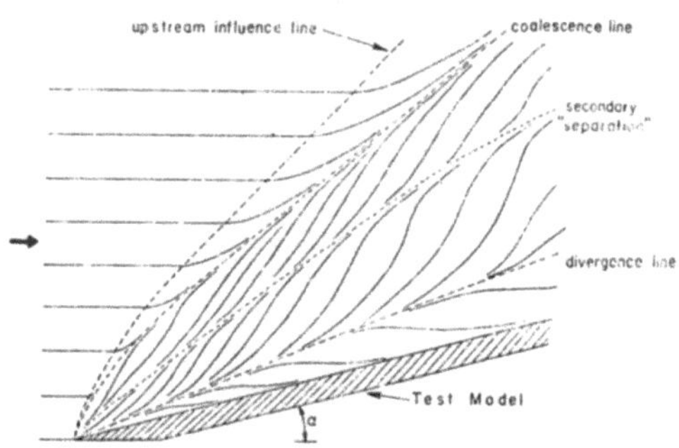

Fig. 7. A Sketch of Surface Patterns for Swept Interaction.

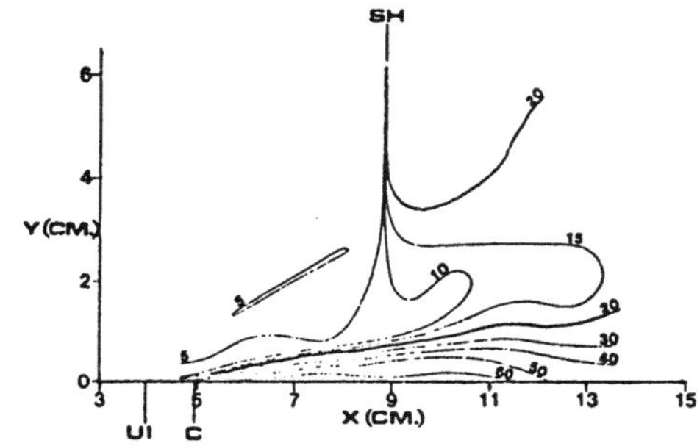

a) Yaw angle contours.

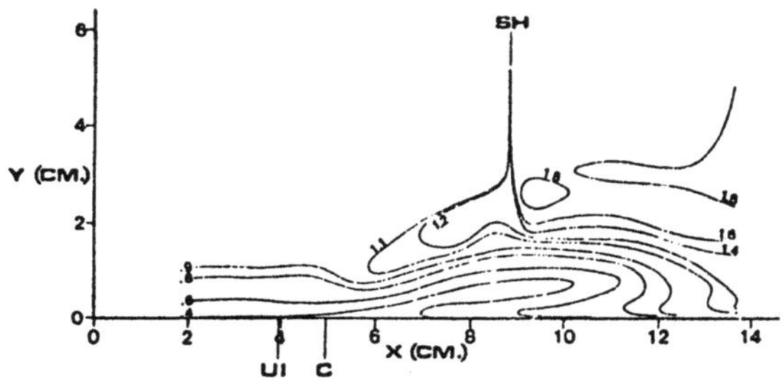

b) Normalized total head contours.

Fig. 8. Contour Plots, in Plane Parallel to Initial Flow, Through the 20° Fin Interaction.

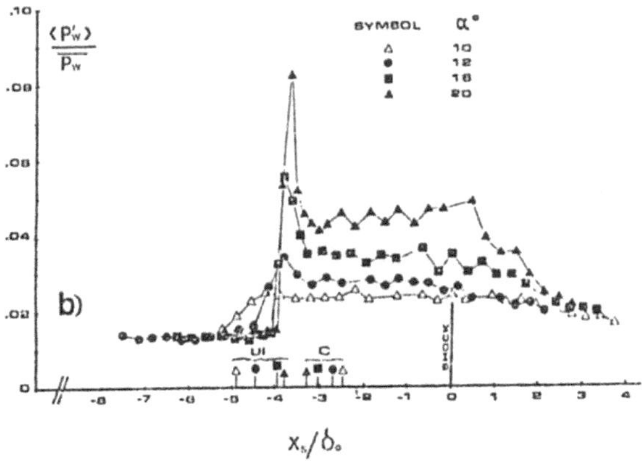

Fig. 9. Distribution of rms Wall Pressure Fluctuation for Several Fin Interactions (normalized by local mean pressure).

Fig. 10. Vapor Screen Photograph Purporting to Show Separation and Reattachment.

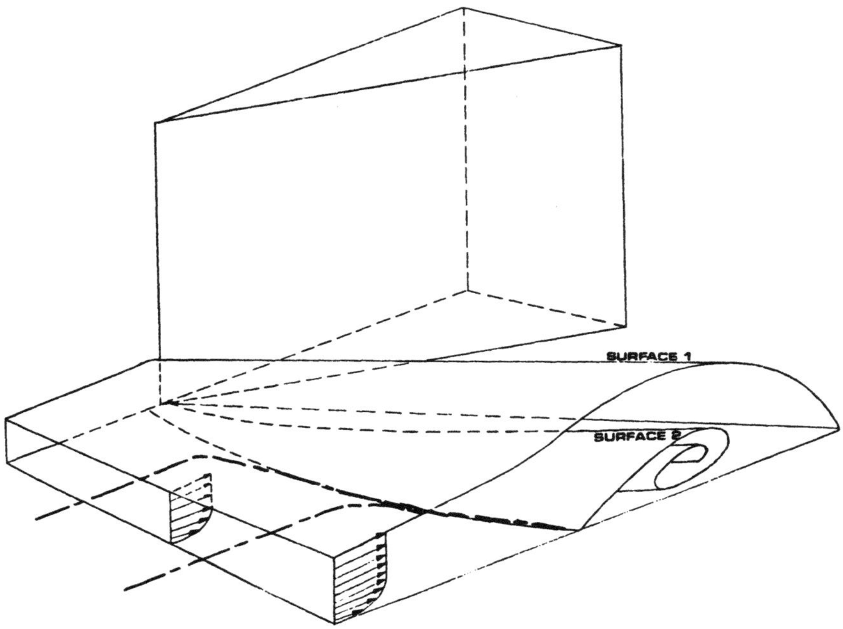

Fig. 11. Flowfield Structure From Computations.

Structure of Flow in Leading-edge Separation Bubbles

M. KIYA

Department of Mechanical Engineering
Hokkaido University
Sapporo, 060 Japan

Summary

This paper describes the turbulence and unsteady structures of the separation zone (separation bubble) at the leading edge of a two-dimensional blunt plate; the growth of the length scales of rolled-up vortices, the structure of large vortices in the reattaching zone of the separated shear layer, the unsteady flow associated with the motion of the large vortices, the low-frequency unsteady flow structure, etc. are presented and discussed. These properties are likely to be similar to those of a wide range of separation bubbles formed behind a salient edge.

Introduction

Flows at high Reynolds numbers with separation and reattachment (separation zones or separation bubbles) are characterized by strong rolled-up vortices formed in a separated shear layer bounding a large-scale recirculating turbulent region and by their interaction with a solid boundary. The extent to which this shear layer differs from a classic mixing layer between two streams is still uncertain. From a practical standpoint the turbulence structure in the reattaching zone of the shear layer is particularly important because the root-mean-square surface pressure and the heat-transfer rate attain a maximum in this zone. Previous works on the separation bubbles are mostly devoted to measurements of Reynolds stresses, velocity spectra and correlations, turbulence intermittency, and discussions of the turbulent energy balance; Bradshaw & Wong [1], Chandrsuda & Bradshaw [2], Eaton & Johnston [3], Troutt et al. [4] for separation bubbles behind backward-facing steps; Ota & Narita [5], Cherry et al. [6], Kiya & Sasaki [7][8] for the leading-edge

F. T. Smith, S. N. Brown, (Eds.)
Boundary-Layer Separation
IUTAM Symposium London 1986

separation bubble of a blunt plate; Ruaerich & Fernholz [9] for a separation bubble behind a normal plate with a long splitter plate. Only a few papers are concerned with the structure and motion of the rolled-up vortices and the unsteady flow structure in separation bubbles (Cherry et al. [6], Kiya & Sasaki [8][10], Troutt et al. [4]). A good deal of work is needed to understand the separation bubble flows well enough to predict their behaviour in general, to elucidate mechanisms of the high r.m.s. pressure and heat-transfer rate (Cherry et al. [6], Hourigan et al. [11], Cooper et al. [12]), and to manipulate them (Parker & Welsh [13], Bhattacharjee et al. [14]).

This paper describes the structure of the large vortices and the unsteady flow in a separation bubble formed at the leading edge of a blunt plate with right-angled corners, mainly based on results of Kiya & Sasaki [8][10] and Cherry et al. [6]. The geometry of flow is illustrated in figure 1. The x-axis is taken in the longitudinal direction along a side of the plate of thickness 2H, the y-axis normal to the side and the z-axis normal to the x- and y-axes. The velocity components in the x- y- and z-directions and the pressure are denoted by u, v, w, p which are divided into the time-mean components U, V, W (= 0), P and the fluctuating components u', v', w', p'. The velocity at infinity upstream is U_∞. The distance between the separation edge to the time-mean reattachment line is written as x_R, which will be used as a length scale in what follows.

Experimental results presented in this paper were obtained in a wind tunnel at Reynolds numbers (2.3-3.2) x 10^4. Measurements of the velocity fluctuations were made by single- and X-wire

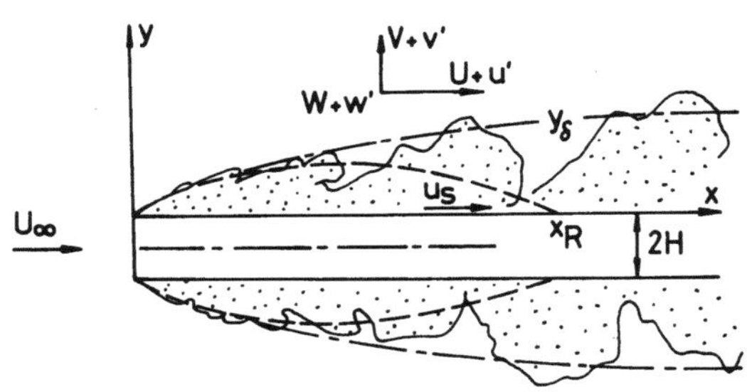

Fig. 1. Configuration of flow and definition of symbols

probes in regions without significant reverse flow and by split film probes in unsteady reverse flow regions. These probes were operated by constant-temperature hot-wire anemometers. Surface-pressure fluctuations were measured by semiconductor strain-gauged pressure transducers. The reattachment length x_R is (9.8-10.1)H.

Turbulence and vortex structures

The separated shear layer differs from a classic plane mixing layer between two streams in that the former is bounded at its low-velocity side by a highly turbulent reverse flow. A part of fluid with turbulence is convected upstream from the reattaching zone and entrained into the rolled-up vortices in the shear layer. This is manifested by the fact that the maximum r.m.s. longitudinal velocity normalized by the actual velocity difference increases all the way from the separation edge to the reattachment line. The same feature is found for a separation bubble behind a normal plate with a long splitter plate (Ruderich & Fernholz [9]) and that behind a backward-facing step (Eaton & Johnston [3]). Accordingly this is likely to be the case for a wide range of reattaching separated shear layers.

A measure of the longitudinal length scale of the rolled-up vortices can be obtained by the integral length scale of the velocity fluctuations u', v', w' and q' multiplied by the free-stream velocity. Here q' is the velocity fluctuation measured by a single hot-wire probe being normal to the xy-plane. As shown in figure 2, all of the four length scales are approximately the same, increasing approximately linearly with the longitudinal distance x. The latter feature is a result of the successive vortex coalescence, so that their being constant downstream of the reattachment line indicates the absence of further vortex coalescence. This is probably due to the effect of image vortices within the plate.

The rolled-up vortices are not two-dimensional but have a spanwise structure. The spanwise length scale of the vortices

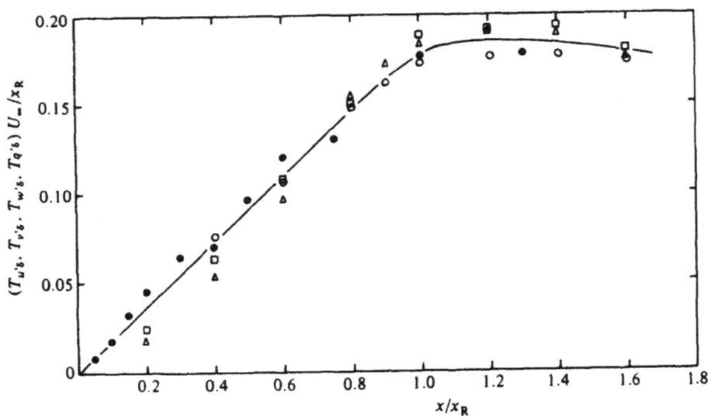

Fig. 2. Integral time scales at edge of shear layer: o, u'; Δ, v'; □, w'; ●, q'(single hot-wire normal to xy-plane)

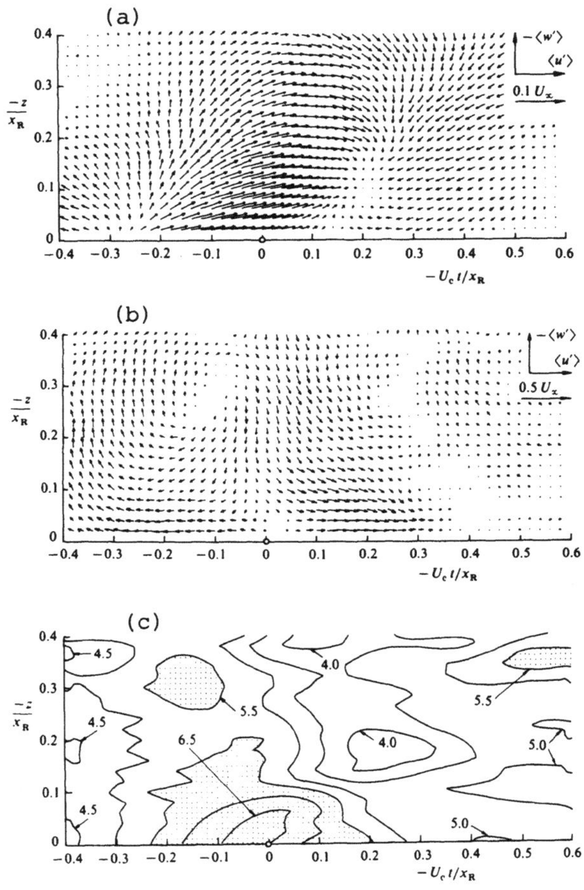

Fig. 3. Space-time (z, t) distribution of fluctuating velocity vector ($\langle u' \rangle$, $\langle w' \rangle$) at (a) edge and (b) centre of shear layer and (c) high-frequency energy $\langle u_T'^2 \rangle$ at centre of shear layer

grows approximately linearly up to the reattachment line and remains unchanged further downstream. This trend is reasonable because vortices of larger cross section are expected to have velocity fluctuations being more correlated in the spanwise direction. At the outer edge of the shear layer in the reattach-

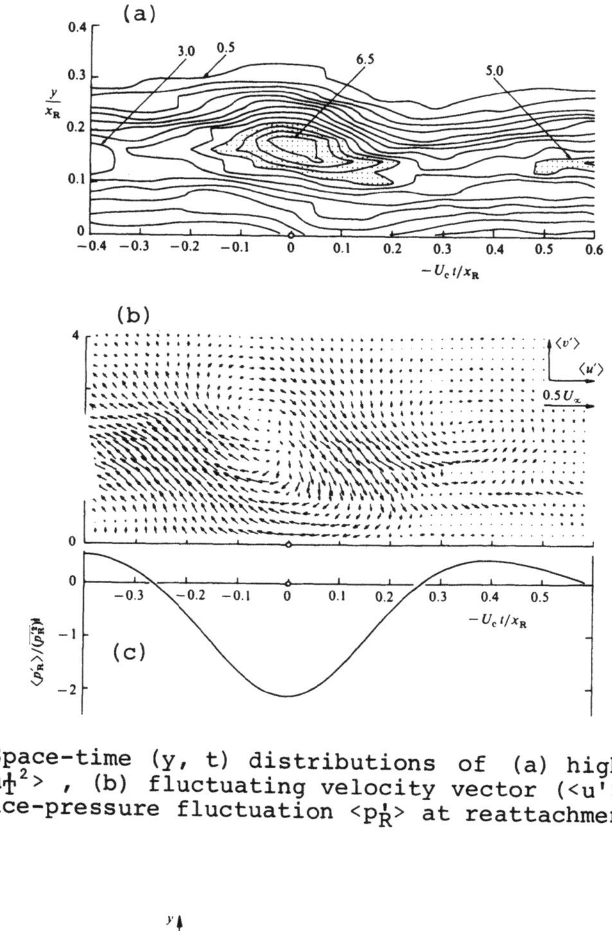

Fig. 4. Space-time (y, t) distributions of (a) high-frequency energy $\langle u_f^2 \rangle$, (b) fluctuating velocity vector ($\langle u' \rangle$,$\langle v' \rangle$) and (c) surface-pressure fluctuation $\langle p_R' \rangle$ at reattachment position

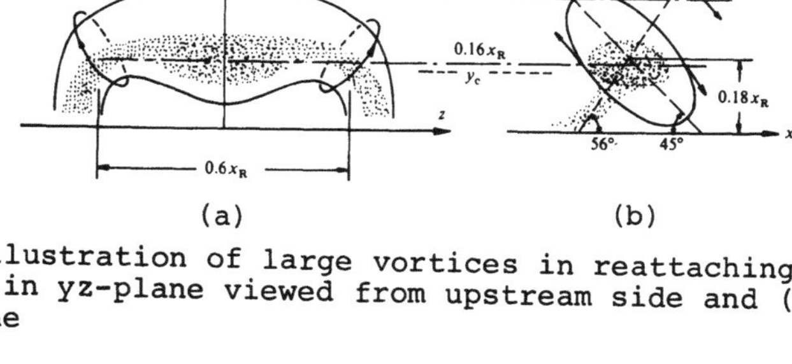

(a) (b)

Fig. 5. Illustration of large vortices in reattaching zone: (a) structure in yz-plane viewed from upstream side and (b) section in xy-plane ing zone, the spanwise length scale is approximately 2.6 times the longitudinal length scale which is the integral time scale multiplied by the local convection velocity.

It may be noted that the growth rate of the vorticity thickness of the shear layer is not linear all the way to the reattachment line (Kiya & Sasaki [8], Ruderich & Fernholz [9]). The initial growth rate of the shear layer is higher than that of the plane mixing layer, being followed by the growth rate which seems to be equal to that of the latter. The integral time scales in figure 2 are expected to reflect this tendency but the experimental uncertainty does not allow one to draw a definite conclusion about this.

The vortices are convected downstream with a velocity U_c which is equal to $(0.5\text{-}0.6)U_\infty$ in the reattaching zone of the shear layer. Downstream of the reattachment line the convection velocity slightly increases with increasing x. This increase is perhaps caused by the turbulent diffusion of vorticity, which reduces the circulation of the central part of the vortices and thus the velocity induced by the image vortices. The vortices are shed downstream from the reattaching zone with a frequency $(0.6\text{-}0.7)U_\infty/x_R$. The vortex shedding is shown by flow-visualization experiments (Cherry et al. [6]), the frequency being determined from the pressure spectrum of the surface pressure in the reattaching zone, that of u' at the edge of the shear layer, and the cross spectra of u' and v' in the shear layer; all these spectra have a broad peak at the above frequency. The shedding frequency and the convection velocity of the vortices give an estimate of the longitudinal distance between two consecutive vortices. This is $(0.8\text{-}0.9)x_R$ in and downstream of the reattaching zone.

The motion of the large vortices produces intense pressure fluctuations in the reattaching zone. A discrete-vortex simulation (Kiya et al. [7]) demonstrates that the fluctuating surface pressure is negative beneath a large vortex, attaining a definite minimum directly below its centre. This has suggested that the structure of the vortices can be obtained by conditionally averaging the fluctuating velocities with respect to the time when the minima lower than a threshold level are detected.

Figures 3 and 4 show the conditionally averaged velocity fluctuations (<u'>, <w'>), (<u'>, <v'>) and the high-frequency turbulent energy $\langle u_T'^2 \rangle$ in the space-time (z, t) and (y, t) domains, which are measured in the reattachment section $x = x_R$. Here u_T' is the velocity fluctuation at frequencies of the order of the Kolmogorov frequency. A region where $\langle u_T'^2 \rangle$ attains a maximum is assumed to be a region of large vorticity, viz. the central part of a vortex. This assumption might be justified by the fact that this high-frequency component of the velocity fluctuation is rotational. A large system of flow rotating in the clockwise sense is seen in figure 4(b). This can be interpreted as a real vortex because $\langle u_T'^2 \rangle$ attains a maximum near the centre of the rotating system, as seen in figure 4(a). The velocity-vector field (<u'>, <w'>) in the (z, t) domain near the edge of the shear layer (figure 3(b)) shows a centre rotating in the clockwise sense at $(-z, -U_c t)/x_R = (0.28, -0.1)$. The high-frequency energy attains a maximum near this centre (figure 3(c)). Moreover, from the symmetry, a centre rotating in the counterclockwise sense should exist at the image point (-0.28, -0.1); these centres constitute a pair of counter-rotating vortices. On the other hand, the vector field near the central part of the shear layer ($y/x_R = 0.32$) has a sink-like centre at $(-z, -U_c t)/x_R = (0, -0.25)$ and a source-like centre at (0.1, 0.2), see figure 3(b).

The most economical interpretation of the space-time distributions in figures 3 and 4 , although it may not be the only one, is that the large vortices in the reattaching zone are hair-pin vortices as depicted in figure 5. The large vortices are the most dominant feature of turbulence in the reattaching zone in that they contribute to most of the Reynolds stresses and the surface-pressure fluctuation. The Reynolds shearing stress is produced near the front and back edges of the vortices while its production is negligible in the central part of the vortices.

Unsteady structure of separation bubbles

The unsteady flow in the separation bubble is dominated by two agents; one is the motion of rolled-up vortices and the other is the low-frequency unsteadiness. The general feature of the unsteady flow can be represented by the reverse-flow intermittency I_r. Figure 6 shows I_r measured near the surface (y/x_R = 0.005) as a function of the longitudinal distance x. The time-mean reattachment position x_R has been defined as a position where $I_r = 0.5$. I_r never becomes unity even in the middle of the separation bubble. A sharp decrease of I_r towards the separation edge suggests the formation of a small secondary separation bubble with the opposite direction of recirculation. The unsteadiness can also be represented by the frequency f_s of switching of the local-flow direction, which is the number of switchingsfrom a forward to reverse flow (or from a reverse to forward flow) per unit time. This attains a maximum of $0.7U_\infty/x_R$ at the reattachment position, which is close to the vortex-shedding frequency.

Firstly the flow unsteadiness associated with the motion of the large vortices in the reattaching zone will be described. A reverse-flow region is defined as a region where the instantaneous longitudinal velocity u_s near the surface is negative; the velocity at the height y/x_R = 0.005 is chosen rather arbitrarily. This velocity is short-time averaged with an

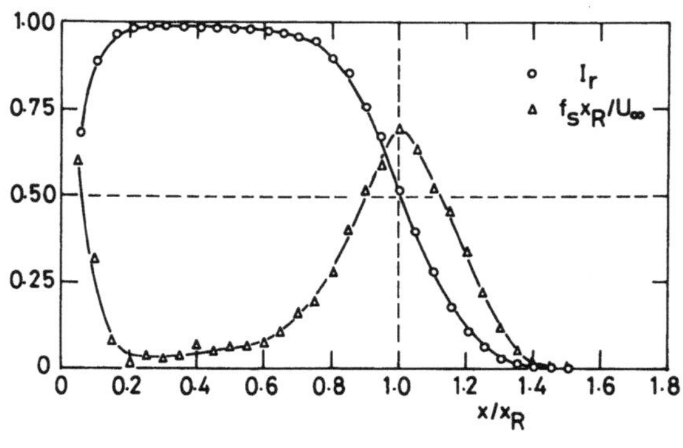

Fig. 6. Reverse-flow intermittency I_r and frequency of swithing of local-flow direction f_s measured at height y/x_R = 0.005

averaging time of about one-tenth of the vortex-shedding period. A boundary between a reverse-flow region and a forward-flow region is described by either $u_s = 0$, $du_s/dt < 0$, t being time, or $u_s = 0$, $du_s/dt > 0$, when u_s is measured at a fixed position x as a function of time. The former is denoted by zero A and the latter by zero B. If the spatial distribution of u_s were measured, zeros A and B would be identified as positions of $u_s = 0$, $du_s/dx > 0$ and $u_s = 0$, $du_s/dt < 0$.

The motion of zeros A and B is obtained in the following way. At a given position x near the reattachment line, the velocity u_s is sampled to find the zeros. The time at which zero A or B is detected is then employed as the synchronization time to phase-average the surface-pressure fluctuation p'_R at the reattachment position. A peak appears in each phase-averaged pressure at a certain time t_P. On the assumption that the peak corresponds to a particular phase of motion of the large vortices, the time t_P is taken as a origin to measure the time when zero A or B appears. This gives the motion of the zeros in the space-time $(x - x_R, t)$ domain, which is shown in figure 7. There exists one reverse-flow region or two of it in the reattaching zone, the number depending on the phase of motion of the large vortices. Zeros B move with a speed of $0.7U_\infty$, which is much greater than the speed of $0.3U_\infty$ of zeros A. The average speed $0.5U_\infty$ is close to the convection velocity of the vortices. The extent to which the motion of zeros A and B are coherent in the spanwise direction is uncertain.

The second unsteadiness has frequencies less than approximately $0.25U_\infty/x_R$, the central frequency being $0.12U_\infty/x_R$ which is approximately one-sixth of the central vortex frequency. The existence of this low-frequency unsteadiness is indicated by (i) a long tail of the autocorrelation of u' near the centre of the shear layer at $x/x_R = 0.2$, (ii) high energy contained in the low-frequency range of the surface-pressure fluctuation at $x/x_R = 0.2$ (figure 8), (iii) a high value of the cross correlation of $0.25U_\infty/x_R$ Hz low-pass filtered surface-pressure fluctuations at two largely separated positions $x/x_R = 0.2$ and 1.0, and (iv) a spatial distribution of the cross correlation of the low-pass

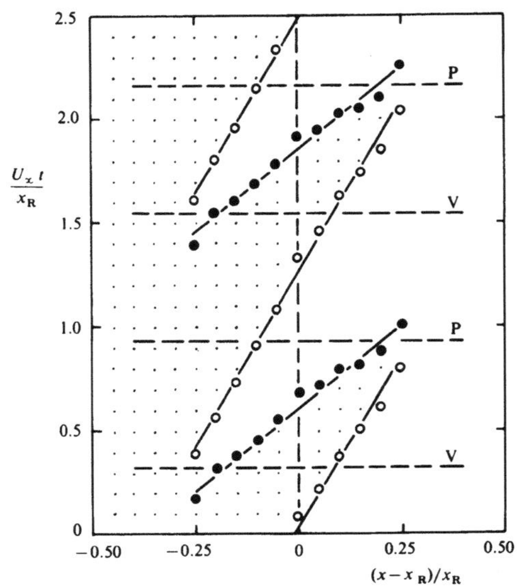

Fig. 7. Motion of zeros A and B of longitudinal velocity near surface in space-time ($x - x_R$, t) domain: o, zero A; •, zero B

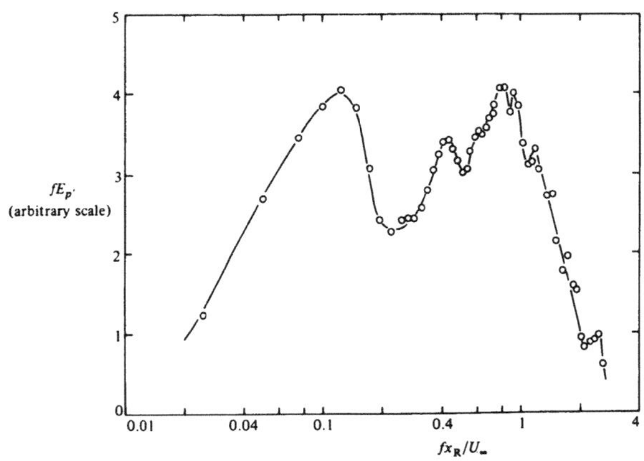

Fig. 8. Surface-pressure spectrum at position $x/x_R = 0.2$

filtered longitudinal velocity fluctuation at various positions in the shear layer with the surface-pressure fluctuation at a fixed position $x/x_R = 0.2$. The feature (iii) indicates that the low-frequency unsteadiness is of large scale in the sense that its effect is felt throughout the separation bubble. Moreover the feature (iv) indicates that the low-frequency

unsteadiness is accompanied by a flapping motion of the shear layer in the first half of the separation bubble.

The flapping motion is confirmed in terms of the longitudinal velocity profiles which are conditionally averaged only when the low-pass filtered surface-pressure fluctuation at x/x_R = 0.2 attains maxima greater than a threshold level and minima lower than another one. This procedure is based on the idea that the flapping motion of the shear layer has a definite phase relation to the surface-pressure fluctuations in the first half of the separation bubble. The flapping motion is found to have an amplitude of the order of $0.005x_R$. This rather small amplitude, however, is sufficient to yield a u' fluctuation of the order of $0.1U_\infty$ in the central part of the shear layer.

The low-frequency unsteadiness is accompanied by shrinkage and enlagement of the separation bubble. The shrinkage and enlargement can be demonstrated by the motion of zeros of the $0.25U_\infty/x_R$ Hz low-pass filtered longitudinal velocity u_{s1} near the surface. There are two kinds of zeros; one is characterized by $u_{s1} = 0$ and $du_{s1}/dt < 0$ (denoted by zero A_1) and the other by $u_{s1} = 0$ and $du_{s1}/dt > 0$ (denoted by zero B_1). In the same way as in determining the zeros A and B, the motion of zeros A_1 and B_1 is obtained with reference to the time when the maxima and minima (denoted by P_1 and V_1) are attained by the low-pass filtered surface pressure p'_{R1} at the reattachment position. The result is shown in figure 9. It is seen that zeros A_1 move upstream while zeros B_1 move downstream. The speed of A_1 is much lower than that of B_1; in other words the shrinkage is swifter than the enlargement, as has been suggested by Eaton & Johnston [3].

The low-frequency unsteadiness also seems to be accompanied by a change in the strength of the large vortices. Figure 10 shows the spectra of the surface-pressure fluctuation p'_R at the reattachment position, which are separately obtained in the shrunk state and in the enlarged state. The spectrum for the shrunk state, for example, is obtained from the waveform during

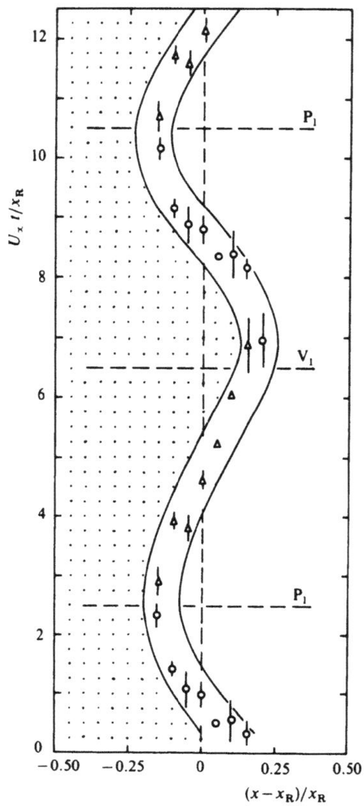

Fig. 9. Motion of zeros A_1 (Δ) and B_1 (o) of longitudinal velocity near surface in space-time ($x - x_R$, t) domain

about a half of the central period of the low-frequency unsteadiness (centred around the time of the maxima P_1 of p'_{R1}). In the same way the spectra for the enlarged state is calculated for the minima V_1. The spectra in figure 10 have a peak at the vortex-shedding frequency $0.6U_\infty/x_R$. The important point is that the spectrum peak is much higher for P_1 than for V_1. This suggests that the large vortices shed during the shrunk state are stronger than those shed during the enlarged state because, the stronger the vortices, the greater is the surface-pressure fluctuation associated with the motion of the vortices.

The low-frequency unsteadiness is also found for separation

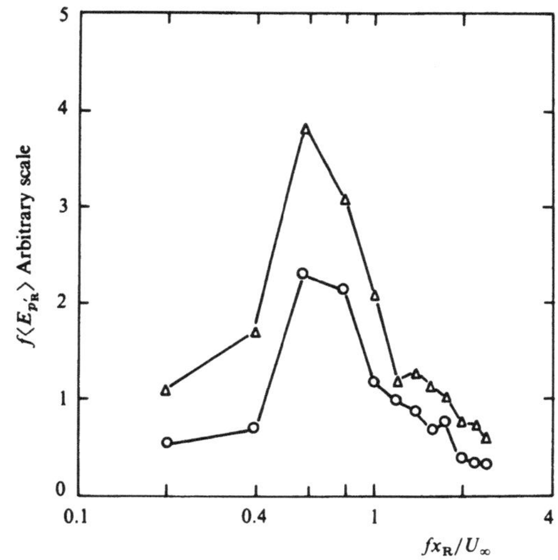

Fig. 10. Pressure spectra for enlarged (Δ) and shrunk (o) states of separation bubble due to low-frequency unsteadiness

bubbles behind a normal plate with a long splitter plate (Cherry et al. [6]), behind a backward-facing step (Eaton & Johnston [3]) and at the leading edge of a blunt circular cylinder (Kiya & Nozawa [15]), and even for the blunt-plate separation bubble with turbulent free streams (Cherry et al. [6]). Accordingly this is probably an inherent feature of separation bubbles in general. An exact mechanism of the low-frequency unsteadiness is uncertain. Eaton & Johnston [3] suggest that this is caused by an instantaneous imbalance between the entrainment from the recirculation zone and the reinjection of fluid in the reattaching zone. Moreover the fair periodicity of the low-frequency unsteadiness in figure 10 seems to indicate a feedback of disturbances from the reattaching zone to the shear layer near the separation line (Rockwell [16]).

Conclusion

This paper has discussed, for the separation bubble at the leading edge of a bluntplate, the growth of the length scales of the rolled-up vortices,the structure of the large vortices in the reattaching zone, their dynamical significance, the unsteady flow associated with the motion of the vortices, and the structure of the low-frequency unsteadiness. It is likely that these properties are basically similar in a wide variety of separation bubbles formed behind a salient edge. Finally it should be emphasized that the surface-pressure fluctuation can be employed as a convenient reference signal to deduce the structure of large vortices in the separation bubble. This is probably the case whenever large strong vortices exist near a solid boundary.

References

1. Bradshaw, P.; Wong, F.Y.F.: The reattachment and relaxation of a turbulent boundary layer. J. Fluid Mech. 52(1972) 113-135.

2. Chandrsuda, C.; Bradshaw, P.: Turbulence structure of a reattaching shear layer. J. Fluid Mech. 110(1981) 171-194.

3. Eaton, J.K.; Johnston, J.P.: Turbulent flow reattachment: An experimental study of the flow and structure behind a downward facing step. Rep.MD-39(1980), Thermosciences Div., Dept. Mech. Engng, Stanford University.

4. Troutt, T.R.; Scheelke, B.; Norman, T.R.: Organized structures in a reattaching separated flow field. J. Fluid Mech. 143(1984) 413-427.

5. Ota, T.; Narita, M.: Turbulence measurements in a separated and reattached flow over a blunt flat plate. ASME J. Fluids Engng 100(1978) 224-228.

6. Cherry, N.J.; Hillier, R.; Latour, M.E.M.P.: Unsteady measurements in a separated and reattaching flow. J. Fluid Mech. 144(1984) 13-46.

7. Kiya, M.; Sasaki, K.: Discrete-vortex simulation of a turbulent separation bubble. J. Fluid Mech. 120(1982) 219-244.

8. Kiya, M.; Sasaki, K.: Structure of a turbulent separation bubble. J. Fluid Mech. 137(1983) 83-113.

9. Ruderich, R.; Fernholz, H.H.: An experimental investigation of a turbulent shear flow with separation, reverse flow, and reattachment. J. Fluid Mech. 163(1986) 283-322.

10. Kiya, M.; Sasaki, K.: Structure of large-scale vortices and unsteady reverse flow in the reattaching zone of a turbulent separation bubble. J. Fluid Mech. 154(1985) 463-491.

11. Hourigan, K.; Welsh, M.C.; Welch, L.W.: Augmented forced convection around a heated flat plate. 23rd AIChE/ASME National Heat Transfer Conference, Denver, Colorado, August 4-7, 1985.

12. Cooper, P.I.; Sheridan, J.C.; Flood, G.J.: The effects of sound on forced convection over a flat plate. Int. J. Heat & Fluid Flow 7(1986) 61-68.

13. Parker, R.; Welsh, M.C.: Effects of sound on flow separation from blunt flat plate. Int. J. Heat & Fluid Flow 4(1983) 113-127.

14. Bhattacharjee, S.; Scheelke, B.; Troutt, T.R.: Modification of vortex interaction in a reattaching separated flow. AIAA Journal 24(1986) 623-629.

15. Kiya, M.; Nozawa, T.: Turbulence structure in the leading-edge separation zone of a blunt circular cylinder. Trans. JSME (in Japanese) (to be published).

16. Rockwell, D.: Oscillations of impinging shear layers. AIAA Journal 21(1983) 645-664.

Computation of Unsteady Separation Using Lagrangian Procedures

L. L. VAN DOMMELEN

FAMU/FSU College of Engineering
Dept. of Mechanical Engineering
Florida State University, Tallahassee

Summary
The interaction between increased knowledge of unsteady separation processes and computational techniques is examined, emphasizing the use of Lagrangian coordinates. Examples include boundary layer solutions of two dimensional unsteady flows, and separation from rotating and translating spheres. Advances in Lagrangian techniques for the Navier-Stokes equations are illustrated by means of high resolution random-walk computations of the impulsively started circular cylinder.

Introduction

The computation of unsteady boundary layer separation (or more precisely, "break-away" [1]) has proven to be difficult for both conventional [2-5] and less conventional [6,7] numerical procedures. Van Dommelen & Shen [8,9] showed that the Lagrangian approach can be a powerful tool to compute such flows.

Lagrangian approaches align the mesh with the fluid particles, rather than choosing a definite spatial location. By following the particles, convective processes can be accurately evaluated, which conventional procedures must estimate indirectly. Since it is believed that unsteady separation processes are dominated by convection in their final stages [8-10], Lagrangian schemes can have a significant advantage.

Yet for the computation to be meaningful, it is not sufficient to evaluate the convection processes accurately. For, though the actual separation is nearly inviscid, the original boundary layer forms through viscous diffusion. Furthermore, the viscous processes must initiate the second Moore-Rott-Sears condition of vanishing vorticity, a necessary condition for separation to occur [1]. Any good procedure must therefore be able to evaluate both the regular viscous diffusion effects and the singular convection accurately.

But the viscous terms are Eulerian in nature, and are difficult to evaluate on a Lagrangian mesh that tends to become highly distorted when seen in an Eulerian coordinate system.

F. T. Smith, S. N. Brown, (Eds.)
Boundary-Layer Separation
IUTAM Symposium London 1986

A number of solutions such as reconnecting the mesh points, relocating them, or restricting their motion have been used (eg. [11]), with varying success. However, Van Dommelen & Shen found that for processes with a limited evolution time, such as unsteady separation, a more simple regular mesh structure gives surprisingly good accuracy. Part of the reason is that away from separation, the mesh cells remain well defined for limited times. Near separation, the mesh cells do become highly distorted but the viscous terms are small and relatively unimportant near the onset of separation.

The evaluation of the viscous terms could destroy the computation regardless: many "finite volume" like approaches write the equations of motion for Eulerian, polygon shaped regions between mesh points. Such discretizations have the volume of the polygon in the denominator, and when this discretisized volume vanishes, the results become indeterminate. The present procedure avoids this, since the volume in the denominator is implicitly evaluated from continuity, rather than from differencing.

But the success of the Lagrangian approach is not merely due to the fact that the viscous terms are small, so that any non-singular evaluation will do. According to the asymptotic descriptions found by Van Dommelen & Shen [8] and Van Dommelen & Cowley [12], the Lagrangian solution remains regular at separation, and the viscous terms are in fact correctly evaluated.

An additional advantage of Lagrangian coordinates is that the fluid drifts toward the regions of separation, so that local resolution increases sizably (cf. [8,12]).

It may seem a simple procedure to solve the Lagrangian equations on a regular mesh structure but actually the variations in the coefficients is such that many ordinary methods do not perform in the usual way. Gauss-Seidel must be modified to account for the large first order terms [9]; multigrid works but results are disappointing [unpublished]. The presently favored approach is a time linearized method which proves to be fairly stable.

Unlike other Lagrangian algorithms, the regular mesh is limited to finite evolution times. Clearly, it is possible to remove this objection by interpolating a complete new mesh periodically. However, in order not to lose the Lagrangian advantage over Eulerian schemes, such a mesh should be solution adaptive; it should be able to resolve the already computed solution. The field of adaptive mesh generation is undergoing rapid growth

[13], and it is hoped that such schemes will be developed in the near future.

For the full Navier-Stokes equations, the pressure gradient is no longer a priori known, as it is for the boundary layer equations. This poses another difficulty for Lagrangian computations, since the velocity field is now related to the vorticity by means of the Laplace equation which is Eulerian in nature. Use of a Lagrangian mesh to solve this equation appears inadvisable, since the convection describes the separation process and must be evaluated accurately.

Dropping the vorticity down on an Eulerian mesh and solving it on that mesh seems in principle possible, but such a mesh would have to be solution adaptive in order that the advantages of the Lagrangian computation are retained. In this paper a new approach is proposed which solves the Laplace equation directly.

Lagrangian Boundary Layer Scheme

The Lagrangian boundary layer equations for axially symmetric flow about a unit sphere are [12]:

$$\dot{u} - w^2 \cot g\, \theta = -\frac{1}{\rho} p_x + D_y{}^2 u \tag{1a}$$

$$\dot{w} + uw \cot g\, \theta = D_y{}^2 w \tag{1b}$$

$$\dot{\theta} = u \qquad D_y\, y = 1 \tag{1c,d}$$

where θ is the polar angle measured from the axis of symmetry and u and w are the meridional and azimuthal velocities respectively.

The y-derivatives are described by

$$D_y = \frac{\sin\phi}{\sin\theta}\left[\frac{\partial\theta}{\partial\phi}\frac{\partial}{\partial\eta} \quad \frac{\partial\theta}{\partial\eta}\frac{\partial}{\partial\phi}\right] \tag{2}$$

where ϕ is the polar position of the particle at time t=0 and η is the radial distance from the wall at that time, (with the usual $\sqrt{\nu}$ boundary layer scaling).

The initial conditions were either taken to be

$$u = \tfrac{3}{2} V \sin\theta \qquad w = 0 \qquad (\eta \neq 0,\ t = 0) \tag{3a,b}$$

corresponding to flow about a sphere in translational motion, or

$$u = 0 \qquad w = \Omega_{inf} \sin\theta \qquad (\eta \neq 0,\ t = 0)\ . \qquad (3c,d)$$

corresponding to a sphere in a rotating external flow. The wall boundary conditions were taken to be

$$u = 0 \qquad w = \Omega_{wall} \sin\theta \qquad (\eta = 0) \qquad (4a,b)$$

The advantage of excluding the combination of translation and a rotating external flow is that the external flow is steady. The sphere itself may perform a spinning motion in addition to a translating one.

The equations of motion may be simplified by introducing the following new variables:

$$U = u \sin\theta \qquad W = w \sin\theta \qquad (5a,b)$$
$$X = 1 - \cos\theta \qquad \xi = 1 - \cos\phi \qquad (5c,d)$$

This has the advantage of replacing the $\sin\theta$ and $\mathrm{cotg}\theta$ terms by algebraic ones. In addition, W corresponds to the azimuthal circulation; a quantity which is invisicidly constant on account of Kelvin's theorem.

To eliminate the singularity caused by the impulsive start, the computational coordinate

$$\beta = \frac{2}{\pi} \operatorname{atan} \frac{\eta}{\delta} \qquad \delta = \delta_0\ [t/(T+t)]^{\frac{1}{2}} \qquad (6a,b)$$

was introduced. Using this scaling, the Lagrangian partial difference equations reduce to ordinary ones at time $t = 0$, which were solved to find the initial conditions.

In the meridional direction, the Lagrangian independent coordinate was stretched by the simple transformation

$$\xi = 2[(1-\varepsilon)(3-2\alpha)\alpha^2 + \varepsilon\alpha] \qquad (6c)$$

where ε was rather arbitrarily taken to be 0.05. This scaling improves the resolution near the axis, which is originally small on account of the transformation from ϕ to ξ.

The equations of motion for U, W and X were individually discretisized following Beam & Warming [14], by time linearizing U, W and X respectively. This implies that the other two dependent variables in each equation were NOT time-linearized; the most recently computed values for these variables were used instead. In order to maintain second order accuracy, a predictor-corrector approach was adopted in which the three equations are solved twice. By using a backward-Euler predictor step, the inaccuracies on the predictor step are damped, while a Crank-Nicolson corrector step leads to second order accuracy for the final results.

It might seem as though linearization of the full system of equations would be preferable, but such an approach was tried and proved to be about 3 times slower, in addition to being significantly less stable numerically.

Rather than solving the difference equations iteratively, they were factored into α and β-operators, again following Beam & Warming [14]. The resulting tridiagonal systems were solved using the procedure described by Van Dommelen [15], which is efficient for vector computers such as the Florida State CYBER 205. For a 144 x 128 mesh, we computed an average computational speed of 126 MFLOPS; the effective machine limit is 200 MFLOPS (linked triads are not very useful for the Lagrangian equations of motion).

For three dimensional flows, the most meaningful quantity to describe the influence of the external flow onto the boundary layer flow is possibly the transpiration velocity.

$$v_\infty = \frac{\partial}{\partial x} \int_0^\infty (u_\infty - u)dy \tag{7}$$

where the integration is to be performed along the lines of constant x. In Lagrangian coordinates, these lines are described as:

$$dy : d\xi : d\eta = 1 : -x_\eta : x_\xi \tag{8}$$

In this study, we evaluated the transpiration velocity simply by doing the integral and taking x-differences.

Although the results will prove to be quite satisfactory, a number of further improvements seem possible on the flowing points: (a) the integration of the transpiration velocity; (b) the discretization of the second order cross derivatives; (c) the stretching of the ϕ-coordinate; (d) the evaluation of the wall shear. In a subsequent study, we hope to be able to elaborate on each of these points.

A Simple Lagrangian Navier-Stokes Scheme

The two-dimensional full vorticity diffusion equation can be written in Lagrangian form as:

$$\dot{\omega} = \nu \nabla^2 \omega \tag{9}$$

Although the Lagrangian boundary layer scheme seems the most promising procedure to integrate this equation, for the results presented here Chorin's Random Walk method [16] was used. His method has several advantages: (a) it is simple; (b) it does not employ any mesh structure at all. In addition, it was found that (c) the scheme appears quite robust.

Using this method, the flow about an impulsively started circular cylinder of unit radius was determined. In order to satisfy the no-slip condition, at each time step new vortex blobs were generated at the wall, an angular distance

$$\Delta\theta = 2\sqrt{\nu\Delta t} \tag{10}$$

apart. The vorticity distribution was taken to be

$$\omega \propto d^2/[r^2+d^2]^2 \tag{11}$$

with r^2 the distance from the center of the blob. The vortices were initially placed a distance from the wall equal to their diameter. It was found that changes in $\Delta\theta$, vortex diameter or initial distance from the wall have little influence on the results obtained.

The remaining difficulty is to find the velocity. Using mirror vortices, the exact solution is given by

$$u_i = \sum_j g_j \frac{y_i - y_j}{[(x-x_i)^2+(y-y_i)^2]^{\frac{1}{2}}} \tag{12a}$$

$$v_i = \sum_j g_j \frac{x_i - x_j}{[(x-x_i)^2+(y-y_i)^2]^{\frac{1}{2}}} \tag{12b}$$

These sums can be evaluated at maximum speed on a vector computer such as the CYBER 205. However, the number of operations is proportional to the square of the number of vortices, so that the final computational time is large.

To improve the situation, the following scheme was developed: First the vortices are enclosed by the smallest possible square. Next this square is recursively bisected into smaller squares each holding fewer vortices. Any square is only bisected further when the number of vortex blobs inside the square is still larger than a chosen number (here taken to be 400). In this way, the vortices are arranged into an adaptive panel structure, with each panel holding in the order of 100 vortices on the average.

The advantage of the panel structure is that the sum over j in the expressions for the velocity components can be replaced by a Laurent series. Provided that the considered vortices j are located in a panel which is at least one panel distant from the panel in which the vortex i is located, the convergence of this series is exponential, so that the number of operations is reduced.

The coding is fairly elaborate, since the vortices must be re-grouped into panels, neighboring panels summed by means of direct evaluation of the original sum and panels further apart by means of the Laurent series. In addition, for efficiency the Laurent series of the larger panels must be found by combination of those of its smaller sub-panel. Available space does not allow a full description here.

Yet it is important to realize that all the steps involved are quite straight forward, provided that a means can be found to address a complex, solution adaptive panel structure. The addressing which was used here is illustrated in Figure 1. To label the panels on any fixed level of subdivision of the original domain, binary numberings of the x-position and the y-positions are used, with one quirk: the symbol "1" is used as binary ZERO, and the symbol "2" as binary ONE. The x- and y-numbers are then combined into a final single identifying base-3 number (of 14 digits) by interleaving the x- and y-digits and filling up with zero's on the right. In this way, each panel is indicated by a single base-3 number, with the following pleasant properties: (a) smaller sub-panels always immediately follow the larger panels in which they are located; (b) neighboring panels can be found by binary subtractions and additions.

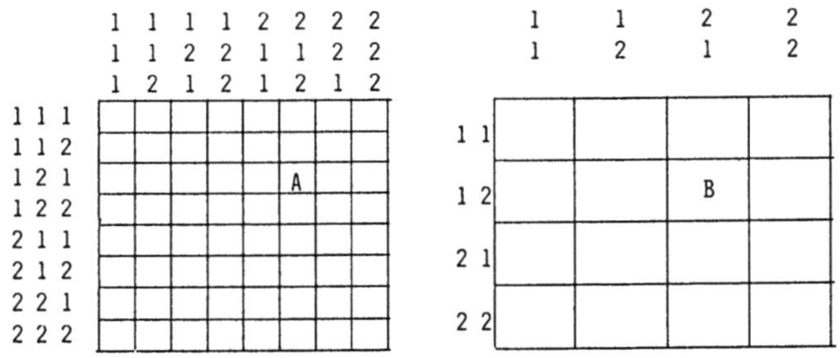

Use the symbol "0" to indicate non-existing digits:

panel A: 2112210000...... (child)
xyxyxyxyxy......

panel B: 2112000000...... (mother)
xyxyxyxyxy......

Fig. 1: Panel labeling scheme.

The effectiveness of the method is illustrated by Table 1, where CYBER 205 times are listed both for vortices equally spaced inside a circle and on the circumference of a circle. Note that the new procedure is both much faster then the original sum and "fast": the CPU time is roughly proportional to the number of vortices. Also note that the best performance is achieved for the most relevant second case, where the vorticity is sparsely distributed.

Vortices:	793	1955	3849	6919	15369	31411	57205
original:	0.06	0.34	1.28	4.05	19.72	81.83*	270.61*
present:	0.06	0.24	0.53	1.18	2.74	6.53	11.68

N:	2000	4000	8000	16000	32000	64000
original:	0.37	1.38	5.40	21.30*	84.70*	338.00*
present:	0.14	0.31	0.65	1.42	3.05	6.45

Table 1: CYBER 205 CPU times for varying number of vortices equally spaced inside a circle (top) and on the perimeter of a circle (bottom).

Results for the Sphere

For the boundary layer flow about the sphere, the first case studied was when the sphere is impulsively given a spinning motion in fluid at rest. Figure 2 shows the lines of particles at constant polar angle Θ in the Lagrangian ρ,η)-plane, for various times. Note that where these lines diverge the particles are accumulating, indicating incipient separation. In this case, this occurs at the meridional plane, where the separation has previously been described, one-dimensionally, by Banks & Zaturska [17] and two dimensionally by Van Dommelen & Cowley [12]. The reason for the drift toward the meridional plane and the subsequent separation is the centripetal acceleration of the fluid caused by the rotational motion.

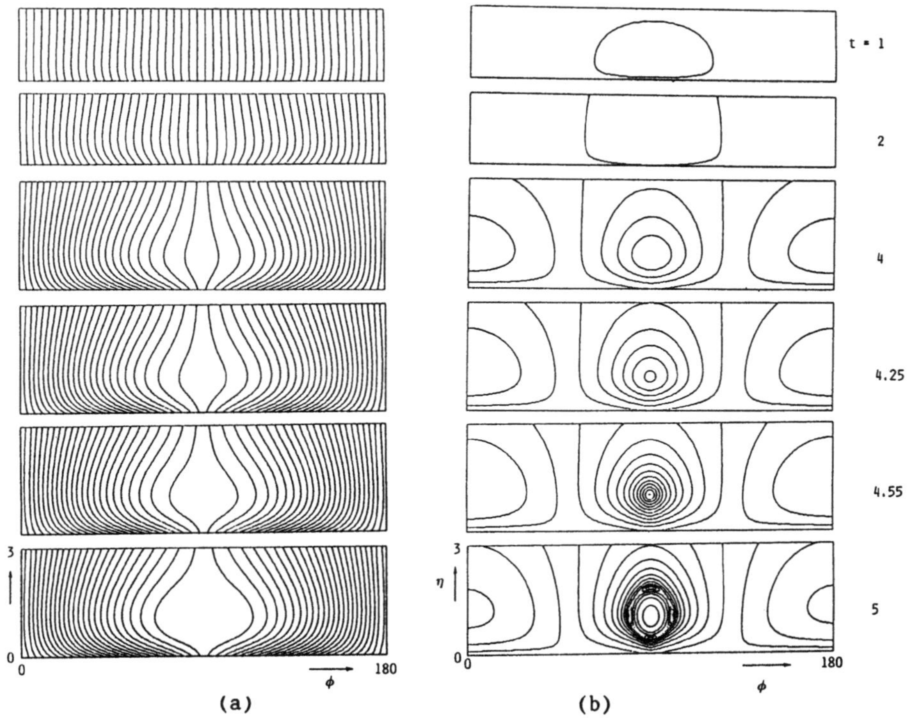

Figure 2: Lines of constant polar angle and particle distortion when the sphere spins and the external flow is at rest (Banks & Zaturska case).

Figure 2b shows lines of constant gradient G, defined as:

$$G = [\Theta_\rho{}^2 + \Theta_\eta{}^2]^{\frac{1}{2}} \quad ;$$

G is a measure of the reciprocal of the particle distortion [12]. Separation is indicated by a zero. Indeed, the first curve in Figure 3 shows Θ_ϕ in the meridional plane, and the curve crosses zero at precisely the separation time found by Banks & Zaturska [17].

The drift toward the meridional plane, and the subsequent separation there can be prevented by giving the external flow an opposite angular rotation. This causes an axial pressure gradient towards the pole, canceling the centripetal acceleration at the wall. However, the boundary layer particles near reversal of the sense of rotation experience little centripetal acceleration and are now driven toward the axis, causing separation to occur there. This separation was previously described one-dimensionally by Bodonyi & Stewartson [18], and two-dimensionally by Van Dommelen & Cowley [12]. The present results are shown in Figures 3 and 4.

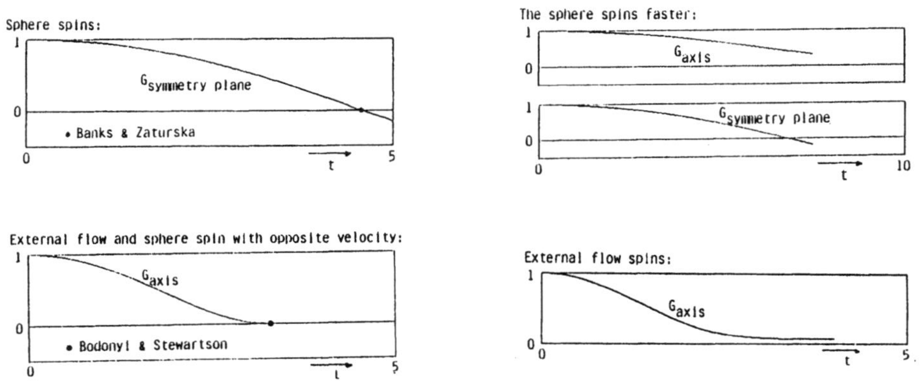

Figure 3: The reciprocal of the particle distortion as a function of time for various cases. The dots denote the separation times found by Bodonyi & Stewartson [18] and Banks & Zaturska [17].

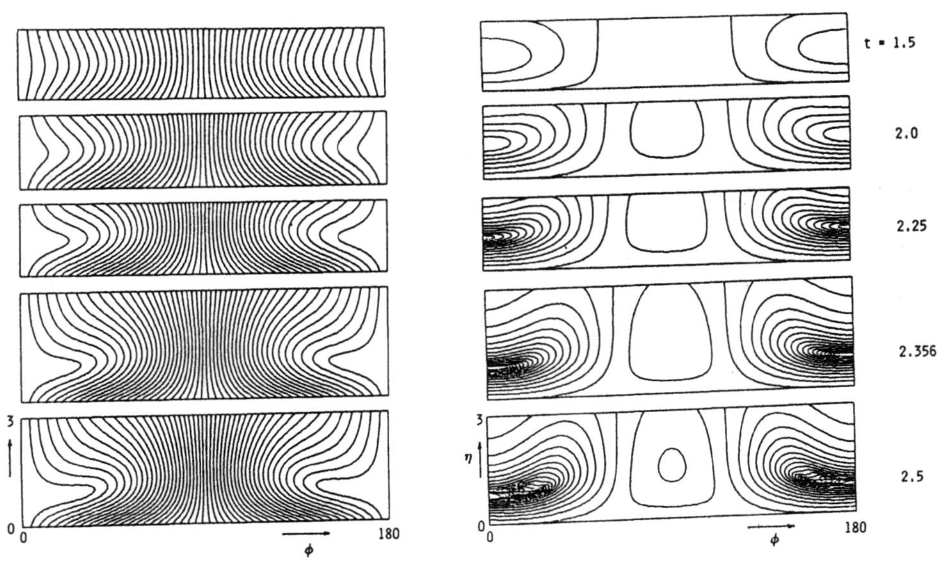

(a) (b)

Figure 4: Lines of constant polar angle and particle distortion when the sphere and external flow spin with opposite velocities (Bodonyi & Stewartson case)

An interesting case arises when the angular velocity given to the sphere is larger than that of the external flow. As an example, we took

$$\Omega_{wall} = \cos 15° \qquad \Omega_{ext} = -\sin 15°$$

Figures 3 and 5 show that the boundary layer particles near the wall are driven toward the meridional plane, while near reversal of rotation they are driven toward the axis. In this case, the separation near the meridional plane occurs first. But the results suggest that it might be possible to adjust the angular velocities such that simultaneous separations would occur at both the meridional plane and the axis.

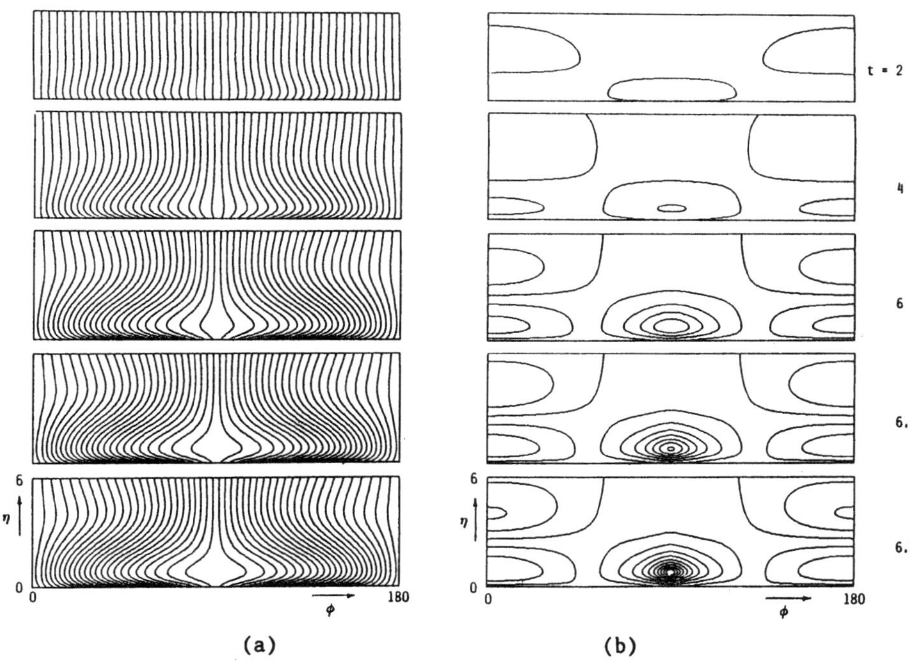

Figure 5: Lines of constant polar angle and particle distortion (intermediate case)

The case of a translating sphere is illustrated in Figure 6; the results are very similar to the translating circular cylinder [8]; the separation structure appears to be as described by Van Dommelen & Cowley [12].

In the case that the external flow is rotating and the sphere is not, according to Van Dommelen [9], inviscid separation at the axis is not possible. This appears to be in agreement with the behaviour of the particle deformation as shown in Figure 3. In this case, there may be little advantage in using a Lagrangian scheme. Our results did in fact suggest that the Eulerian computation by Dennis & Ingham [19] are here more accurate.

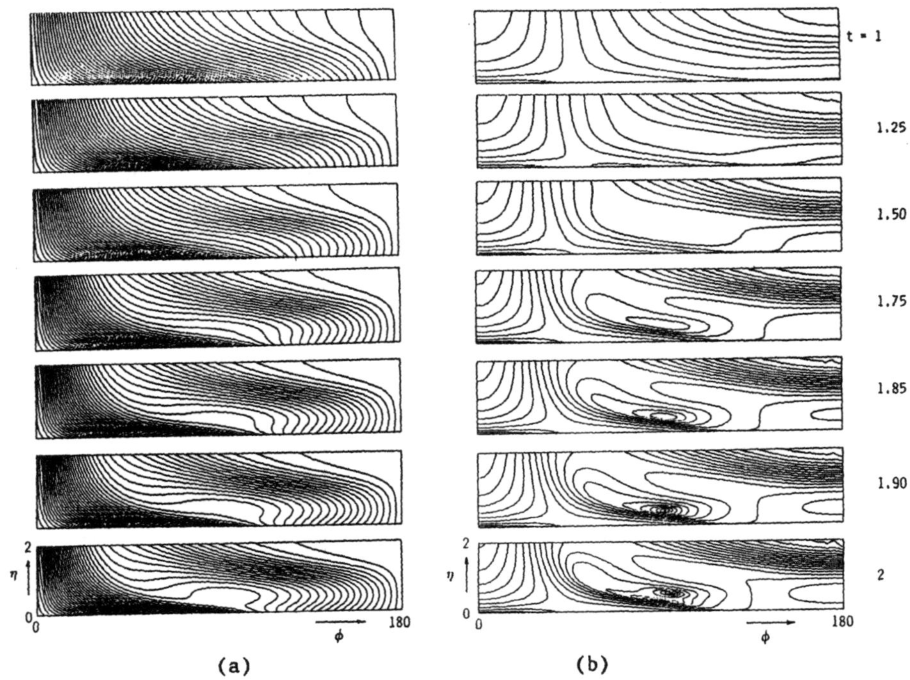

Figure 6: Lines of constant polar angle and particle distortion for the case that the sphere translates.

<u>Results for the Cylinder</u>

Figure 7 shows streamlines for the circular cylinder, at a Reynolds number R_D = 550. The arrows indicate the eddy length according to Loc [20] and the vortex separation (0.55 D) according to Lourenco & Krothapalli [21]. Figure 8 shows the lift and drag coefficients; though the computed drag compares reasonably with other methods, it appears that the random walk exaggerates the actual asymmetry of the flow.

The randomness introduced by the random walk procedure makes interpolation of the vorticity field quite difficult, yet there seems little evidence of the separation process described by Van Dommelen & Shen [8]. This disappointing result may have several reasons; (a) the randomness does certainly obscure the smaller scale phenomena; (b) according to the boundary layer results, the separation becomes only clearly evident late in the evolution, and Reynolds number 550 is without doubt much too low for the boundary layer equations still to be truly meaningful at that time; (c) there is still little known about the properties of unsteady separation.

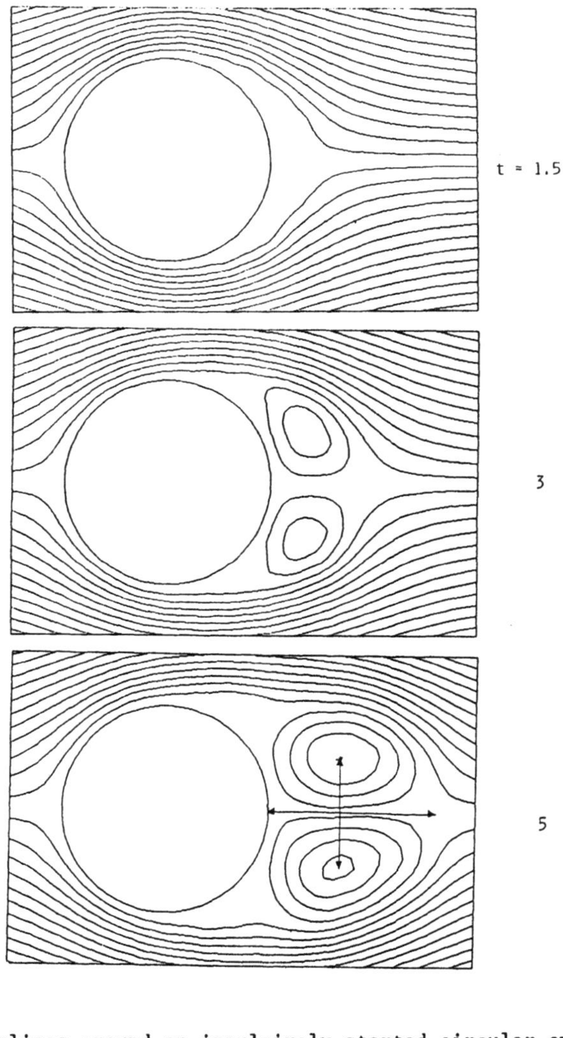

Figure 7: Streamlines around an impulsively started circular cylinder.

Acknowledgements

Various parts of this research were supported by the U.S. Dept. of Energy, the Supercomputer Computations Research Institute, NASA Ames Center, and by the Florida State University through access of its CYBER 205.

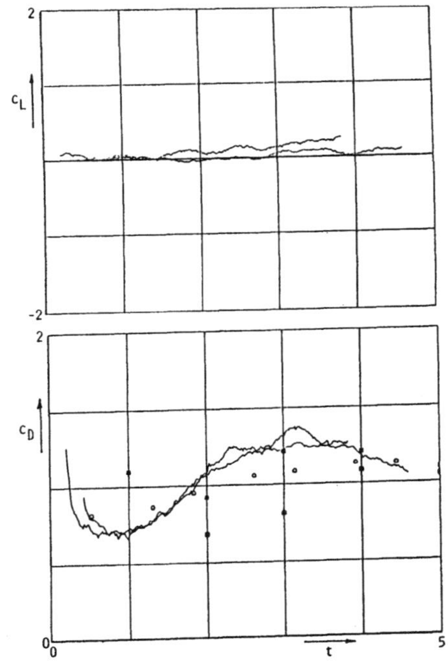

Figure 8: Computed transient lift and drag. Squares are from Loc [20] and circles from circles: Thoman & Szewczyk [22].

References

1. Sears, W.P. & Telionis, D.P., "Boundary Layer Separation in Unsteady Flow", SIAM J. Appl Math, Vol. 28, 1975, pp. 215-235.

2. Ece, M.C., Walker, J.D.A. & Doligalski, T.L., "The boundary layer on an impulsively started rotating and translating cylinder." Phys. Fluids, Vol. 27, 1984, pp. 1077-1089.

3. Cebeci, T., "Unsteady separation," in "Proc. Symp. Num. Phys. Aspects Aerod. Flows," pp. 265-277. Springer-Verlag 1982.

4. Cebeci, T., "Unsteady boundary layers with an intelligent numerical scheme": private communication 1984.

5. Wang, K.C., "On the current controversy about unsteady separation," in "Proc. Symp. Num. Phys. Aspects Aerod. Flows," pp. 279-291, Springer-Verlag 1982.

6. Ingham, D.B., "Unsteady separation," J. Comp. Phys., to appear.

7. Cowley, S.J., "Computer extension and analytic continuation of Blasius' expansion for impulsive flow past a circular cylinder," J. Fluid Mech., Vol. 135, 1984, pp. 389-405.

8. Van Dommelen, L.L. & Shen, S.F., "The genesis of separation", "Proc.

Symp. Num. Phys. Aspects Aerod. Flows," pp. 293-311. Springer-Verlag 1982.

9. Van Dommelen, L.L., "Unsteady Separation", Ph.D. thesis, Cornell University, 1981.

10. Elliot, J.S., Cowley, S.J. and Smith, F.T., "Breakdown of boundary layers: (i) on moving surfaces (ii) in semi-similar unsteady flow; (iii) in fully unsteady flow," Geophys. Astrophys. Fluid Dyn., Vol. 25, 1983, pp. 77-138.

11. Fritts, M.J., "Two-dimensional Lagrangian fluid dynamics using triangular grids", in "Finite Difference Techniques for Vectorized Fluid Dynamics Calculations", (D.L. Book, Ed.), pp. 98-116. Springer-Verlag 1981.

12. Van Dommelen, L.L. & Cowley, S.J., "Lagrangian numerical and analytical description of unsteady boundary layer separation. Part 2", J. Fluid Mech., in preparation.

13. Thompson, Joe E., Warsi, Z.U.A. & Mastin , C.W., "Numerical grid generation", Elsevier, 1985.

14. Beam, R.M. & Warming, R.F., "An implicit factored scheme for the compressible Navier-Stokes equations", AIAA J., Vol. 16, 1978, pp. 393-402.

15. Van Dommelen, L.L., "Vectorized Lagrangian Computation of Unsteady Separation", AIAA paper 85-1492-CP, 7th Comp. Fluid Dyn. Conf., Cincinnati, Ohio, Aug. 1985.

16. Chorin, A.J., "Numerical Study of Slightly Viscous Flow", J. Fluid Mech., Vol. 57, 1973, pp. 785-796.

17. Banks, W.H.B. & Zaturska, M.B., "The collision of unsteady laminar boundary layers", J. Engineering Math., Vol. 13, 1979, pp. 193-212.

18. Stewartson, K., Simpson, C.J. & Bodonyi, R.J. "The unsteady boundary layer on a rotating disk in a counter rotating fluid. Part 2", J. Fluid Mech. Vol. 121, 1982, pp. 507-515.

19. Dennis, S.C.R., & Ingham, D.B., "The boundary layer on a fixed sphere on the axis of an unbounded rotating fluid", J. Fluid Mech, Vol. 123, 1982, pp. 219-236.

20. Loc, T.P., "Numerical analysis of unsteady secondary vortices generated by an impulsively started circular cylinder", J. Fluid Mech., Vol. 100, 1980, pp. 111-128.

21. Lourenco, L.M., and Krothapalli, A., "Application of PIDV to the study of the flow past a circular cylinder" Proceedings of Third International Symposium on Application of Laser Anemometry to Fluid Mechanics, Lisbon, Portugal, July 1986.

22. Thoman, D.C., & Szewczyk, A.A., "Time dependent viscous flow over a circular cylinder", Phys. Fluids. Suppl. II, 1969. pp. 76-86.

Session 2

On Turbulent Boundary-Layer Separation

VLADIMIR V. SYCHEV

(Moscow)

1. The purpose of this paper will be an analysis of a turbulent boundary layer separation from a smooth surface using the method of matched asymptotic expansions for $Re \to \infty$.

As an initial equation system we shall consider the Reynolds equations for averaged flow functions without resorting to any of closure hypotheses. Let us begin with the investigation of the structure of the turbulent boundary layer under a pressure gradient. Classical laws that determine the velocity profile in its inner and outer parts are the Prandtl's law of the wall and the von Kármán's velocity-defect law, respectively. Many authors beginning from Millikan [1] used the assumption that the validity ranges of both laws overlap each other and the velocity profile in the area of their overlapping is logarithmic. The Yajnik's paper [2] seems to be a first attempt of a systematic approach to the problem of using the method of matched asymptotic expansions for averaged flow functions in a turbulent boundary layer at $Re \to \infty$. When treating, as before, the boundary layer structure as a double-deck one and performing a formal matching of asymptotic expansions (in their main terms) for regions where wall and velocity-defect laws are valid he found that their thicknesses are of the order of $\ln Re/Re$ and $1/\ln Re$, respectively.

This double-deck structure of the turbulent boundary layer does not contain a region where internal shear stresses, pressure gradient, and inertia forces are of the same order of magnitude as $Re \to \infty$, i.e. the region that is precisely the boundary layer, as defined by Prandtl. Its absence makes it

F. T. Smith, S. N. Brown, (Eds.)
Boundary-Layer Separation
IUTAM Symposium London 1986

impossible to explain, among other things, a flow separation under the action of adverse pressure gradient since the flow in the velocity-defect region is not subjected (to the first approximation) to the action of shear stresses and the flow in the wall region is not subject to pressure gradient (the latter should, then, have unrealistically high value of the order of $Re/\ln^3 Re$).

In 1956, using experimental observations Coles [3] introduced his law of the wake that suitably relates the law of the wall and the velocity-defect law. According to this new law the velocity profile in the boundary layer depends substantially on pressure gradient and varies in such a manner that when approaching the separation point it assumes the shape of the profile in a wake (Fig.1). The law of the wake may be considered as an indication that the boundary layer structure is not a double-deck one and that actually there is no overlap domain for the wall and velocity-defect laws (even for the flow without pressure gradient) and, therefore, it is necessary to introduce at least one more intermediate zone.

Let us show now that the necessity of introducing such an intermediate zone (where pressure and friction inertia forces are of the same order of magnitude) may be justified theoretically using the method of matched asymptotic expansions.

For this purpose consider first of all the wall-layer part of the flow. According to the law of the wall governing parameter for the flow here is dynamic or wall-friction velocity $u^*=(\tau_w/\rho)^{1/2}$, where τ_w is the wall friction and ρ is the density of fluid. As usual, let us consider the ratio of some characteristic value of friction velocity u_0^* to characteristic value of the velocity in outer inviscid flow as a small parameter

$$\varepsilon = u_0^*/U_\infty \ . \tag{1.1}$$

It is not difficult to estimate the orders of magnitudes for different terms of the Reynolds equations assuming nondimensional averaged values of longitudinal velocity and shear stress components in the wall region as values on the order of ε

and ε^2 , respectively. We shall use a nondimensional set of independent and dependent variables choosing the characteristic size L of a body as a unit for length, free-stream velocity U_∞ for velocity, and doubled pressure head ρU_∞^2 for pressure differences and shear stress measurements.
The Reynolds number is defined as

$$Re = U_\infty L/\nu \quad . \tag{1.2}$$

The Reynolds equations for a two-dimensional flow written in a nondimensional form are as follows

$$\begin{gathered} u\frac{\partial u}{\partial x} + v\frac{\partial u}{\partial y} + \frac{\partial p}{\partial x} = Re^{-1}\left(\frac{\partial^2 u}{\partial x^2} + \frac{\partial^2 u}{\partial y^2}\right) + \frac{\partial \tau_{xx}}{\partial x} + \frac{\partial \tau_{xy}}{\partial y} \\ u\frac{\partial v}{\partial x} + v\frac{\partial v}{\partial y} + \frac{\partial p}{\partial y} = Re^{-1}\left(\frac{\partial^2 v}{\partial x^2} + \frac{\partial^2 v}{\partial y^2}\right) + \frac{\partial \tau_{xy}}{\partial x} + \frac{\partial \tau_{yy}}{\partial y} \\ \frac{\partial u}{\partial x} + \frac{\partial v}{\partial y} = 0 \quad . \end{gathered} \tag{1.3}$$

In the wall layer viscous and turbulent shear stresses are of the same order of magnitude, so that

$$Re^{-1}\frac{\partial u}{\partial y} \sim \tau_{xy} \quad .$$

Making use of the above estimates for velocity and shear stress we can find that the wall layer thickness is

$$\delta = 0[1/(\varepsilon Re)] \tag{1.4}$$

and the corresponding terms in the first momentum equation (1.3) are

$$Re^{-1}\frac{\partial^2 u}{\partial y^2} \sim \frac{\partial \tau_{xy}}{\partial y} \sim \varepsilon^3 Re \quad .$$

Let us assume now that the dimensionless pressure gradient $\frac{\partial p}{\partial x}$ is of the order of unity; based on the above order estimates the contribution of inertia forces, e.g. $u\frac{\partial u}{\partial x}$, will be of the order of ε^2 . Therefore, using equations (1.3) we can write the following asymptotic expansions of the functions in the wall region of the boundary layer*:

* The occurrence of terms with $\ln \varepsilon$ in these expansions (as is shown later) is associated with the effect of the induced pressure gradient.

$$u=\varepsilon\left(u_1^+ + \frac{1}{\varepsilon^3 \mathrm{Re}} u_2^+ + \frac{\ln\varepsilon}{\varepsilon \mathrm{Re}} u_3^+ + \frac{1}{\varepsilon \mathrm{Re}} u_4^+ + \cdots\right)$$
$$v=\mathrm{Re}^{-1}\left(v_1^+ + \frac{1}{\varepsilon^3 \mathrm{Re}} v_2^+ + \frac{\ln\varepsilon}{\varepsilon \mathrm{Re}} v_3^+ + \frac{1}{\varepsilon \mathrm{Re}} v_4^+ + \cdots\right)$$
$$p=p_1^+ + \varepsilon^2 \ln\varepsilon\, p_2^+ + \varepsilon^2 p_3^+ + \cdots \tag{1.5}$$
$$\tau_{xy}=\varepsilon^2\left(\tau_1^+ + \frac{1}{\varepsilon^3 \mathrm{Re}} \tau_2^+ + \frac{\ln\varepsilon}{\varepsilon \mathrm{Re}} \tau_3^+ + \frac{1}{\varepsilon \mathrm{Re}} \tau_4^+ + \cdots\right)$$
$$\tau_{xx}= \varepsilon^2 \pi_1^+ + \cdots \quad , \qquad \tau_{yy}=\varepsilon^2 \sigma_1^+ + \cdots$$

Here, independent variables of the order of unity are

$$x \ , \ y^+ = \varepsilon \mathrm{Re}\, y \ . \tag{1.6}$$

The behavior of all functions at $y^+ \to \infty$ may be found from the equations obtained by substituting (1.5) and (1.6) in (1.3). Thus, the system of equations for functions of the first approximation makes it possible to find that at $y^+ \to \infty$

$$\tau_1^+ = g_1(x) - \frac{h_1(x)}{y^+} + \cdots$$
$$u_1^+ = h_1(x)\ln y^+ + f_1(x) + \cdots \tag{1.7}$$
$$p_1^+ = p_1^+(x)$$

Here $g_1(x)$ is the dimensionless wall shear stress; $h_1(x)$ and $f_1(x)$ are arbitrary integration functions. Note that the first of the expansions (1.7) uses the assumption that the viscosity contribution to internal friction in the outer part of the wall sublayer vanishes as an inverse of y^+ . This assumption corresponds to the logarithmic velocity law [2].

If the equations for functions of the second approximation are taken into account it is not difficult to see that

$$\tau_2^+ = p_1^{+\prime}(x)\, y^+ - h_2(x)\ln y^+ + g_2(x) + \cdots , \quad y^+ \to \infty \tag{1.8}$$

and from the equations for next expansion terms we have

$$\tau_3^+ = p_2^{+\prime}(x)\, y^+ + \cdots \tag{1.9}$$
$$\tau_4^+ = h_1(x) h_1'(x) y^+ \ln^2 y^+ + \cdots , \qquad y^+ \to \infty$$

Using the asymptotic representations for τ_1^+ and τ_2^+ at $y^+ \to \infty$ we find that the corresponding expansion terms (1.5) for τ_{xy} become of the same order of magnitude at $y^+ = O(\varepsilon^3 Re)$. The fourth term of this expansion will be of the same order of magnitude if $\varepsilon \ln \varepsilon = O(1)$. Note that then, according to (1.5), (1.7), the longitudinal velocity component u at $y^+ = O(\varepsilon^3 Re)$ becomes of the order of unity. Therefore, without any loss of generality we can assume that

$$\varepsilon = 1 / \ln Re \tag{1.10}$$

which corresponds to the result of [2] mentioned above.

Thus, we see that expansion (1.5) for τ_{xy} ceases to be valid in the region with a lateral dimension of the order of $\varepsilon^3 Re \delta$ or ε^2.
Let us assume now that the regions of invalidity of expansions for u and τ_{xy} coincide, i.e. corresponding expansion terms of (1.5) also become the same order of magnitude at $y^+ = O(\varepsilon^3 Re)$ or $y = O(\varepsilon^2)$. Then, the main terms of asymptotic representations of u_2^+ and u_4^+ at $y^+ \to \infty$ should be as follows

$$\begin{aligned} u_2^+ &= h_2(x) y^+ \ln y^+ + \cdots \\ u_4^+ &= h_4(x) y^+ \ln^3 y^+ + \cdots \end{aligned} \tag{1.11}$$

The first of these expressions corresponds to (1.8), and the second one to (1.9).

Let us introduce the flow region with independent variables of the order of unity

$$x \, , Y = y / \varepsilon^2 = y^+ / \varepsilon^3 Re \; . \tag{1.12}$$

Using asymptotic expansions (1.5), equations (1.7)-(1.11) and conditions of matching of the solutions in regions $y^+ = O(1)$ and $Y = O(1)$ we can find that the expansions for functions u, v, p, τ_{xy} in region $Y = O(1)$ can be written as

$$\begin{aligned} u &= u_1 + \varepsilon \ln \varepsilon \, u_2 + \varepsilon u_3 + \cdots \\ v &= \varepsilon^2 v_1 + \varepsilon^3 \ln \varepsilon \, v_2 + \varepsilon^3 v_3 + \cdots \\ p &= p_1 + \varepsilon^2 \ln \varepsilon \, p_2 + \varepsilon^2 p_3 + \cdots \\ \tau_{xy} &= \varepsilon^2 \tau_1 + \varepsilon^3 \ln \varepsilon \, \tau_2 + \varepsilon^3 \tau_3 + \cdots \end{aligned} \tag{1.13}$$

Substituting these expansions in Reynolds equations (1.3) rewritten in dimensionless independent variables (1.12) we shall obtain a set of boundary layer equations for the main terms of (1.13) in the form

$$
\begin{gathered}
u_1\frac{\partial u_1}{\partial x}+v_1\frac{\partial u_1}{\partial Y}+\frac{\partial P_1}{\partial x}=\frac{\partial \tau_1}{\partial Y}\\
\frac{\partial P_1}{\partial Y}=0\quad,\quad \frac{\partial u_1}{\partial x}+\frac{\partial v_1}{\partial Y}=0\ .
\end{gathered}
\tag{1.14}
$$

Boundary conditions at $Y=0$ are those of matching with the solution for the viscous wall-layer region. Using (1.5),(1.7), (1.10), (1.12) we find that they are as follows:

$$
\begin{gathered}
u_1(x,0)=h_1(x)\ ,\quad v_1(x,0)=0\ ,\quad P_1(x)=p_1^+(x)\ ,\\
\tau_1=g_1(x)+\left[P_1'(x)+h_1(x)h_1'(x)\right]Y+\cdots,\quad Y\to 0\ .
\end{gathered}
\tag{1.15}
$$

Boundary conditions at $Y\to\infty$ are governed by the conditions of matching with the solution for the velocity-defect region. Hence, it is necessary to consider the flow in the latter one. A longitudinal velocity component here is equal, to the first approximation, to the velocity $U_e(x)$ of an outer potential flow, and the turbulent shear stress effect becomes important only in the second approximation. The relative thickness of the velocity-defect region is known to be a value of the order of ε [2]. Therefore, for the velocity-defect region dimensionless independent variables of the order of unity are

$$
x\ ,\ \bar{y}=y/\varepsilon=\varepsilon Y\ . \tag{1.16}
$$

Expansions for dependent variables can be represented in the form

$$
\begin{aligned}
u&=U_e(x)+\varepsilon U_1+\varepsilon^2\ln\varepsilon\, U_2+\varepsilon^2 U_3+\cdots\\
v&=\varepsilon V_1+\varepsilon^2\ln\varepsilon\, V_2+\varepsilon^2 V_3+\cdots\\
p&=P_1+\varepsilon^2\ln\varepsilon\, P_2+\varepsilon^2 P_3+\cdots\\
\tau_{xy}&=\varepsilon^2 T_1+\varepsilon^3\ln\varepsilon\, T_2+\varepsilon^3 T_3+\cdots
\end{aligned}
\tag{1.17}
$$

Having substituted these expansions in (1.3) we find that

$$P_1' = P_e'(x) = -U_e(x)U_e'(x) \quad , \quad V_1 = -\bar{y}\, U_e'(x)$$

$$U_e \frac{\partial U_1}{\partial x} + U_e' U_1 - \bar{y} U_e' \frac{\partial U_1}{\partial \bar{y}} = \frac{\partial T_1}{\partial \bar{y}} \ . \tag{1.18}$$

We can assume that asymptotic expansions (1.17) become invalid in the region of $y = O(\varepsilon^2)$, i.e. at $\bar{y} = O(\varepsilon)$. If this is the case then the condition

$$U_1 = \frac{F_1(x)}{\bar{y}} + \cdots \quad , \quad \bar{y} \to 0 \tag{1.19}$$

should be valid.
Substituting this expression in (1.18) and requiring that

$$T_1 = G_1(x) + \cdots \qquad \bar{y} \to 0 \tag{1.20}$$

i.e. friction τ_{xy} should remain the order of ε^2 , we obtain the equation

$$U_e F_1' + 2U_e' F_1 = 0$$

and integrating this equation we find that

$$F_1(x) = C_1 / U_e^2(x) \tag{1.21}$$

where C_1 is an arbitrary constant.
Using the continuity equation for further terms of expansions (1.17) we can see that

$$\frac{\partial V_2}{\partial \bar{y}} = 0 \quad \text{and} \quad V_2 = H_2(x) \tag{1.22}$$

and from (1.19)

$$V_3 = -F_1'(x) \ln \bar{y} + \cdots \ , \qquad \bar{y} \to 0 \ . \tag{1.23}$$

To match expansions(1.17) and (1.13) for function v we rewrite (1.23) using independent variable Y :

$$V_3 = -F_1'(x) \ln \varepsilon - F_1'(x) \ln Y + \cdots$$

and since the term of the order of $\varepsilon^2 \ln \varepsilon$ in expansion of v for intermediate region (1.13) is absent we have to assume that

$$H_2(x) = F_1'(x) \ . \tag{1.24}$$

Considering further terms of expansions (1.17) and (1.13) we can find that at $\bar{y} \to 0$

$$U_2 = \frac{F_2(x)}{\bar{y}^2} + \cdots , \quad U_3 = -\frac{F_2(x)}{\bar{y}^2}\ln\bar{y} + \frac{F_3(x)}{\bar{y}^2} + \cdots \tag{1.25}$$

$$T_2 = \frac{G_2(x)}{\bar{y}} + \cdots , \quad T_3 = -\frac{G_2(x)}{\bar{y}}\ln\bar{y} + \frac{G_3(x)}{\bar{y}} + \cdots$$

where

$$G_2 = F_1 F_1' - (3F_2 U_e' + F_2' U_e) ,$$
$$G_3 = -G_2 - F_2 U_e' - F_1(F_1' - H_3) - (3F_3 U_e' + F_3' U_e) . \tag{1.26}$$

Here, functions $F_2(x)$, $F_3(x)$, $H_3(x)$, as well as $G_1(x)$ in (1.20) remain arbitrary.

Higher approximations will not be considered here because for the further discussion it is important to clear up only the boundary conditions that the functions of system (1.14) have to satisfy at $Y \to \infty$. Corresponding matching conditions, as it follows from the above expressions, are

$$p_1'(x) = p_e'(x) = -U_e(x)U_e'(x)$$

$$u_1 = U_e(x) + \frac{F_1(x)}{Y} - \frac{F_2(x)}{Y^2}\ln Y + \frac{F_3(x)}{Y^2} + \cdots$$

$$v_1 = -U_e'(x)Y - F_1'(x)\ln Y + H_3(x) - F_2'(x)\frac{\ln Y}{Y} + \tag{1.27}$$

$$+ \left[-F_2'(x) + F_3'(x)\right]\frac{1}{Y} + \cdots$$

$$\tau_1 = G_1(x) - \frac{G_2(x)}{Y}\ln Y + \frac{G_3(x)}{Y} + \cdots , \quad Y \to \infty .$$

2. The problem of a self-induced separation of the turbulent boundary layer from a smooth surface was treated for the first time in [4]. According to the initial assumption underlying the asymptotic theory developed in that paper the limiting state of the flow field in the scale of a body dimension, when the Reynolds number (Re) tends to infinity, is sought in the class of potential flows with free streamlines

(the Kirchhoff's scheme). Nondimensional values of pressure gradient along the body surface and free-streamline curvature in the vicinity of the separation point are determined by the expressions

$$\frac{\partial p}{\partial x} = \begin{cases} \dfrac{K_0}{\sqrt{-x}} + O(1) , & x \to -0 \\ 0 , & x > 0 \end{cases} \qquad (2.1)$$

$$\varkappa = \frac{K_0}{U_{oo}^2 \sqrt{x}} + \varkappa_0 + O(\sqrt{x}) , \quad x \to +0 .$$

(Here, $\varkappa_0$ is the body surface curvature in the separation point $x=0$; U_{oo} is the velocity on a free streamline). The value of the positive constant K_0 related to the position of the separation point remains finite at $Re \to \infty$ (does not tend to zero as in case of a laminar flow) i.e. the limit solution does not meet the Brillouin-Villat condition. This initial assumption is in full agreement with numerous experimental results according to which the turbulent boundary layer separation point is located at a finite distance downstream of the laminar separation point. It also corresponds to the calculations of Tshahalis and Telionis [5], that were performed using the pressure distribution obtained experimentally in the well-known paper of Schubauer and Klebanoff [6], and showed that inertia terms of eqs (1.14) behave proportionally to $|x|^{-1/2}$ when approaching the separation point ($x \to 0$). Eqs (1.14) were used as a basis for the analysis in [4]. For the sake of convenience they are written in von Mises variables

$$u\frac{\partial u}{\partial x} + \frac{\partial p}{\partial x} = u\frac{\partial \tau}{\partial \psi} , \quad \frac{\partial Y}{\partial \psi} = \frac{1}{u} , \quad \frac{\partial p}{\partial \psi} = 0 \qquad (2.2)$$

with indices of the approximation number just as the equation for the velocity vector component v being omitted.* It was shown in [4] as a result of an asymptotic analysis of the Reynolds equations that the turbulent boundary separation is always associated with its interaction with the outer flow. Approaching the interaction region ($x \to -0$) the solution of eq. (2.2) under pressure gradient behaves singularly. E.g., the boundary layer thickness (i.e., the region where $y=O(\varepsilon^2)$)

---

* Note that in [4] the value that is notified as ε^2 throughout this paper was used as a small parameter ε .

increases according to the logarithmic law; an intense fluid deceleration takes place in its inner portion and the velocity profile assumes the form of the velocity profile in a wake near the point of its breakdown [7]. These results are in good agreement with experimental data (Fig.1).

The solution is no longer valid in the region with the longitudinal dimension $|x| = O(\varepsilon^{4/3})$, where streamline slopes and self-induction pressure gradient become governing, i.e. there is a boundary layer-outer flow interaction. The boundary-layer structure in the interaction region becomes a multilayer one (Fig.2). In its main part (II) where $\Psi = O(1)$ the velocity profile and turbulent stresses remain along streamlines, and corresponding asymptotic expansions are as follows:

$$u = u_0(\Psi) + \varepsilon^{2/3} u_1^*(x^*, \Psi) + \cdots, \quad x^* = \varepsilon^{-4/3} x$$
$$\tau = \tau_0(\Psi) + \cdots, \quad Y = -(6\alpha_0)^{-1} \ln \varepsilon^2 + Y_0^*(x^*, \Psi) + \cdots \tag{2.3}$$
$$p = p_{00} + \varepsilon^{2/3} p_1^*(x^*, \Psi) + \cdots, \quad p_{00} = (1 - U_{00}^2)/2$$
$$\alpha_0 = \text{const} > 0 .$$

A high self-induced pressure gradient $\frac{\partial p}{\partial x} = O(\varepsilon^{-2/3})$ leads to that a variation of the turbulent shear stress has no effect on the solution for main expansion terms. It is obtained by substituting (2.3) in the initial Reynolds equations or, which is equivalent, in eq. (2.2):

$$p_1^* = p^*(x^*), \quad u_1^* = -\frac{p^*(x^*)}{u_0(\Psi)}$$
$$Y_0^* = \int \frac{d\Psi}{u_0} + G^*(x^*) + \Phi(x^*)\Theta(x^*), \quad u_0(\infty) = U_e(0) = U_{00} . \tag{2.4}$$

Here, functions $G^*(x^*)$ and $\Phi(x^*)$ determine the effect of the boundary layer displacement and zero streamline shape behind separation point; $\Theta(x^*)$ is the Heaviside function. The solutions (2.3), (2.4) are invalid for region III (Fig.2), where $\Psi = O(\varepsilon^{1/3})$. This is a slow flow region that reveals a nonlinearity of initial equations, but just as in region II, the turbulent stress effect is absent in the main terms, so that here the Bernoulli integral is valid.

The flow in this region determines a basic contribution to the effect of the boundary layer displacement which makes it possible to find pressure distribution $p^*(x^*)$ in the interaction region without any closure conditions. The corresponding solution is

$$\psi = \varepsilon^{1/3}\psi^*, \quad u = \varepsilon^{1/3} u^*(x^*, \psi^*) + \cdots$$

$$\tau = \tau^*(x^*, \psi^*) + \cdots, \quad Y = Y^*(x^*, \psi^*) + \cdots \quad (2.5)$$

$$p = p_\infty + \varepsilon^{2/3} p^*(x^*)$$

where

$$u^{*2} + 2p^* = H(\psi^*), \quad \frac{\partial Y^*}{\partial \psi^*} = \frac{1}{u^*} \quad . \quad (2.6)$$

The Bernoulli function

$$H(\psi^*) = a_0^2 \psi^{*2} + 2B^*, \quad B^* = \text{Const} \quad (2.7)$$

is found from matching with the boundary layer solution at $x \to -0$ and then it follows from (1.6) that

$$Y^* = \frac{1}{2a_0} \ln\left(\frac{u^* + a_0\psi^*}{u^* - a_0\psi^*}\right) + \Phi(x^*)\Theta(x^*) \quad . \quad (2.8)$$

Solution matching used for regions II and III yields

$$u_0 = a_0\psi + \cdots, \quad \tau_0 = b_0 + \cdots, \quad \psi \to 0$$
$$\tau^* = b_0 + \cdots, \quad \psi^* \to \infty, \quad b_0 = const \quad . \quad (2.9)$$

Also a relation for functions $p^*(x^*)$ and $G^*(x^*)$ is found to be

$$G^* = -\frac{1}{2a_0} \ln\left(-\frac{p^* - B^*}{2a_0^2}\right) \quad . \quad (2.10)$$

Other relationships connecting functions $p^*(x^*)$, $G^*(x^*)$ and $\Phi(x^*)$ and closing the problem for regions II and III are determined using the small-disturbance theory for the flow in region I having lateral dimension of the order of longitudinal dimension of the interaction region.

The problem just formulated is in some sense analogous to

that of the laminar boundary layer separation from a downstream moving surface [8]. The reason for this is that the interaction mechanism in both cases is due to the displacement effect of the nonlinear region of the flow field where shear stresses are negligible, and the form of the Bernoulli function is that of (2.7). On the basis of the results of papers [9], [10] it is possible to assume the constant B^* to be equal to zero. Then it turns out that the flow separation does not occur in the interaction region, but at a distance downstream of it. If $B^*=0$ the integral of the small-disturbance theory of potential flow together with (2.10) yields the following equation

$$\frac{U_{\infty}^2}{2a_0}\frac{p^{*\prime}(x^*)}{p^*(x^*)}=-2K_0\sqrt{x^*}\,\theta(x^*)+\frac{1}{\pi}\int_{-\infty}^{\infty}\frac{p^*(t)+2K_0\sqrt{-t}\,\theta(-t)}{t-x^*}dt\,. \qquad (2.11)$$

This equation can be solved numerically and the solution corresponds to an unseparated flow in region III, i.e. $\phi(x^*)=0$. In view of a high pressure gradient in the interaction region the effect of turbulent shear stresses, according to [4] becomes appreciable (in main terms) in thin layer IV only, where

$$\begin{aligned}\psi&=\varepsilon^{5/3}\psi^0\,, \qquad u=\varepsilon^{1/3}u^0(x^*,\psi^0)+\cdots\\ \tau&=\varepsilon^{2/3}\tau^0(x^*,\psi^0)+\cdots\,, \qquad Y=\varepsilon^{4/3}Y^0(x^*,\psi^0)+\cdots\end{aligned} \qquad (2.12)$$

These functions satisfy the boundary layer equations with pressure gradient determined by eqs. (2.5), (2.11) and a set of boundary conditions for matching with adjacent regions that are not presented here. Note solely that as a result of matching of solutions for regions III and IV it follows that at $\psi^0\to\infty$

$$u^0=\sqrt{-2p^*(x^*)}+O(\psi^{0\,-1/2})\,, \qquad \tau^0=S^0(x^*)\psi^{0\,1/2}+\cdots \qquad (2.13)$$

From (2.10), (2.11) at $x^*\to\infty$ we obtain

$$\begin{aligned}p^*&=-2K_0\sqrt{x^*}\exp\left[-\gamma_0 x^{*3/2}+o(1)\right]\\ \sigma^*&=\frac{\gamma_0}{2a_0}x^{*3/2}-\frac{1}{4a_0}\ln x^*+\frac{1}{a_0}\ln\left(\frac{a_0}{\sqrt{K_0}}\right)+o(1)\\ \gamma_0&=\frac{8K_0a_0}{3U_{\infty}^2}\end{aligned} \qquad (2.14)$$

i.e. the pressure tends to zero by the exponential law and, therefore, the value of velocity u^* also diminishes according to (2.6), (2.7), (2.13). Thus, the interaction results in a further flow deceleration in regions III and IV, but it does not cause a boundary layer separation. Hence it is necessary to introduce a new characteristic region B (Fig.2) at some distance downstream of the interaction region which continues the nonlinear boundary layer portion and where the action of the inner shear stress becomes significant due to a flow deceleration. It is this region where a separation will occur. The exponential nature of a decrease of the pressure $p^*(x^*)$ at $x^* \to \infty$ shows that in region B the value $p - p_{oo}$ in its order of magnitude will be less than in the interaction region, i.e.

$$p - p_{oo} = \sigma \varepsilon^{2/3} \hat{P}_0 + \cdots \qquad (2.15)$$

where $\sigma = \sigma(\varepsilon) \to 0$ at $\varepsilon \to 0$. Resorting now to asymptotic representations (2.14) and to solutions (2.6)-(2.8) we can obtain orders of magnitude for variables in region B and present here the solution as follows*

$$x = \gamma_0^{-2/3} \Delta_0 + \Delta_1 X \;, \qquad \psi = \varepsilon^{1/3} \sqrt{\sigma}\, \hat{\psi}$$

$$u = \varepsilon^{1/3} \sigma^{1/2} \hat{U}_0 (X, \hat{\psi}) + \cdots, \quad \tau = \hat{T}_0 (X, \hat{\psi}) + \cdots \qquad (2.16)$$

$$Y = \hat{Y}_0 (X, \hat{\psi}) + \cdots, \quad p = p_{oo} + \varepsilon^{2/3} \sigma \hat{P}_0 (X, \hat{\psi}) + \cdots$$

where

$$\Delta_0 = \varepsilon^{4/3} \left[(\ln \sigma^{-1})^{2/3} + \frac{2}{9} (\ln \ln \sigma^{-1}) (\ln \sigma^{-1})^{-1/3} \right] ,$$

$$\Delta_1 = \varepsilon^{4/3} (\ln \sigma^{-1})^{-1/3} .$$

Substituting (2.16) in Reynolds equations or Eqs (2.2) and equating convective and turbulent shear stress terms we come to boundary layer equations

$$\hat{U}_0 \frac{\partial \hat{U}_0}{\partial X} + \frac{\partial \hat{P}_0}{\partial X} = \hat{U}_0 \frac{\partial \hat{T}_0}{\partial \hat{\psi}} \;, \quad \frac{\partial \hat{P}_0}{\partial \hat{\psi}} = 0 \;, \quad \frac{\partial \hat{Y}_0}{\partial \hat{\psi}} = \frac{1}{\hat{U}_0} . \qquad (2.17)$$

*Note that matching of the solutions in region B with the solution for region III should be performed using p, ψ as independent variables since these solutions have not overlap domain in original variables x, ψ.

Then we find for a small parameter σ the transcendent equation

$$\sigma\left(\ln\sigma^{-1}\right)^{1/3}=\varepsilon^{2/3}. \qquad (2.18)$$

From matching of solutions in regions B and III it follows that at $X\to-\infty$

$$\hat{U}_0=(-\hat{P}_0)^{1/2}(a_0^2\varepsilon^2+2)^{1/2}+\cdots;\quad \hat{Y}_0=\frac{1}{2a_0}\ln\left(\frac{F_0+a_0\varepsilon}{F_0-a_0\varepsilon}\right)+\cdots$$
$$\hat{P}_0=-2K_0\gamma_0^{-1/3}\exp\left(-\frac{3}{2}\gamma_0^{2/3}X\right)+\cdots;\quad \varepsilon=\hat{\Psi}/\sqrt{-\hat{P}_0}. \qquad (2.19)$$

The flow in sublayer IV at $x^*\to\infty$ ($p^*\to 0$) should also be matched with the solution for region B.

Eqs. (2.13), (2.14) indicate that at $x^*\to\infty$ the longitudinal component of the velocity vector continues to decrease both in region IV and region III.

According to (2.12) the cross dimension of this region is $Y=O(\varepsilon^{4/3})$ and turbulent stresses are $\tau=O(\varepsilon^{2/3})$, while values Y and τ in region B are of the order of unity. Therefore, at $x^*\to\infty$ there will be an expansion of region IV and an increase of turbulent stresses. Thus, when a pressure gradient increase ahead of the interaction region ($x\to-0$) results in a simultaneous decrease of turbulent shear stress and turbulent sublayer thickness [4], a reverse situation takes place, i.e. both values increase as pressure gradient tends to zero downstream of the interaction region. These results may be formally obtained as matching conditions of the solutions for regions IV and B (not described here).

At the same time when the separation point is approached a decrease of the turbulent shear stress on the wall (more precisely, on the upper boundary of the viscous wall layer) must occur, i.e. when

$$x^*\to\infty,\quad \tau^0(x^*,0)=\tau_w^*(x^*)\to 0$$

The available experimental data also indicate that as the separation point is approached a rise of the turbulent friction takes place within the boundary layer. Fig.3 gives corresponding measurement results from [6].

Returning now to a system of equations and boundary conditions for region B let us note that its outer boundary conditions (at $\hat{\psi} \to \infty$) are obtained by matching with the solution in region A (Fig.2) that is a continuation of the main part of the boundary layer II with fixed profiles of velocity and turbulent shear stresses.

In addition, the solution (2.17) should satisfy tangency condition at $\hat{\psi} = 0$ and conditions of matching with the solution in the region downstream of region B which is connected with the fact that at a certain $X = X_S$ velocity $\hat{U}_0(X_S, 0)$ vanishes, and a zone of reverse flow originates behind this point. We shall not present here these boundary conditions. Note only that they can be obtained by treating the flow in a mixing layer between the main part of the separated boundary layer and the zone of the reversal flow caused by a suction effect of the mixing layer.

As a result, a whole relationship system can be obtained that describes the flow in region B where a separation of the turbulent boundary layer occurs. This system is unclosed just as the initial system of Reynolds equations. When attempting to construct a closure condition interrelating functions $\hat{U}_0$ and $\hat{T}_0$ (and/or their derivatives) then one must expect that it will be a rather complicated one since it is necessary to take into account the whole set of boundary conditions that provide matching of the solution in B with those in adjacent regions.

An important feature of the flow in region B is that the pressure distribution $\hat{P}_0(X)$ here is not prescribed and should be found as a part of the solution. At the same time the known function is a change of the displacement thickness or, which is just the same, the streamline slope

$$\vartheta = \lim_{\hat{\psi} \to \infty} \left(\frac{\partial \hat{Y}_0}{\partial X} \right) = \frac{3}{4} \frac{\gamma_0^{2/3}}{a_0}$$

at the outer boundary of region B.

Thus, the above analysis shows that a turbulent boundary layer separation in contrast to a laminar separation is not localized in the interaction region and represents a rather complicated gradual transition that is close to a laminar separation on a downstream moving surface. This kind of flow behavior fully corresponds to available experimental data [11].

References

1. Millikan C.B. A critical discussion of turbulent flows in channels and circular tubes. Proc. 5th Int. Congr. Appl. Mech., Wiley 1939, pp.386-392.
2. Yajnik K.S. Asymptotic theory of turbulent shear flows. J.Fluid Mech. 42 (1970) 411-427.
3. Coles D. The law of the wake in the turbulent boundary layer. J.Fluid Mech. 1 (1956) 191-226.
4. Сычев В.В., Сычев Вик.В. О турбулентном отрыве. Ж.Вычисл. Мат.и Мат.Физ. 20 (1980) 1500-1512.
5. Tsahalis D.Th., Telionis D.P. On the behavior of turbulent boundary layers near separation. AIAA J. 13 (1975) 1261-1262.
6. Schubauer G.B., Klebanoff P.S. Investigation of separation of turbulent boundary layer. NACA Rep. 1030, 1951, 20 p.
7. Сычев Вик.В. О разрушении плоского ламинарного следа. Уч.Зап.ЦАГИ 9 (1978) 6, 9-16.
8. Сычев Вик.В. Асимптотическая теория нестационарного отрыва. Изв.АН СССР, МЖГ (1979) 6, 21-32.
9. van Dommelen L.L., Shen S.F. An unsteady interactive separation process. AIAA J. 21 (1983) 358-362.
10. Сычев Вик.В. К асимптотической теории ламинарного отрыва на подвижной поверхности. Прикл.Мат.и Мех. 48 (1984) 247-253.
11. Simpson R.L. A review of some phenomena in turbulent flow separation. Turbulent boundary layers (ed. Weber H.E.) N-Y, ASME, 1979, pp.1-14.

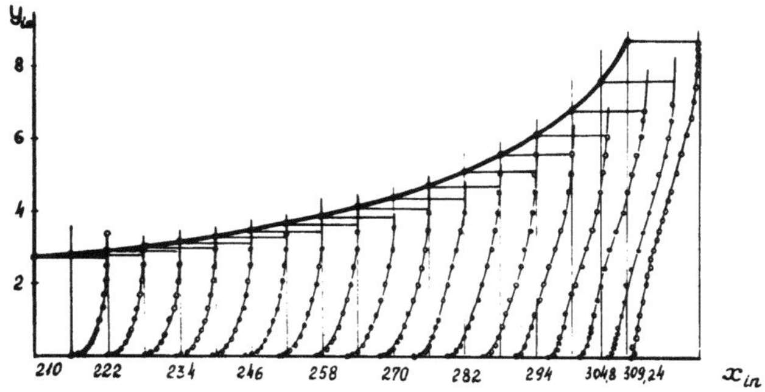

Fig. 1

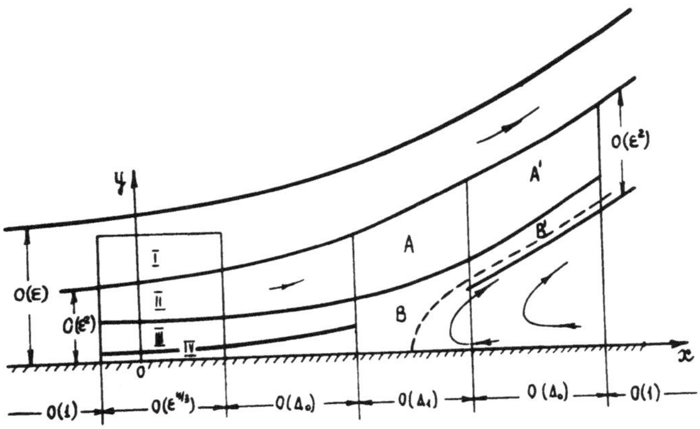

Fig. 2

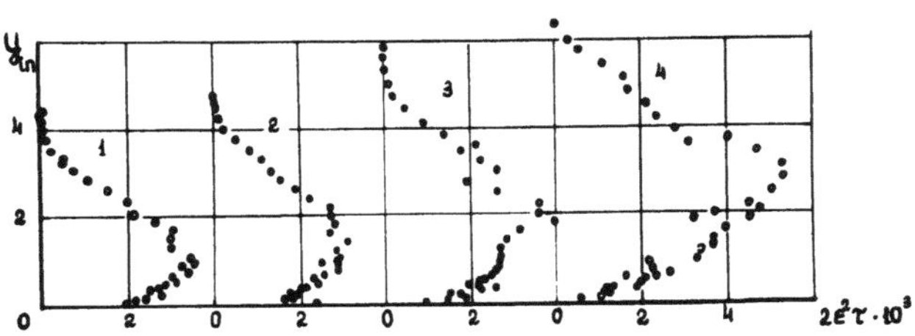

Fig. 3

Experiments on Hydrodynamic Instability of Boundary Layers with Separation

A.V. DOVGAL, V.V. KOZLOV, O.A. SIMONOV

Institute of Theoretical and Applied Mechanics,
Siberian Branch USSR Academy of Sciences,
Novosibirsk, USSR

Summary

The survey of the experimental results devoted to the instability and transition to turbulence in flows with laminar boundary layer separation is given. The development of infinitesimal oscillations in two- and three-dimensional separation regions is discussed.

Introduction

To date a significant success has been achieved in studies of steady laminar boundary layer separation. The problem has been solved numerically for Navier-Stokes equations ([1-3] among others), for more simple approaches (see in [4]) and using the asympthotic methods in solving the hydrodynamic equations [5-11].

By the way the shear layer starting at the point of separation is usually unstable (the mean velocity profile with the inflection point is typical for separation). The instability results in transition to turbulence in separated flow so the transitional type of separation is formed. Just this class of separation is usually met with rather than the laminar one. Under certain conditions the reattachment of turbulent separated layer may occur and the transitional separation bubble appears. The lack of understanding of the transition process in separated flows compels researchers to use the empirical models of the transition in the calculations [12-16]. Empirical data on the formation of transitional separation bubbles have been obtained in numerous experiments ([17-19], for example) for different experimental conditions. The discrepancy of the experimental results is obviously due to the

F. T. Smith, S. N. Brown, (Eds.)
Boundary-Layer Separation
IUTAM Symposium London 1986

fact that the structure of such flows essentially depends on the transition process in separation bubble. The transition in turn is defined by several experimental conditions which in the above studies were not adequately controlled. The sensitivity of the calculation results to the transition model has been stressed in [16]; also the remarkable effect of the transition position in separation region on the structure of the flow has been observed in [20]. The importance of investigation of the laminar-turbulent transition in separated flows has been emphasised in [21].

The experimental results [22-24] demonstrated that the well known concept of instability of shear layers to small amplitude oscillations is applicable to separated flows. In [22, 23] vortical disturbances in the region of boundary layer separation were observed while in [24] it was shown that the turbulization of separated flow occurs because of amplification of infinitesimal oscillations. As regards theoretical studies the conclusion of the common nature of the separation and the instability has been done in [25]. The calculations of linear instability of separated flows are those of [26, 27].

In this paper the experimental studies of flow instability in the regions of boundary layer separation and associated phenomena are discussed. The experiments were performed in the low turbulence wind tunnel, the main results have been obtained by means of the hot-wire measurements. Artificial excitation and control of instability waves governing the transition process were adopted.

2-D Separation

The first case under the consideration is boundary layer separation from the smooth surface. The experiments have been performed on the model of a straight wing [28]. The scheme of the flow is shown in Fig.1. Here is also given the distribution of pressure on the test surface (U_∞=5.7 m/s). It can be noted that in the region of adverse pressure gradient the flow remains laminar and only downstream there appear natural oscillations leading to turbulence and to the reattachment of the separated flow.

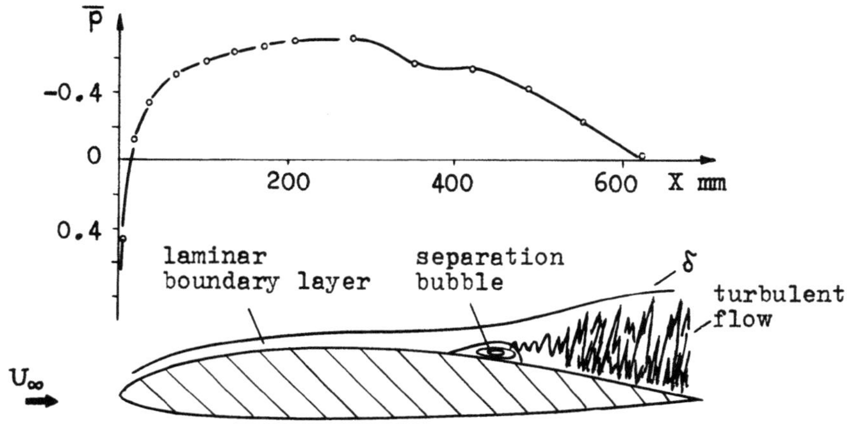

Fig.1

The vibrating ribbon technique was used for excitation of controlled two-dimensional vortical disturbances in the boundary layer upstream of separation. Amplification of instability waves excited in such a way is shown in Fig.2. Eddy oscillations evolving in the pre-separation region while being transformed into separation flow disturbances, amplify in the separation bubble. Thereafter they are observed to strongly influ-

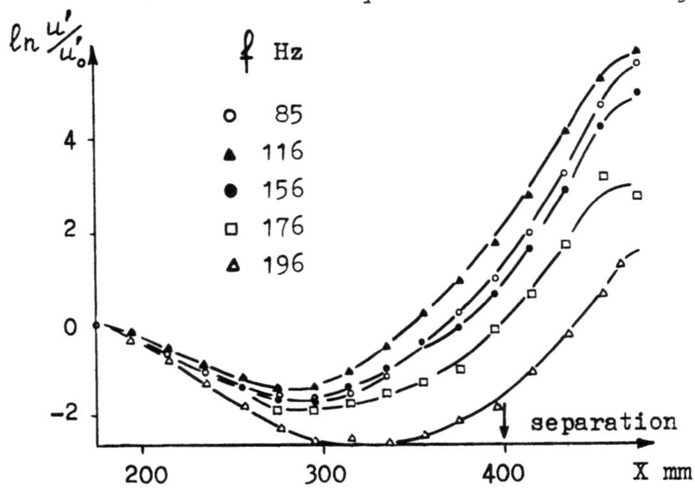

Fig.2

ence the flow structure in the adverse pressure gradient region. The change in the mean velocity flow field caused by the excitation is illustrated in Fig.3. Essential is the fact that the velocity profiles filling doesn't result from the flow turbulization but takes place in a laminar layer whereas the laminar-turbulent transition occurs much later. The mean flow distortion exceeds the local oscillation amplitude more than by an order, which is surely due to the rearrangement of the separation bubble as a whole.

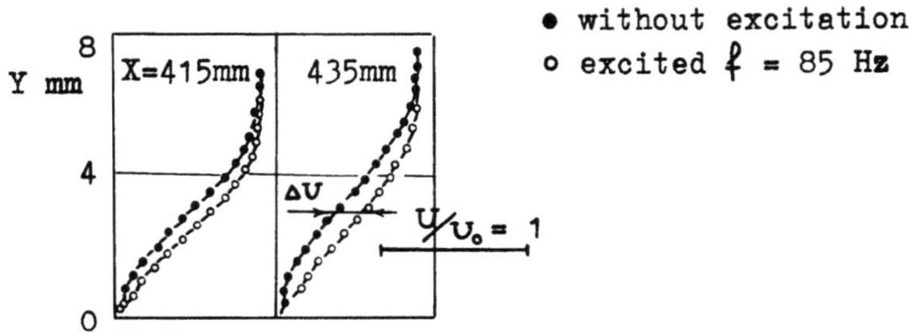

Fig.3

Nonlocal nature of the mean flow-oscillations interaction is obviously seen in Fig.4 where the effect of "point source" excitation is shown [29] . The solid lines represent amplitudes of oscillations in the wave packet while the dashed ones - the values of mean velocity distortion. The mean flow profiles are distorted not only in the region of the wave packet propagation but also outside where the flow remains undisturbed.

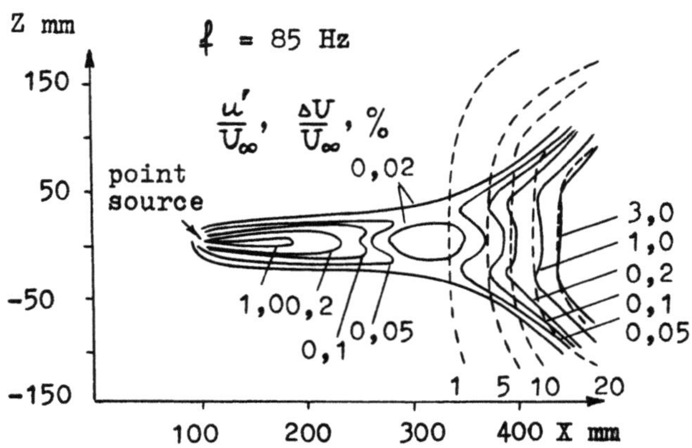

Fig.4

The disturbances amplifying in separated region may be generated by some mechanisms. Acoustic excitation is among them. The influence of the external sound field on separation from the smooth surface has been studied in [30]. The scheme of the experiment remained the same as in [28] with the exception that instead of the vibrating ribbon the dynamic loudspeaker was used to generate acoustic oscillations of the amplitude of the order of 100 dB. It was shown that when generating acoustic flow oscillations in the boundary layer in the region of the adverse pressure gradient there appeared two-dimensional vortical waves with the excitation frequency. Those excited in the vicinity of minimum pressure point (i.e. $x \simeq 250$ mm $\div$ 300 mm) are the most significant downstream for instability and transition to turbulence in separation bubble.

The following considerations [31, 32] are focused on the instability of separation bubbles of transitional type forming on fractured surfaces (Fig.5). Laminar boundary layer, separation bubble and turbulent flow downstream of the reattachment are shown by figures 1÷3. In the case 5a two-dimensional oscillations in separation region were excited using the loudspeaker while in 5b by means of vibrating ribbon.

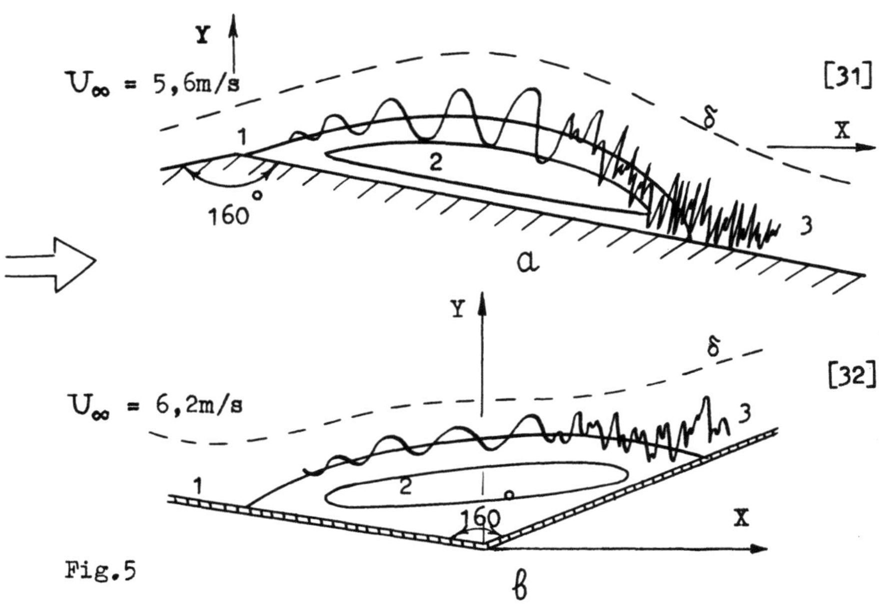

Fig.5

The typical profile of intensity of monochromatic oscillations in separation bubble is given in Fig.6. The position of the intermediate maximum approximately coincides with the inflexion in the mean velocity distribution. The lower maximum is in the reversed flow but associated with the upper one and also travels in the freestream direction.

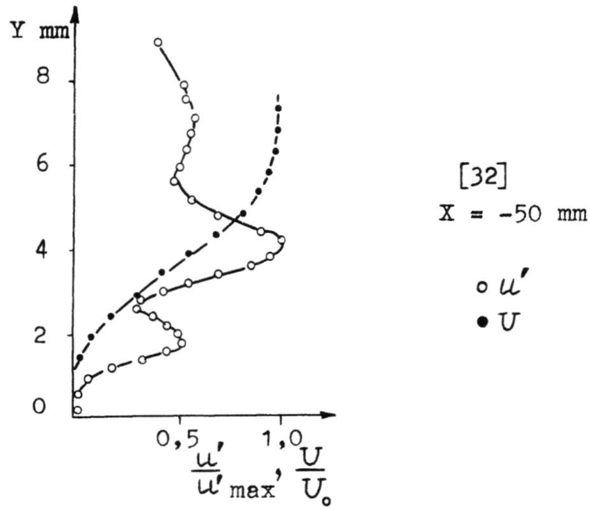

Fig.6

The experimental results of [31, 32] showed the independence of disturbances amplification in separation bubble on their intensities for small amplitude oscillations that enabled one to introduce the coefficients of spatial amplification $-\alpha_i$. The frequency ranges of instability for both cases are given in Fig.7. Note that amplification rates of small amplitude oscillations do not depend on their intensities despite the distortion of the mean flow distribution which in turn defines the stability of the shear layer. The discrepancy is avoided when taking into account that the mean flow profile in the vicinity of inflexion point is not affected by excitation but only shifts in Y-direction (Fig.8). So the stability of separated flow as a whole is determined by stability characteristics of a mixing layer near the boundary of separation bubble.

The effect of sound excitation on the mean flow structure is even more pronounced that in the experiments on the wing model.

Rearrangement of the flow structure caused by excitation of instability waves in separation region produces the change in static pressure distribution on the surface (Fig.9). The pressure distributions with excitation significantly differs from that without one.

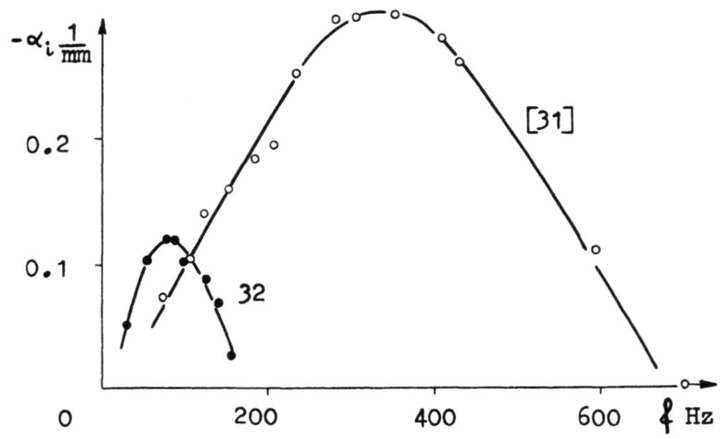

Fig.7

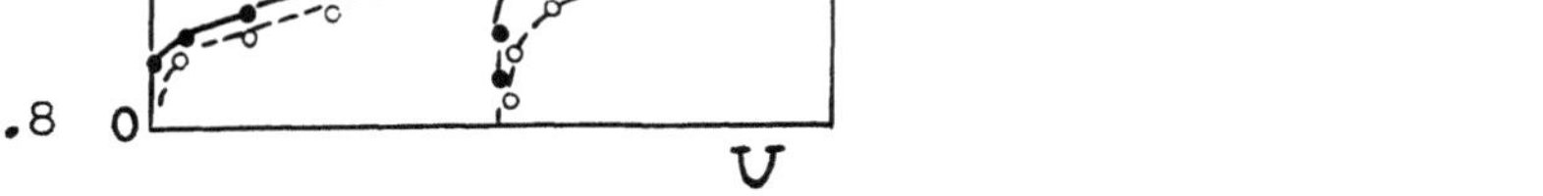

Fig.8

The above effects associated with the instability of separated flow have been also observed in the experiments devoted to the leading-edge separation [33]. The symmetrical airfoil was examined. Acoustic oscillations of the external flow were excited by means of the loudspeaker. When switching the acoustic field to the definite frequency range in the laminar separation region there appears a Tollmien-Schlichting wave, i.e. in this region the acoustic disturbances are transformed into vortical ones as was already mentioned.

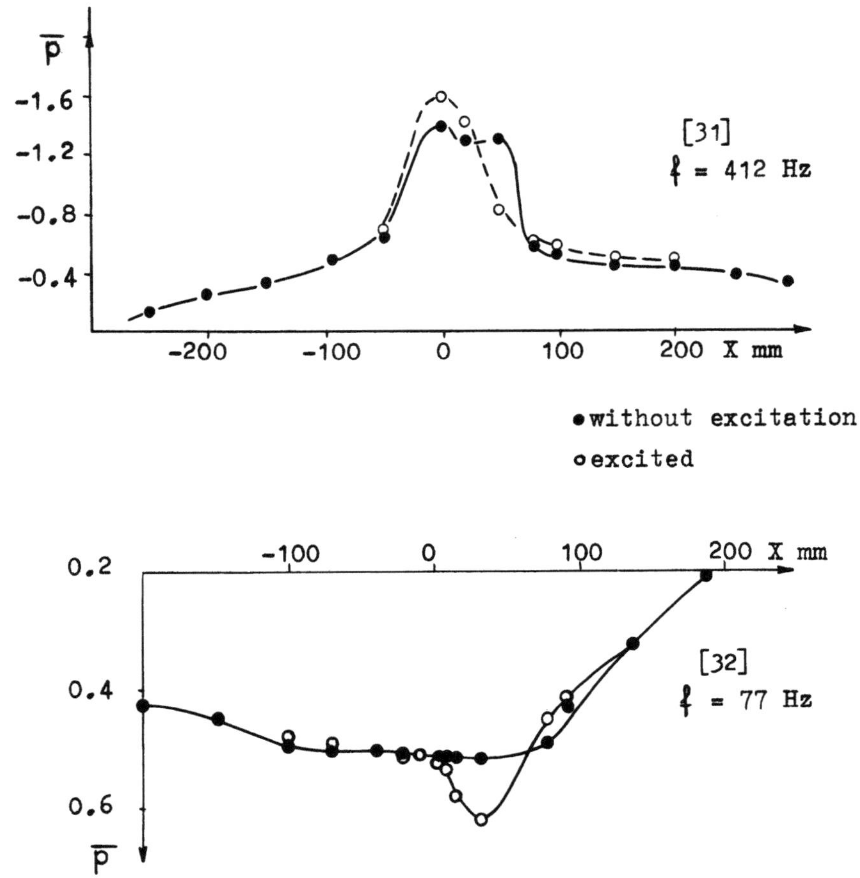

Fig.9

Like in the abovementioned references this leads to the change in the mean flow structure: the boundary layer separation vanishies and the flow attaches to the upper surface of the airfoil (Fig.10).

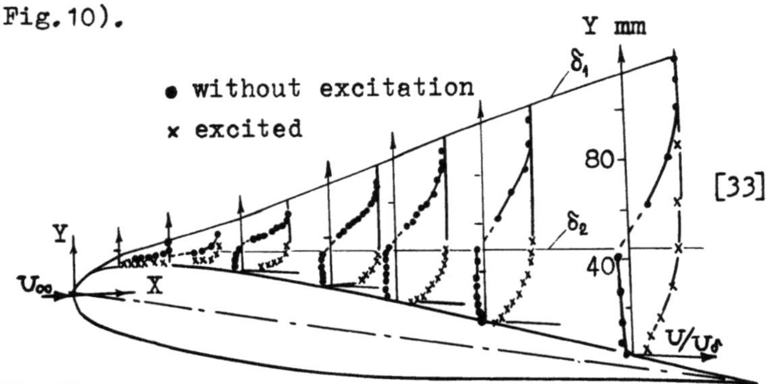

Fig.10

Some more details concerning the experiments summarized one can find in [34]. The stability of the separation from the backward-facing step in plane channel also has been studied in [35].

3-D Separation

The three-dimensional separation in general is a too complicated problem for theoretical and experimental efforts. The reasonable way to understanding the transition process in three-dimensional separated flows surely consists in the studying of the flows with rather simple spatial structure in order to clarify the mechanisms of instability governing the transition process.

The one of them is boundary layer separation which occurs in the region of adverse pressure gradient on the swept wing with infinite span. The separation in this case may be classified as "quasi-three-dimensional": the separation bubble is two-dimensional (when neglect the boundary layer thickening in the span direction), at the same time inside the bubble the cross-flow is presented. The experiments [36] were devoted to the investigation of the development of small amplitude oscillations in such a flow.

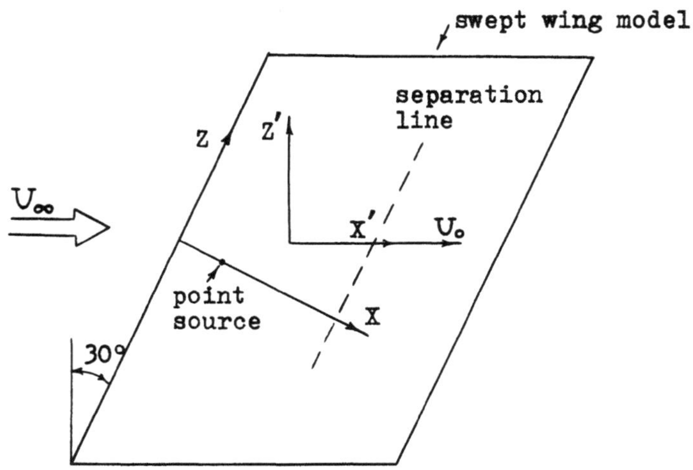

Fig.11

The experiments have been performed on the 30°-swept wing model with a chord equals to 700 mm placed at zero incidence. Free stream velocity U_∞ was equal to 5.7 m/s (Fig.11). Mean flow and oscillations were measured with the use of single hot-wire probe. The hot-wire at each point was placed perpendicular to the local mean flow direction U, which was defined by turning the probe in XZ-plane. Correspondingly all the results concerning the oscillations which are discussed below are given for the component in the local mean flow direction. Harmonic in time boundary layer oscillations were excited by means of the "point source" with the coordinates X=105 mm, Z=0.

The mean flow structure without artificial excitation is given in Fig.12. The position of separation point X_s is approximately 400 mm. Downstream the natural oscillations amplify which cause the transition (X≃ 580 mm) and the reattachment of turbulent flow. The pressure distribution is compared with the one at the same unswept model.

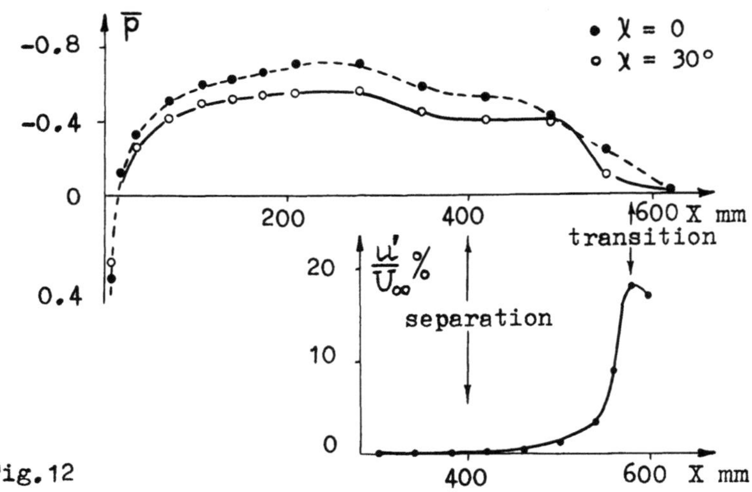

Fig.12

The mean flow distribution in the boundary layer and separation bubble is shown in Fig.13 where U_x, and U_z, components in the section Z=0 are presented. Here U_o is the external velocity at the edge of the viscous layer. All over the tested X-range under given experimental conditions the directions of $\vec{U_o}$ and

$\vec{U}_\infty$ were the same. Inside the boundary layer and separation bubble the mean flow direction differs from the inviscid one (at the separation point the surface streamline is oriented towards Z-direction).

The measurements at different Z-positions showed the 10% non-uniformity of the boundary layer thickness along the range $\Delta Z = 150$ mm in the vicinity of separation point. Neverseless it didn't result in noticeable difference in oscillation amplitudes in the transverse direction; also the transition line was parallel to Z-axis.

Spatial wave-packet of harmonic in time oscillations was excited in the boundary layer by means of the point source. The excitation frequency equal to 81 Hz was chosen from the instability range of the flow in adverse pressure gradient region. The strength of the excitation was limited in order to deal with small amplitude oscillations ($u'/U_\infty \leq 1\%$). The amplitude and phase (Φ) profiles of disturbances in the boundary layer and separation bubble are presented in Fig.14. One can see that the maximum intensity of oscillations is far enough from the wall and associated with the inflexion point in the mean velocity distribution.

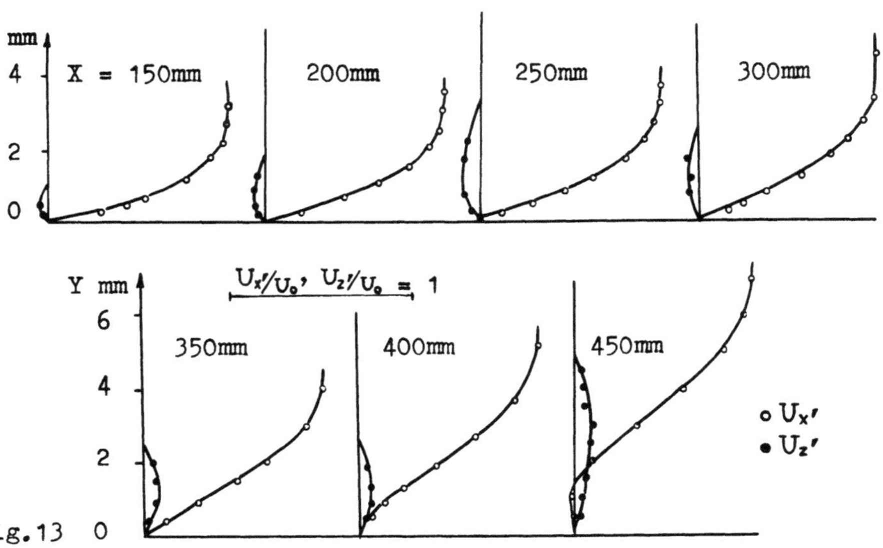

Fig.13

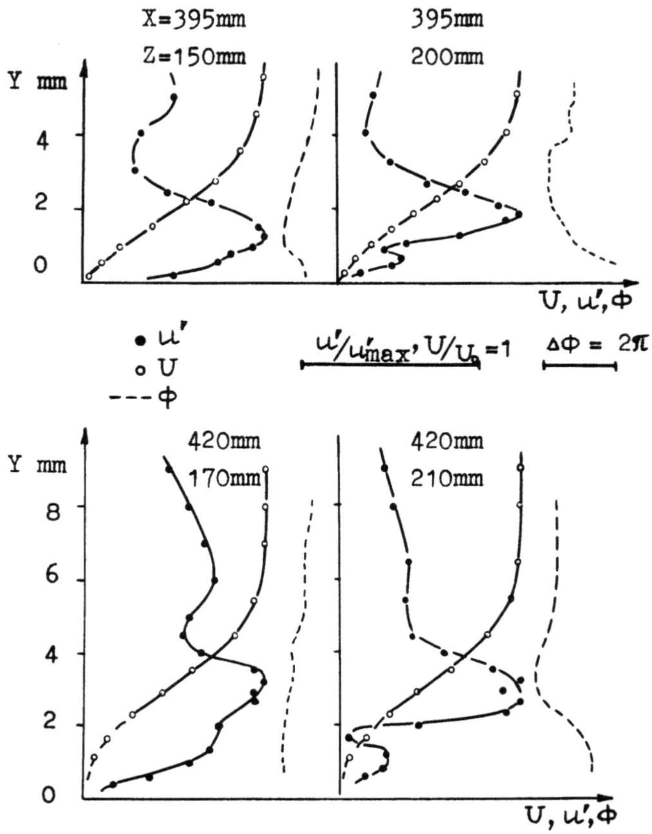

Fig.14

The spreading of the wave-packet in XZ-plane is illustrated in Fig.15 and 16. In Fig.15 the lines of constant amplitudes of oscillations are given (the intensity expressed in per cent of the free stream velocity is shown by figures). Fig.16 shows the constant phase lines in a 2π -interval. The data have been obtained in the two sections at different distances from the wall: $U/U_\infty=0.6$ (in the proximity of the inflexion point) and 0.1 - where the main- and crossflow components are of the same order. The results show that the wave packet propagates approximately in the external flow direction (the boundaries of the disturbed flow are stretched with $\vec{U}_0$). In the region of the positive pressure gradient the spreading of the packet increases, i.e. the difference in amplification rates between components of wave spectrum becomes larger. According to the amplitude and phase distributions in the separation bubble one may suppose that the most amplified components are those with the fronts nearly parallel to Z-axis.

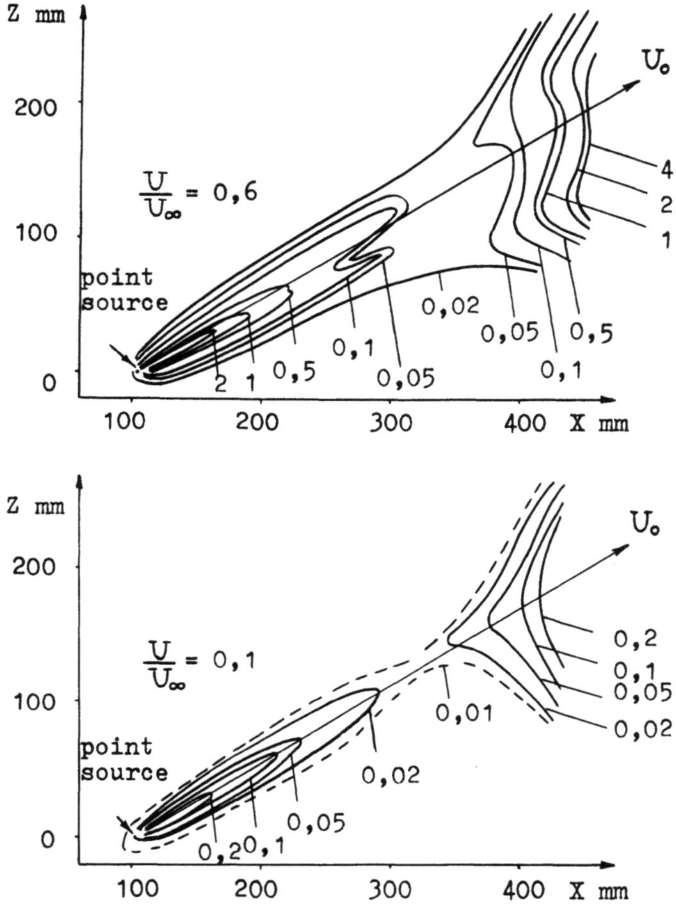

Fig.15

In other respects despite the crossflow component the given distributions are similar to those obtained at the straight wing [29] for the same airfoil and experimental conditions (Fig.17). Thus, in the experiments discussed considerable three-dimensionality inside the boundary layer at the adverse pressure gradient region and in the separation bubble have no significance for the development of laminar flow disturbances. Similar to that in two-dimensional separation the most important is the mean flow profile near the inflexion in the outer part of the layer (in separation bubble - the flow structure near its boundary) where the crossflow component is small enough.

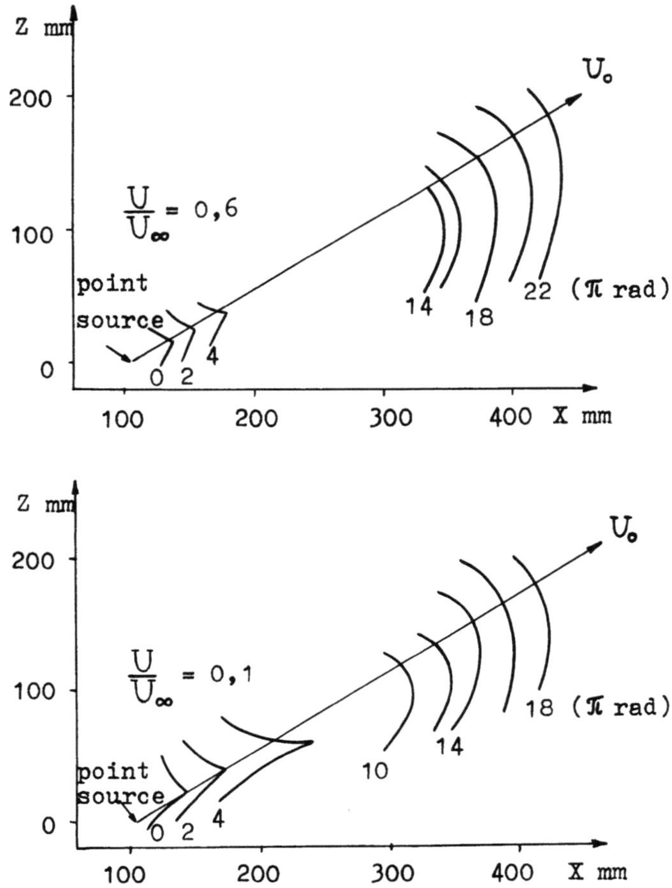

Fig.16

The spreading of the wave-packet as it was told about corresponds to the difference in amplification rates of the spectral components forming the packet. As a result from the initial wide spectrum generated by the point-source in the separation bubble the most amplified harmonics predominate. The one with the front parallel to Z-axis was introduced by means of vibrating ribbon placed in the section X=105 mm; the frequency of the oscillations was the same as for the point source excitation.

The smoke-wire visualization of instability waves excited in such a way is represented in the photo.

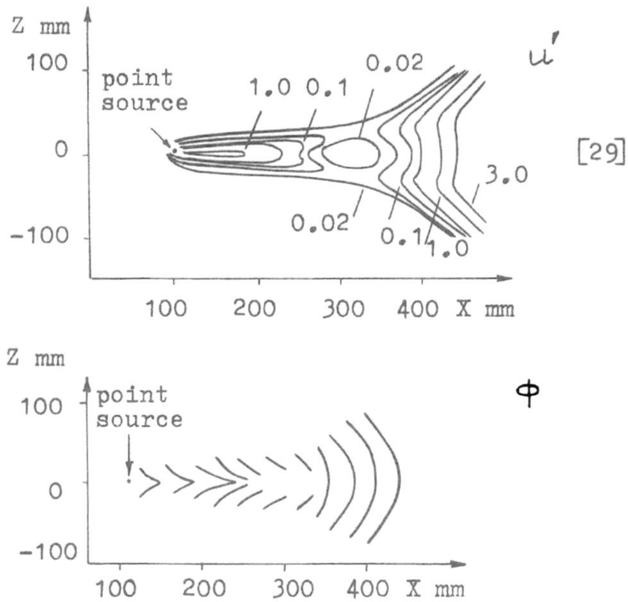

Fig.17

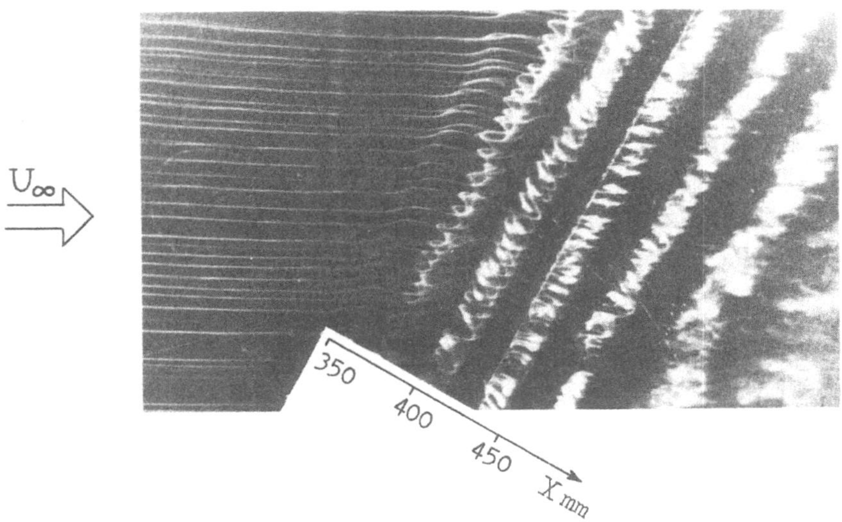

The spatial growth of the amplitude and the phase of the oscillations in the direction perpendicular to the wave front is shown in Fig.18. The measurements were carried out in the vicinity of the maximum intensity (U/U_∞=0.4). The phase velocity is equal to $0.47U_\infty$. The amplification rates for the vibrating ribbon and the point source excitation are nearly the same. The coincidence indirectly confirms that the oscillations with the front parallel to Z-direction belong to the spectral range of the highest amplification.

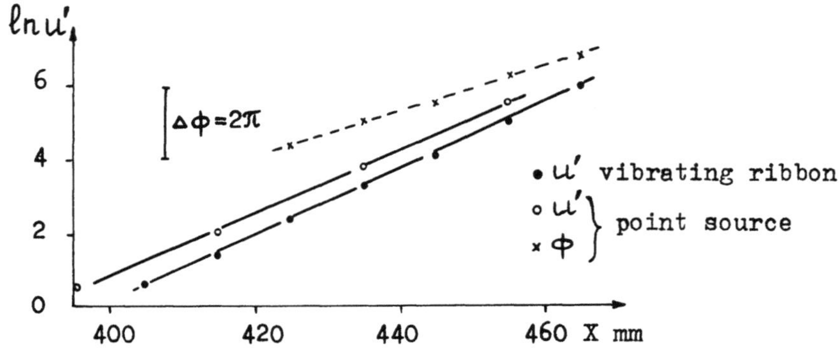

Fig. 18

Now separated flow at a surface with a fracture similar to that at the swept wing is considered. Such a separation was examined on the model which was the same as in [31] but turned at the angle of 30° in XZ-plane (Fig. 19). The experimental technique in this case remained the same as in the swept wing experiments.

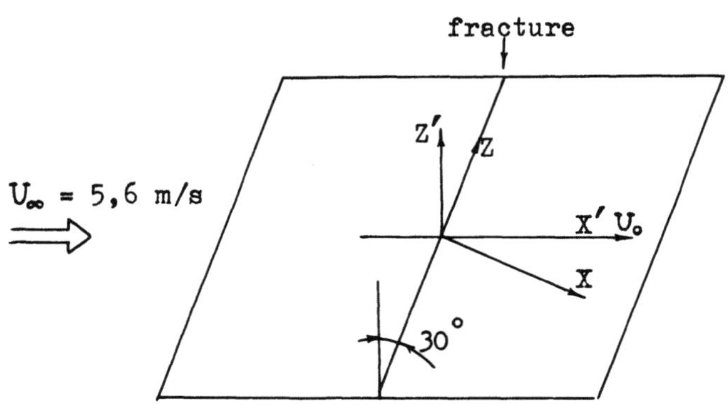

Fig. 19

The scheme of the flow is given in Fig.20. Fig.21 represents the frequency spectrum of disturbances in the transition region. One may distinguish two energy-containing parts: amplifying disturbances with the central frequency f_c equal to 230 Hz that are associated with the instability and the low-frequency oscillations caused by unsteadiness (but not insta-

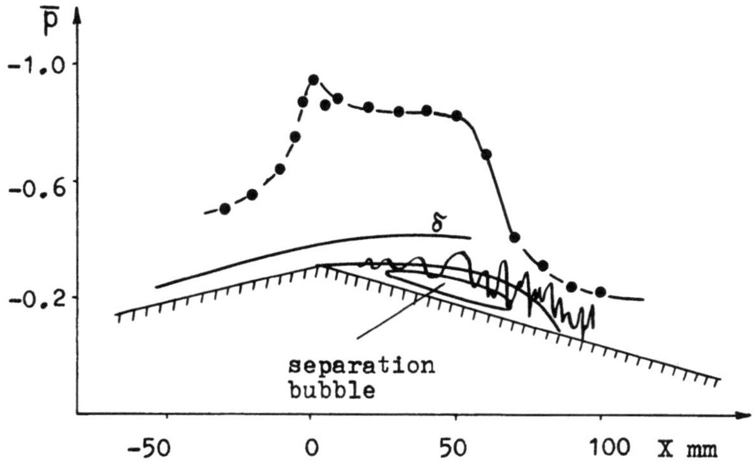

Fig.20

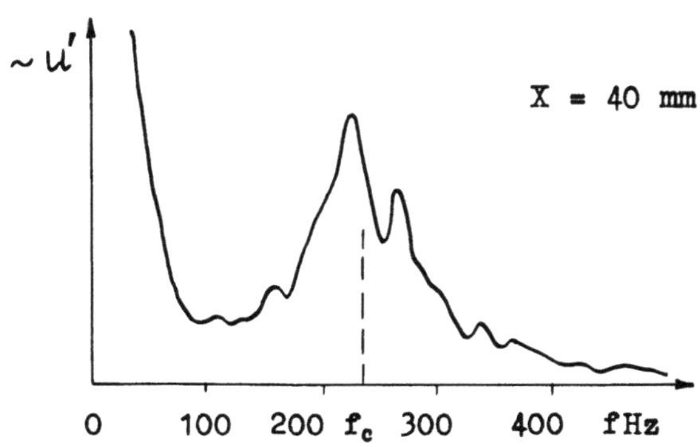

Fig.21

bility) of the separation bubble. Their contributions to the total intensity of oscillations are shown in Fig.22. Evidently the transition to turbulence in separated flow results from amplification just of the higher frequency oscillations.

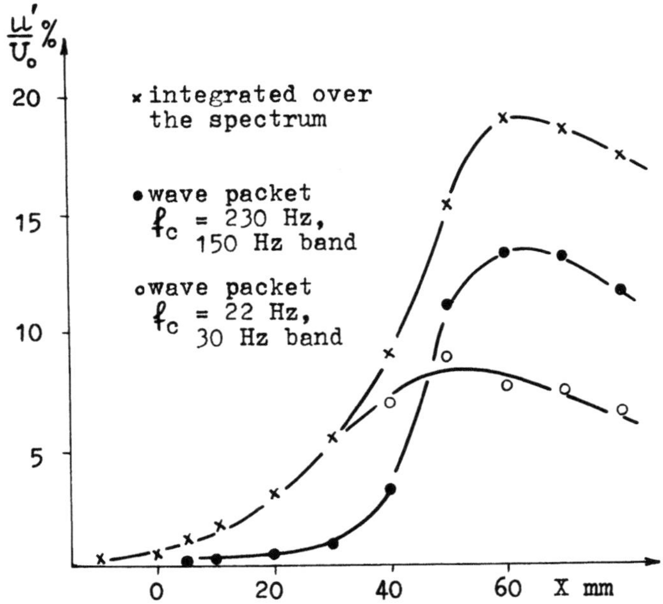

Fig.22

The disturbance profiles in the region of transition together with the mean flow components are given in Fig.23. Note that the relative magnitude of the crossflow component $U_{z'}/U_0$ here is greater than at the swept wing and is equal to 0.35 whereas at the wing it didn't exceed 0.10. Maximum intensity of the disturbances similar to that in the above considerations is located near the boundary of the separation bubble and is associated with the inflexion in the mean flow profile.

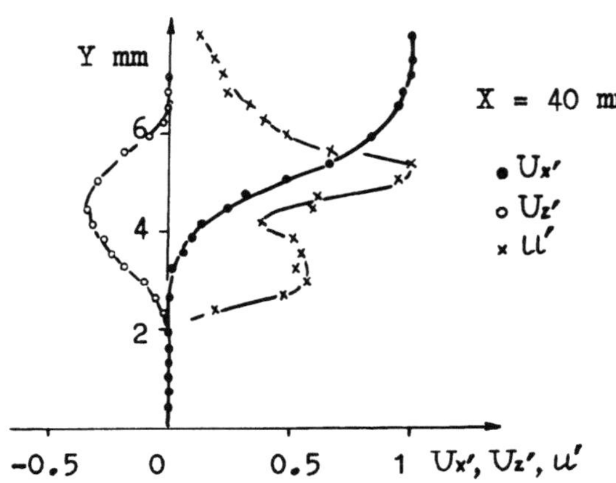

Fig.23

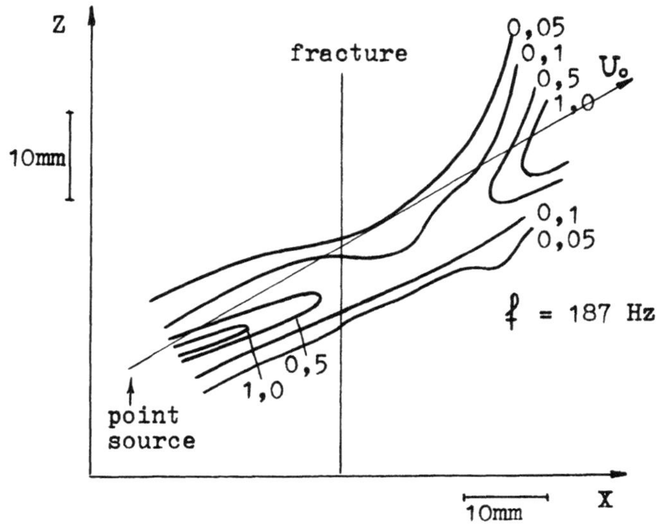

Fig.24

Despite the crossflow component that is much more than at the swept wing the wave-packet of the oscillations in this case also propagates approximately in the external flow direction. Fig.24 shows the distribution of oscillations amplitude in the wave-packet; the data are given for the maximum in the profile of disturbances.

The last example under discussion is the boundary layer separation from the body of revolution at incidence. There are a lot of results known concerning the separation from axisymmetrical bodies. In particular the thorough studying of separation from the cylindrical body with the hemisphere nose was undertaken in [37] where the different conditions were examined by means of vizualization technique.

When Reynolds number and free stream turbulence level are small enough three-dimensional separation takes place at the nose of such a body which is followed by transition to turbulence in separated flow. All the phenomena mentioned above that are associated with the instability of separated layer remain here qualitatively the same. Sensitivity of separation bubble to the external acoustic disturbances is shown in Fig.25 for $Re=1.6x10^4$. Under the excitation from the definite frequency range amplifying vortical disturbances were observed in the separation bubble which forcing the transition resulted in shortening of the separated flow. The mean flow rearrangement manifests itself in the distortion of velocity profiles and the pressure distribution on the surface,

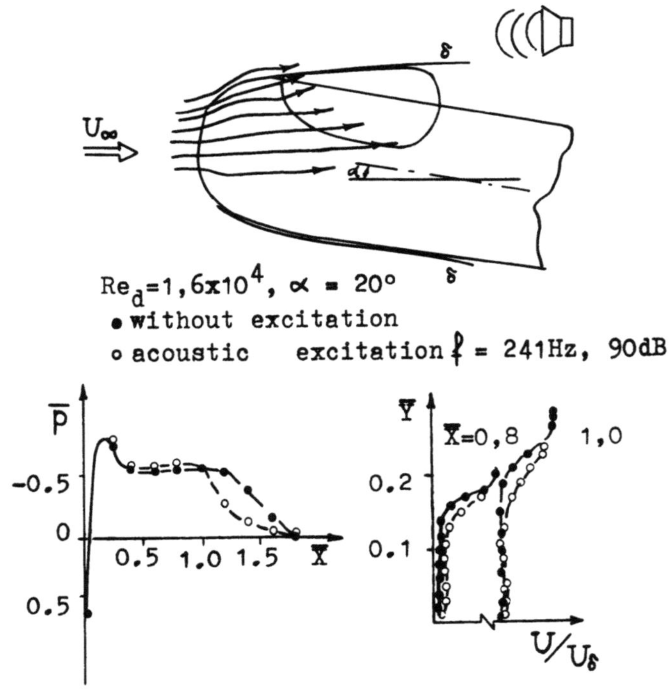

Fig.25

Conclusion

The above considerations are devoted mainly to the instability of transitional separation bubbles with respect to small amplitude oscillations. Despite the "sensitivity" of the mean flow to the disturbances their amplification doesn't depend on the initial amplitude when it remains small enough. Thus properly formulated linear problem should adequately describe this stage of transition in separated flows.

Among other aspects of the transition problem is the receptivity of separated flows to external disturbances. The recent work [38] is devoted to the correspondent theoretical analysis. The experimental results summarized in this paper demonstrate the transformation of acoustic oscillations into instability waves of separated flows and the efficiency of their influence on the mean flow characteristics.

References

1. Briley, W.R.: Journ. of Fluid Mech., 47 (1971).
2. Leal, L.G.: Journ. of Fluid Mech., 59 (1973).
3. Martsinkovsky, V.; Shkadov, V.Ya. Izv. AN SSSR, Mekh. Zhid. i Gaza, 1 (1985).
4. Williams, J.C.: Ann. Rev. Fluid Mech., 9 (1977).
5. Neiland, V.Ya.; Ruban, A.I.; Sychev, V.V. In: Modeli v mekhanike sploshnoy sredi, Novosibirsk, 1979.
6. Ruban, A.I.; Sychev, V.V. Uspekhi Mekhaniki, 2 (1979), N°4.
7. Stewartson, K.: Adv. Appl. Mech., 14 (1974).
8. Smith, F.T.: Journ. of Fluid Mech., 79 (1977).
9. Smith, F.T.; Brighton, P.W.M.; Jackson, P.S.; Hunt, J.C.R. Journ. of Fluid Mech., 113 (1981).
10. Diesperov, V.N.: Prikl. matem. i mekhan., 47 (1983), N°3.
11. Fomina, I.G.: Uchenie zapisky TsAGI, 14 (1983), N°5.
12. Horton, H.P. Aeronautical research council, England, 1967, CP 1073.
13. Roberts, W.B.: AIAA Journal, 18 (1980).
14. Crimi, P.; Reeves, B.L. AIAA Journal, 14 (1976).
15. Briley, W.R.; McDonald, H. Journ. of Fluid Mech., 68 (1975).
16. Vatsa, V.N.; Carter, J.E. AIAA Journal, 22 (1984).
17. Gaster, M.: AGARD CP, 1966, N°4.
18. Gault, D.E.: NASA TN, 1955, N°3505.
19. Mueller, T.J.; Batill, S.M. AIAA Journal, 20 (1982), N°4.
20. Chapman, D.R.; Kuehn, D.M.; Larson, H.K. NACA Rep., 1958, N°1356.
21. Chang, P.K.: Control of flow separation. Washington etc., 1976.

22. Rannaher, J.: Kurzfassung in machinenbantechnic, 1969, N°1.

23. Arena, A.V.; Mueller, T.J. AIAA Journal, 18 (1980).

24. Dovgal, A.V.; Zanin, B.Yu. In: Neustoichivost do- i sverhzvukovikh techeny, Novosibirsk, 1982.

25. Ryzhov, O.S.: In: Proc. of IUTAM-Symp. on Lam.-Turb. Transition, Springer-Verlag, 1985.

26. Taghavi, H.; Wazzan, A.R. Phys. of Fluids, 17 (1974).

27. Van Inger, J.V.: AGARD CP, 1975, N°168.

28. Dovgal, A.V.; Kozlov, V.V.; Kosorygin, V.S.; Ramazanov, M.P.: Dokl. AN SSSR, 258 (1981), N°1.

29. Gilev, V.M.; Dovgal, A.V.; Kozlov, V.V. Preprint ITAM SO AN SSSR, 1984, N°6-84,

30. Dovgal, A.V.; Kozlov, V.V. Izv. AN SSSR, Mekh. Zhid. i Gaza, 1983, N°2.

31. Dovgal, A.V.; Kozlov, V.V. Dokl. AN SSSR, 280 (1983), N°6.

32. Dovgal, A.V.: In: Proc. of IUTAM-Symp. on Lam.-Turb. Transition, Springer-Verlag, 1985.

33. Kozlov, V.V.: Prikladnaya matem. i tekh. phizika, 1985, N°2.

34. Kozlov, V.V.: In: Proc. of IUTAM-Symp. on Lam.-Turb. Transition, Springer-Verlag, 1985.

35. Kozlov, V.V.; Ramazanov, M.P. Izv. SO AN SSSR. Seria tekh. nauk, 1984, N°16, vip.3.

36. Dovgal, A.V.; Kozlov, V.V.; Simonov, O.A. Preprint ITAM SO AN SSSR, 1985, N°34-85.

37. Bippes, H.; Turk, M. DFVLR-FB, 1984, 84-20.

38. Goldstein, M.E.: Journ. of Fluid Mech., 145 (1984).

Session 3

The Brillouin – Villat Condition in the Theory of Separation

Oleg S. RYZHOV

Computing Center of the USSR Academy of Sciences, Moscow

Summary

The incompressible boundary layer is considered in the region located ahead of the streamline coming off the surface of a smooth body or a corner point on its contour. The potential of the external irrotational velocity field is provided by the free-streamline flow theory. As for the initial magnitude of the skin friction, its order is supposed to vary within broad limits both remaining finite and taking on great values. A similarity parameter is introduced that incorporates the Reynolds number, the absolute value of the first coefficient from the external pressure gradient expansion and the initial skin friction. The flow pattern evolution is traced and discussed as the gradual increase of the similarity parameter and the accompanying negative-to-positive sign reversal of the pressure gradient take place.

The External Potential Flow

Let us choose a nondimensional system of units for measuring both independent variables and desired fluid parameters where the generatrix curvature radius at the separation point, the external potential flow velocity and the density of fluid are the basic reference scales. We suppose the s- and n-axes of the curvilinear orthogonal co-ordinates to be directed along and normal to the body contour, respectively. Denote by u' and v' the corresponding components of the velocity perturbation vector, by p' the excess pressure in the external potential flow region. According to the theory of free-streamline flows of ideal incompressible fluid the complex velocity has a singularity in the vicinity of the separation point. In the linear approximation the Bernoulli integral reads $u'=-p'$, whence [1]

$$w'=p'+iv'=ib_{1/2}z^{1/2}+\ldots\ , \qquad z=s+in\ . \tag{1.1}$$

F. T. Smith, S. N. Brown, (Eds.)
Boundary-Layer Separation
IUTAM Symposium London 1986

In the limit $\arg z \to 0$ the excess pressure $p' \to 0$, if otherwise $\arg z \to \pi$, then

$$p' \to -b_{1/2}(-s)^{1/2} + \ldots \quad (1.2)$$

The constant $b_{1/2}$ is defined by the global pattern of the flow which can be a cavitating one. In the last case the value of $b_{1/2}$ varies in compliance with the model at hand though these variations are considerably small [2]. Within the framework of any local theory aimed at construction of the velocity field near the point of separation the constant $b_{1/2}$ should be chosen so as to take account of specific situations for flows around bodies of different shapes. The Brillouin - Villat condition $b_{1/2} = 0$ implies that the free-streamline curvature is the same as that of the body at the point of separation and there is no singularity in the pressure gradient. It is generally accepted that the Brillouin - Villat condition governs the separating flow past the bluff body behind which the fluid is either at rest or in slow recirculating motion [2]. In other regimes the free-streamline curvature and the pressure gradient are infinite according to (1.1) and (1.2) at the separation point and their signs depend on the sign of $b_{1/2}$.

The pressure of a corner point on the body is as a rule regarded to be a reason for separation and large-scale breakaway with a stagnant zone extending far beyond the vicinity of that point. As a matter of fact the crucial factor for the flowpattern under consideration to be realised is not only the existence of the corner point but also the presence in front of it of a point fixed by the Brillouin - Villat condition. If there is the point for which $b_{1/2} = 0$ on the smooth forepart of the body contour then it gives rise to a free streamline.

In Fig. 1a the circular cylinder is shown as a typical smooth bluff body. Crosses on its front mark the points where the free streamlines leave the surface in accord with the Brillouin - Villat condition [1, 2]. The pressure gradient in this case remains favourable everywhere over the wetted part of the cylinder up to the separation points where it vanishes.

The streamline breakaway shown in Fig. 1b occurs if we cut off the cylinder and consider the remaining short symmetric arc with both edges located ahead of the Brillouin - Villat points. The pressure gradient is favourable over the whole arc and its magnitude near the edges depends essentially on how far they are from the Brillouin - Villat points.

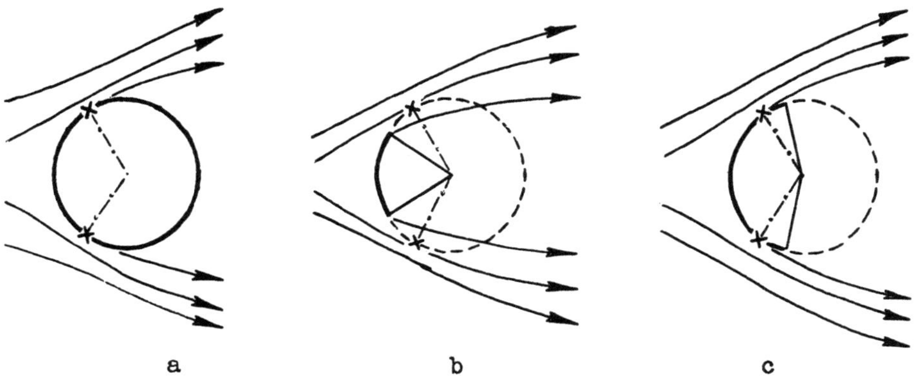

Fig. 1. Flow breakaway from circular arcs of various length

Fig. 1c presents the flow pattern with the free streamlines coming off the long arc at some distance from the edges. Since this arc contains obviously the Brillouin - Villat points its tips are submerged into a stagnant zone where a slow recirculating motion may have some other singularities.

For the flows in Figs. 1a-c the sign and the absolute value of the coefficient $b_{1/2}$ serve as characteristics of relative position of the arc edges and the Brillouin - Villat points. So, $b_{1/2} < 0$ for the short arc (Fig. 1b) and $b_{1/2} > 0$ at the edges of the long one (Fig. 1c). The velocity fields around both types of arcs are typical for bodies with a jump in the slope of their contours.

The Linear Region of Pre-Separated Flow

Squire [3] was probably the first to give convincing arguments in favour of why the separation of viscous incompressible flow from a smooth body obeys the limiting form $b_{1/2} \rightarrow 0$ of the

Brillouin - Villat condition as the Reynolds number $R \to \infty$. Later Sychev [4] showed that with viscosity taken into account the constant $b_{1/2} > 0$ is of order $R^{-1/16}$ if at the beginning of the region of the pre-separated boundary layer the reduced skin friction $\lambda \sim 1$. Sychev's theory was developed by Smith [5] who calculated numerically the solution of the basic nonlinear boundary-value problem. The separation of viscous flow from the corner point of an airfoil was treated by Ackerberg [6] and Ruban [7], both authors assuming that $b_{1/2} < 0$ and $|b_{1/2}| \sim \lambda \sim 1$. The last conjecture holds true not for all shapes: as for the short arc in Fig. 1b it means that the edges of this particular body are some finite distance apart from the Brillouin - Villat points.

The Prandtl boundary layer with the singularity (1.2) in the external pressure gradient was studied by Brown & Stewartson [8] for the related problem on the laminar flow past a plate at incidence with the angle $\alpha = b_{1/2} \sim R^{-1/16}$ $(\lambda \sim 1)$. A more comprehensive analysis was given by Messiter & Enlow [9]. Our purpose is to provide a unified treatment of flow regimes which are realized around different bodies in dependence on $b_{1/2}$ and λ.

In the viscous sublayer adjacent to the solid surface let us perform the affine transformation

$$s = \lambda^{-2} \Delta x_1 \quad , \quad n = R^{-1/2} \lambda^{-1} \Delta^{1/3} y_3 \tag{2.1}$$

of both the co-ordinates and normalize accordingly

$$\psi = R^{-1/2} \lambda^{-1} \Delta^{2/3} \psi_3 \quad , \quad p' = \Delta^{3/2} p_1 \tag{2.2}$$

the streamfunction and the excess pressure. In the new variables the Prandtl equation reads

$$\frac{\partial \psi_3}{\partial y_3} \frac{\partial^2 \psi_3}{\partial x_1 \partial y_3} - \frac{\partial \psi_3}{\partial x_1} \frac{\partial^2 \psi_3}{\partial y_3^2} = \frac{\partial^3 \psi_3}{\partial y_3^3} - \Delta^{5/6} \frac{d p_1}{d x_1} \tag{2.3a}$$

$$p_1 = -\operatorname{sign} b_{1/2} (-x_1)^{1/2} + \ldots \qquad (2.3b)$$

We suppose hereinafter that being expressed in terms of the Reynolds number $R \to \infty$ the ratio $\Delta = |b_{1/2}|/\lambda \to 0$ as a result of at least one of the two conditions $|b_{1/2}| \to 0$ or $\lambda \to \infty$.

Insofar as Δ is a small parameter, the solution of the equation (2.3a) may be sought for in the form of the asymptotic sequence

$$\psi_3 = \frac{1}{2} y_3^2 + \Delta^{5/6} (-x_1)^{1/2} f^{(1/2)}(\xi) + \ldots, \qquad \xi = (-x_1)^{-1/3} y_3 . \qquad (2.4)$$

The function $f^{(1/2)}$ satisfies the linear differential equation

$$\frac{d^3 f^{(1/2)}}{d\xi^3} - \frac{1}{3}\xi^2 \frac{d^2 f^{(1/2)}}{d\xi^2} + \frac{1}{2}\xi \frac{d f^{(1/2)}}{d\xi} - \frac{1}{2} f^{(1/2)} = \frac{1}{2} \operatorname{sign} b_{1/2}$$

and the boundary conditions $f^{(1/2)} = df^{(1/2)}/d\xi = 0$ as $\xi = 0$ which follow from the no-slip requirements at the wall. Besides, the exponentially growing, as $\xi \to \infty$, terms are to be excluded from the desired solution. The properties of the function $f^{(1/2)}$ were examined by Brown & Stewartson [8]. Using them, one can get, in particular

$$\psi_3 = \frac{1}{2} y_3^2 + \Delta^{5/6} \Big\{ \operatorname{sign} b_{1/2} \Big[-\frac{2\Gamma(1/6)}{3\Gamma(2/3)} y_3^2 +$$

$$+ 2^{1/3} 3^{1/3} \Gamma(1/3) (-x_1)^{1/6} y_3 - (-x_1)^{1/2} + \ldots \Big] \Big\} + \ldots \qquad \text{as } y_3 \to \infty. \qquad (2.5)$$

For the most of the boundary layer the longitudinal coordinate x_1 and the pressure p_1 remain the same as in (2.1) and (2.2), respectively, while the lateral co-ordinate y_2 and the stream-function ψ_2 are introduced by $n = R^{-1/2} \lambda^{-1} y_2$, $\psi = R^{-1/2} \lambda^{-1} \psi_2$. In these variables the Prandtl equation

$$\frac{\partial \psi_2}{\partial y_2} \frac{\partial^2 \psi_2}{\partial x_1 \partial y_2} - \frac{\partial \psi_2}{\partial x_1} \frac{\partial^2 \psi_2}{\partial y_2^2} = \Delta \frac{\partial^3 \psi_2}{\partial y_2^3} - \Delta^{3/2} \frac{dp_1}{dx_1} \qquad (2.6)$$

where p_1 is prescribed by means of (2.3b) contains also the small parameter Δ. The asymptotic sequence (2.5) dictates the form

$$\psi_2 = \psi_{21}^{(0)} + \Delta(\psi_{22}^{(0)} + \dots) + \Delta^{3/6}(-x_1)^{1/6}(\psi_{23}^{(1/6)} + \dots) +$$

$$+ \Delta^{3/2}(-x_1)^{1/2}(\psi_{24}^{(1/2)} + \dots) + \dots \tag{2.7}$$

of the expansion which stands for the solution of (2.6). The functions $\psi_{21}^{(0)}$, ..., $\psi_{24}^{(1/2)}$ involved in (2.7) depend on y_2 only and can be derived from the results of Messiter & Enlow [9]. The reduced skin friction $\tau_w = (\partial^2\psi_3/\partial y_3^2)_{y_3=0}$ is defined as follows

$$\tau_w = 1 - \Delta^{5/6} \operatorname{sign} b_{1/2} \frac{\Gamma(1/6)\Gamma(1/3)}{2\, 3^{1/3}\pi^{1/2}\Gamma(2/3)}(-x_1)^{-1/6} + \dots \tag{2.8}$$

being valid for arbitrary values of $b_{1/2}$ and λ , that is for different regimes of the free-streamline flows past smooth bodies as well as arcs exemplified in Figs. 1b,c (or past bodies with corner points).

The Similarity Parameters

As one can see from (2.8) the correction term on the right-hand side becomes of the order 1 if $-x_1 \sim \Delta^5$. This length scaling is inherent in the nonlinear region where the above analysis fails. In accord with this remark let us put

$$x_1 = \Delta^5 \bar{x}, \quad y_3 = \Delta^{5/3}\bar{y}, \quad \psi_3 = \Delta^{10/3}\bar{\psi}, \quad p_1 = \Delta^{5/2}\bar{p} \tag{3.1}$$

in the viscous wall sublayer and reduce the Prandtl equation to the form

$$\frac{\partial\bar{\psi}}{\partial\bar{y}}\frac{\partial^2\bar{\psi}}{\partial\bar{x}\partial\bar{y}} - \frac{\partial\bar{\psi}}{\partial\bar{x}}\frac{\partial^2\bar{\psi}}{\partial\bar{y}^2} = \frac{\partial^3\bar{\psi}}{\partial\bar{y}^3} - \frac{d\bar{p}}{d\bar{x}} \tag{3.2}$$

where the terms with the small parameter Δ are absent.

Since the similar variable $\xi = (-\bar{x})^{-1/3}\bar{y}$ remains invariant with respect to the affine transformation (3.1) we derive, with (2.4) in mind, the requirements

$$\bar{\psi} \to \frac{1}{2}\bar{y}^2 + (-\bar{x})^{1/2} f^{(1/2)}(\xi) + \dots, \quad \bar{p} \to -\operatorname{sign} b_{1/2}(-\bar{x})^{1/2} + \dots \tag{3.3}$$

which should be posed as $\bar{x} \to -\infty$, $\xi = const$. In order to obtain the boundary condition for the equation (3.2) as $\bar{y}\to\infty$ one needs to know the solution for the main part of the boundary layer in the nonlinear region. Following with minor alterations the reasoning of the previous section we have

$$\psi_2 = \psi_{21}^{(0)} + \Delta\psi_{22}^{(0)} + \Delta^2 A(\bar{x})\frac{d\psi_{21}^{(0)}}{dy_2} + \dots \qquad (3.4)$$

The displacement thickness A can be chosen arbitrarily except for its behaviour $A \to \operatorname{sign} b_{1/2}\, 2^{1/3}3^{1/3}\Gamma(1/3)(-\bar{x})^{1/6} + \dots$ at infinity $\bar{x}\to-\infty$ that is determined from the matching of the asymptotic expansions (2.7) and (3.4). Matching now the streamfunction $\bar{\psi}$ with ψ_2 we get the constraint

$$\bar{\psi} \to \frac{1}{2}\bar{y}^2 - \operatorname{sign} b_{1/2}\frac{2\Gamma(1/6)}{3\Gamma(2/3)}\bar{y}^{3/2} + A(\bar{x})\bar{y} + \dots \quad \text{as } \bar{y}\to\infty \qquad (3.5)$$

at the outer edge of the viscous wall sublayer.

As far as the pressure gradient in (3.2) is concerned, we need to examine the potential motion where the lateral co-ordinate $n = \lambda^{-2}\Delta^6\tilde{y}$ has the same scaling as the longitudinal one. It is convenient to analyse the potential flow directly in terms of the components $u' = \Delta^4\tilde{u}$, $v' = \Delta^4\tilde{v}$ of the disturbed velocity and the excess pressure $p' = \Delta^4\tilde{p}$. As it follows from the matching of these functions with the analogous ones for the most of the boundary layer the conventional slip condition

$$\tilde{v} = -K\frac{dA}{d\bar{x}} \ , \quad K = R^{-1/2}|b_{1/2}|^{-8}\lambda^9 \quad \text{as } \tilde{y}=0, \ \bar{x}<0 \qquad (3.6)$$

for an inviscid fluid holds.

The equation (3.2), the initial conditions (3.3) and the boundary conditions (3.5) and (3.6) comprise only two similarity parameters $\operatorname{sign} b_{1/2}$ and K which depend on the type of flow under consideration. One should distinguish between four cases: 1) $b_{1/2}<0$, $K\to0$; 2) $b_{1/2}<0$, $K\sim1$; 3) $b_{1/2}$ changes its sign from negative to positive, $K\to\infty$; 4) $b_{1/2}>0$, $K\sim1$. In spite of the fact that the velocity field in the linear region of pre-separated flow admits the unified mathematical treatment independently of the sign of $b_{1/2}$ and the magnitude of K, the further analysis has specific features in each of the cases pointed out.

The Nonlinear Region of Pre-Separated Flow

When $b_{1/2}<0$ and $K\to 0$ we have for the complex velocity $\tilde{w}=\tilde{p}+i\tilde{v}$ a problem on the flow past the original body insomuch as $\tilde{v}(\bar{x},0)=0$ owing to (3.6). The excess pressure is determined then by its limiting value at infinity. Thus $\bar{p}=-\operatorname{sign} b_{1/2}(-\bar{x})^{1/2}+\ldots=(-\bar{x})^{1/2}+\ldots$, while the displacement thickness A in the boundary condition (3.5) remains unknown. It is to be found in the course of integration of (3.2) which starts with the distribution (3.3) for the streamfunction as $\bar{x}\to-\infty$.

It is known from the Prandtl boundary-layer theory [10] that the influence of the initial data is essential over a distance estimated as some few displacement thicknesses. That is why one can in the first approximation bypass the initial data at all when dealing with the asymptotic behaviour of the solution as $\bar{x}\to 0-$. Ackerberg was the first to make use of this simple idea as the basis of his work [6]. Following him let us divide the flow region under consideration where the longitudinal and lateral co-ordinates are scaled by (3.1) into two sublayers. As a result we have the velocity field consisting of four decks.

To meet the no-slip requirements at the wall the viscous shear stresses must be taken into account. Allowing for the fact that the small parameter Δ does not enter the coefficients of (3.2) we arrive at a conclusion concerning the similar form

$$\bar{\psi}=2^{1/2}(-\bar{x})^{5/8}f(\eta),\quad \eta=2^{-1/2}(-\bar{x})^{3/8}\bar{y} \tag{4.1}$$

of the asymptotic solution in the lowest of four decks. The substitution of (4.1) into (3.2) yields the nonlinear differential equation

$$\frac{d^3f}{d\eta^3}-\frac{5}{4}f\frac{d^2f}{d\eta^2}+\frac{1}{2}\left(\frac{df}{d\eta}\right)^2+1=0\ .$$

Bearing in mind the properties of its solution let us write down the asymptotic representation

$$\bar{\Psi} = 2^{-1/3} a_0 \bar{y}^{-5/3} + 2^{1/6} a_3 (-\bar{x})^{3/8} \bar{y}^{2/3} + 2^{1/3} a_4 (-\bar{x})^{1/2} \bar{y}^{1/3} + \dots \quad (4.2)$$

for the streamfunction as $-\bar{x} \ll 1$ is fixed and $\eta \to \infty$. The coefficients $a_0 = 1.950718$, $a_3 = -1.577568$, $a_4 = 9(5a_0)^{-1}$ involved have been computed by Ackerberg [6].

The form of asymptotics (4.2) dictates the existence of a sublayer with the properties much akin to those of the main part of the boundary layer. Indeed, in the new intermediate region

$$\bar{\Psi} = \bar{\Psi}_0 + (-\bar{x})^{3/8} \bar{\Psi}_1 + (-\bar{x})^{1/2} \bar{\Psi}_2 + \dots \quad (4.3)$$

Except for the limiting properties as $\bar{y} \to 0$ and $\bar{y} \to \infty$ the function $\bar{\Psi}_0$ can be chosen arbitrarily while the other functions $\bar{\Psi}_1$ and $\bar{\Psi}_2$ are given as follows

$$\bar{\Psi}_1 = C \frac{d\bar{\Psi}_0}{d\bar{y}}, \quad \bar{\Psi}_2 = \left[\mathcal{D} - 2^{2/3} 3 \left(\frac{3}{5}\right)^2 \frac{1}{a_0^2 \bar{y}^{1/3}}\right] \frac{d\bar{\Psi}_0}{d\bar{y}}$$

$$\mathcal{D}(\bar{y}) = \int_0^{\bar{y}} \left[2^{2/3} \left(\frac{3}{5}\right)^2 \frac{1}{(a_0 y^{4/3})^2} - \left(\frac{d\bar{\Psi}_0}{dy}\right)^{-2} \right] dy$$

where the constant $C = 2^{1/2} 3 a_3 (5a_0)^{-1}$. Since $\bar{\Psi}_1 \to C\bar{y}$ and $\bar{\Psi}_2 \to \mathcal{D}(\infty)\bar{y}$ as $\bar{y} \to \infty$, the desired asymptotic distribution of the displacement thickness according to (4.3) is

$$A = C(-\bar{x})^{3/8} + \mathcal{D}(\infty)(-\bar{x})^{1/2} + \dots$$

The reduced skin friction $\tau_w = 2^{-1/2} \lambda (-\bar{x})^{-1/8} d^2 f(0)/d\eta^2$ corresponding to the solution at hand leads to the estimation $-S \sim |b_{1/2}|^6 \tau_w^{-8}$, which for $K \sim 1$ and $\tau_w \sim \lambda$ is equivalent to the first of the scalings (3.1) ensuing from all other reasonings. However, as $K \to 0$ the local magnitude of the skin friction, increasing in the nonlinear region, gradually surpasses by far its initial value λ. The preliminary growth of the skin friction ensures the following occurence of the interaction between the boundary layer and the external potential flow.

Let us evaluate the coordinate $\bar{x}$ where the aforementioned process starts. As far as the pressure is constant across all the three decks constituting the boundary layer in the case under consideration, we conclude that $\bar{p} \sim \tilde{p} \sim \tilde{v}$. The latter quantity can be found from (3.6) with A being expressed by the first term on the right-hand side of (4.4). The resulting estimation $\bar{p} \sim K(-\bar{x})^{-5/8}$ must be of the same order as the initial excess pressure $(-\bar{x})^{1/2}$. Hence $-\bar{x} \sim R^{-4/9} |b_{1/2}|^{-64/9} \lambda^8$. The ultimate relation

$$-s \sim R^{-4/9} |b_{1/2}|^{-10/9} \qquad (4.5)$$

does not comprise $\lambda \ll R^{1/8} |b_{1/2}|^{8/9}$, which is in full accord with the above assertion that the asymptotic properties of flow as $\bar{x} \to 0-$ are quite independent of the starting velocity profile (as $\bar{x} \to -\infty$). In order to describe the developing interaction process one can apply the solution by Ruban [7] constructed on assumption that $\Delta \sim 1$. This appears to be a simplification to some extent for the linear part of pre-separated region is omitted from the analysis.

The Interactive Boundary Layer Leaving the Edge of an Arc

Let $K \sim 1$ whence $|b_{1/2}| \sim R^{-1/16} \lambda^{9/8}$. Under this condition the interacting boundary layer comes into play without giving rise to the foregoing nonlinear region with the known pressure distribution. The longitudinal extent of the free-interaction zone obeys the estimate (4.5) though $\lambda \sim R^{1/18} |b_{1/2}|^{8/9}$ now. As for the sign of $b_{1/2}$, the possibility of its positive-to-negative reversal is admitted. The passage of $b_{1/2}$ through zero implies that the boundary layer becomes liable to the effect of the adverse pressure gradient. Examining $b_{1/2} > 0$ let us however confine ourselves to the regimes when separation occurs at the arc edges (or the corner point of a body).

We align the s-axes behind the separation point with the free streamline and put $\tilde{z} = \bar{x} + i\tilde{y}$. The moving fluid is divided from that being at rest (in compliance with the Kirchhoff scheme for an ideal inviscid flow) by the shear layer. Then according to Ruban's arguments [7] the pressure $\bar{p} = 0$ throughout this layer as $\bar{x}>0$. Remembering (3.6) we pose for the analytic function $\tilde{w}=\tilde{p}+i\tilde{v}$ the Hilbert problem in the upper semi-plane with the following data

$$\mathrm{Im}\,\tilde{w}=-K\frac{dA}{d\bar{x}} \quad \text{as } \bar{x}<0\,, \qquad \mathrm{Re}\,\tilde{w}=0 \quad \text{as } \bar{x}>0 \tag{5.1}$$

on the axes $\tilde{y}=0$. Besides, we must, owing to (3.3), satisfy the limiting condition $\tilde{w}\to i\,\mathrm{sign}\,b_{1/2}\,\tilde{z}^{1/2}$ as $|\tilde{z}|\to\infty$ in spite of that the boundary value $\mathrm{Im}\,\tilde{w}\to\frac{1}{6}\mathrm{sign}\,b_{1/2}\,2^{-2/3}3^{-2/3}\times\Gamma(1/3)K(-\bar{x})^{-5/6}$ as $\bar{x}\to-\infty$. In the class of functions with such a growth at infinity the problem would be of the zero index to ensure the uniqueness of a solution if $|\tilde{w}|\sim|\tilde{z}|^m$ with $1/2<m\leqslant 3/2$ as $|\tilde{z}|\to 0$. The behaviour of $\tilde{w}$ stated is quite permissible from all points of view since it would imply weakening or even smoothing out a singularity that stems in the flow of ideal fluid from neglect of viscous stresses. However, it is impossible to meet the requirements both at infinity and origin without introducing an additional constraint for the limit $\tilde{w}\to i\,\mathrm{sign}\,b_{1/2}\,\tilde{z}^{1/2}$ as $|\tilde{z}|\to\infty$ is not in line with the asymptotics of $\mathrm{Re}\,\tilde{w}$ and $\mathrm{Im}\,\tilde{w}$ as given by (5.1).

We proceed from the estimates fixing only the growth of $\tilde{w}$ in both singular points. Conventional methods of the theory of functions of a complex variable then give

$$\tilde{w}=\frac{K}{\pi i}\,\tilde{z}^{3/2}\int_{-\infty}^{0}\frac{dA/d\xi}{(-\xi)^{3/2}(\xi-\tilde{z})}\,d\xi\,. \tag{5.2}$$

However, as follows from (5.2), the coefficient of the leading $\tilde{z}^{1/2}$-term in the asymptotic expansion $\tilde{w}$ at infinity is strictly speaking different from $i\,\mathrm{sign}\,b_{1/2}$. To make it equal to the prescribed value it is necessary to put the constraint

$$\int_{-\infty}^{0}\frac{dA/d\xi}{(-\xi)^{3/2}}\,d\xi=\mathrm{sign}\,b_{1/2}\,\frac{\pi}{K} \tag{5.3}$$

on the desired displacement thickness A. As a result the excess pressure in the region $\bar{x}<0$ of the boundary layer is expressed by

$$\bar{p} = (-\bar{x})^{1/2}\left[\operatorname{sign} b_{1/2} + \frac{K}{\pi}\int_{-\infty}^{c} \frac{dA/d\xi}{(-\xi)^{1/2}(\xi-\bar{x})}\,d\xi\right] . \tag{5.4}$$

The boundary-value problem for the Prandtl equation (3.2) is now formulated completely (if we allow for the no-slip conditions on the body surface) but the relation (5.3) is an additional one and, thus, should be verified. If (5.3) was not fulfilled then the application of (5.4) for the pressure would imply that its singularity is not smoothed out in the zone of the interacting boundary layer.

The Linear Interaction

The simplification of the problem at hand is of principal meaning for clarifying the issue on whether the additional condition (5.3) is properly imposed or must be abandoned. To this effect we proceed to treat the passage of $b_{1/2}$ through zero. Suppose that the similarity parameter $K \to \infty$ due to $|b_{1/2}| \to 0$. Let be $b_{1/2} = k b_{1/2}^{c}$ where k is some positive number and substitute $|b_{1/2}^{c}|$ for $|b_{1/2}|$ in all above transformations of both the independent variables and desired functions. The equation (3.2) is invariant with respect to this replacement and preserves its form, on the contrary, in (3.3), (3.5) we have $k \operatorname{sign} b_{1/2}^{c}$ instead of $\operatorname{sign} b_{1/2}$ and on the right-hand side of (3.6) the modified similarity parameter $K^{0} = R^{-1/2}|b_{1/2}^{c}|^{-8}\lambda^{9}$ emerges. Let k be chosen in such a way that $K^{0} = 1$. Hence the normalization $|b_{1/2}^{c}| = R^{-1/16}\lambda^{9/8}$ arises with the obvious consequence $k = R^{1/16}|b_{1/2}|/\lambda^{9/8} = K^{-1/8}$, i.e. $k \sim 1$ if $K \sim 1$, but $k \to 0$ as $K \to \infty$. The longitudinal extent of the interactive boundary layer region is, as distinct from (4.5), expressed in terms of the Reynolds number and the initial magnitude of the skin friction. The thickness of the viscous wall sublayer is estimated as $h \sim R^{-5/8}\lambda^{-3/4}$.

In the boundary-value problem modified in this fashion the small parameter k enters the relations (3.3), (3.5) instead of the interaction condition (3.6). The problem can be linearized if we put

$$\bar{\Psi} = \frac{1}{2}\bar{y}^2 + k\bar{\Psi}'(\bar{x},\bar{y}), \quad \bar{p} = k\bar{p}'(\bar{x}), \quad A = k\bar{A}'(\bar{x})$$

its final formulation for the disturbed streamfunction $\bar{\Psi}'$ in appropriately taken variables being the same as presented by Brown & Stewartson [8] in connection with the evaluation of the antisymmetric part of the pressure over the plate at incidence. Let us designate the angle of incidence by α. Insofar as the reduced angle is $\bar{\alpha} = \alpha R^{1/16} \lambda^{-9/8}$ it corresponds to $k \, \mathrm{sign} b^c_{1/2}$. We are now in a position to make use of the solution from [8], which defines the pressure distribution for $\bar{x} < 0$ in the following way

$$\frac{d\bar{p}'}{d\bar{x}} = \mathcal{D}_0 \int_0^\infty \frac{\exp(\gamma^{3/4}\bar{x}\zeta)\exp[I(\zeta)]}{\zeta^{1/2}(1+\zeta^{8/3})} d\zeta \tag{6.1a}$$

$$\mathcal{D}_0 = \mathrm{sign} b^c_{1/2} \frac{1}{2\pi^{1/2}} \gamma^{3/8}, \quad \gamma = 3^{2/3}\Gamma^{-1}(1/3) \tag{6.1b}$$

$$I = \frac{2}{3\pi} \int_0^\infty \frac{\sigma^{1/3} \ln(\sigma+\zeta)}{1 - 3^{1/2}\sigma^{4/3} + \sigma^{8/3}} d\sigma . \tag{6.1c}$$

Assuming in (6.1a) the co-ordinate $\bar{x} \to 0-$ we obtain the finite value of $d\bar{p}'/d\bar{x}$ what is in line with the exponent $m =$ = 1 in the asymptotic estimate of $|\tilde{w}|$ as $|\tilde{z}| \to 0$. The variation of the displacement thickness along the contour of the body is stated through

$$\frac{d^2\bar{A}'}{d\bar{x}^2} = -\mathcal{D}_0 \int_0^\infty \frac{\exp(\gamma^{3/4}\bar{x}\zeta)\zeta^{5/6}\exp[I(\zeta)]}{1+\zeta^{8/3}} d\zeta \tag{6.2}$$

with the constants $\mathcal{D}_0$, γ and the integral I given by (6.1b) and (6.1c), respectively. The value of $d^2\bar{A}'/d\bar{x}^2$ ensuing from (6.2) in the limit $\bar{x} \to 0-$ is finite whence we get the same exponent m = 1.

Thus, according to the solution of the linearized problem the singularity of the type $(-\bar{x})^{1/2}$ in the pressure distribution that originates in the flow of ideal fluid from the neglect of shear stresses is smoothed out by the interacting boundary layer. As for the integral condition (5.3) with K^0 = 1 substituted for K on the right-hand side, its direct verification based on (6.1a-c) and (6.2) is nevertheless rather difficult. If, however, we suppose that this condition fails then the pressure must due to (5.4) be singular at the edges of the arc in contradiction with the solution under consideration.

Making use of the asymptotic properties of the solution with $b_{1/2}>0$ we can, after Brown & Stewartson [8], points out stagnation points in the external inviscid flow at a distance

$$-s = k^2 R^{-7/8} \lambda^{-1/4} \frac{\cos^2 \pi/8}{2\gamma^{3/2}}$$

from the edges of the arc. In the limiting case $k \to 0$ under consideration these points vanish under the influence of viscosity. If otherwise $b_{1/2}<0$ the stagnation points do not appear in the external flow.

Remarks on Separation from the Smooth Part of the Contour

On condition that $b_{1/2}>0$ the integration of the nonlinear Prandtl equation (3.2) with the pressure prescribed by the Hilbert integral (5.4) must give the value of $k=k_s$ for which the skin friction at the edges of the arc becomes zero for the first time. Though calculations to determine k_s are not carried out yet the desired quantity can be approximately estimated as being equal to 0.4. When $k>k_s$ the edges are submerged into the stagnation zone. The growth of k is limited by the value $k=k_*>k_s$ which is associated with separation from the closed contour (smooth body).

In this case the local flow pattern of the velocity field is provided by the Sychev's [4] theory. Its main distinction from the exposed above lies in the fact that the Hilbert integral for the induced pressure is replaced by the Schwarz integral

$$\bar{p} + k(-\bar{x})^{1/2} H(-\bar{x}) = \frac{1}{\pi} \int_{-\infty}^{\infty} \frac{dA/d\xi + k\xi^{1/2} H(\xi)}{\bar{x}-\xi} d\xi$$

where H is a step-function. As a result we arrive at the nonlinear eigen-value problem. The independent numerical investigations by Smith [5] and Korolev [11] have pointed firmly to the existence of a unique solution with $k = k_* = 0.42-0.44$ what exceeds slightly the magnitude of k_s estimated approximately as 0.4.

When k varies in the range $k_s < k < k_*$ the slow recirculating motion in the stagnation zone can comprise some additional singularities disposed around the arc edges. Their emergence is caused by the stiff Hakkinen & Rott [12] condition with consequences examined by Smith [13]. From the mathematical point of view the main interest is to elucidate how the Prandtl-equation solution with the Hilbert integral for the induced pressure arproaches the limiting form when the pressure is expressed through the Schwarz integral.

References

1. Thwaites, B. (ed.). Incompressible aerodynamics. Oxford: Clarendon Press 1960.

2. Gurevich, M.I. Teor. struy ideal. zhid. Moskwa: Nauka 1979.

3. Squire, H.B.: Phil. Mag. 17 (1934) 1150.

4. Sychev, V.V.: Izv. Akad. Nauk SSSR. Mekh. Zhid. i Gaza (1972), N 3, 47.

5. Smith, F.T.: Proc. Roy. Soc. A356 (1977) 433.

6. Ackerberg, R.C.: J. Fluid Mech. 44 (1950) 211.

7. Ruban, A.I.: Sci. J. TSAGI 5 (1974), N 2, 44.

8. Brown, S.N. & Stewartson, K.: J. Fluid Mech. 42 (1970) 561.

9. Messiter, A.F. & Enlow, R.L.: SIAM J. Appl. Math. 25 (1973) 655.

10. Nickel, K. In: Ann. Rev. Fluid Mech., vol. 5. Palo Alto: Ann. Rev. Inc. 1973.

11. Korolev, G.L.: Sci. J. TSAGI 11 (1980), N 2, 27.

12. Hakkinen, R.J. & Rott, N.: AIAA J. 8 (1965) 1553.

13. Smith, F.T.: J. Fluid Mech. 131 (1983) 219.

Velocity Profiles for Two-dimensional Turbulent Separated Flows

R.C. LOCK

Department of Aeronautics,
The City University,
London*

1. Summary and Introduction

In any 'integral' method for predicting the development of a turbulent boundary layer, an important constituent is the specification of the streamwise velocity profiles of the mean flow. For example, in the widely-used entrainment methods of Green et al[1,2], the definition of the shape factor $H_1=(\delta-\delta^*)/\theta$ that appears in the entrainment equation ought to be directly dependent on the shape of the velocity profiles; and the fact that neither of these methods makes any immediate use of such information, but relies instead on cruder empiricism, forms a serious limitation to their range of validity. This is particularly true when attempts are made to extend these methods to deal with flows approaching, or beyond, separation; where the definitions of H_1 used in Refs. 1 or 2 become increasingly misleading.

In the present paper, a three-parameter family of velocity profiles is suggested which gives reasonable agreement with experimental data for separated flows induced by adverse pressure gradients on smooth surfaces; and from it an approximate formula for H_1 is derived, as a function of the shape factor $H(=\delta^*/\theta)$, the momentum thickness Reynolds number Re and an additional parameter χ. The profiles are an extension of those suggested by Cross[6] for attached flows, which in turn are a generalisation of the well-known Coles profiles. Further improvements are incorporated to extend the definition of the logarithmic law of the wall down into the viscous sub-layer, and to remove the discontinuity in slope of the profiles at the outer edge of the boundary layer which is otherwise present.

2. A three-parameter family of velocity profiles for incompressible flow

We consider the turbulent boundary layer over a smooth surface, and take axes Ox and Oy along and normal to the surface; the corresponding components of mean velocity are U and V. If δ is the thickness of the boundary layer and U_e is the value of U at its outer edge, we can define non-dimensional quantities $\eta = y/\delta$ and $u = U/U_e$.

If τ_w is the surface shear stress and $c_f = \tau_w/(\frac{1}{2}\rho\, U_e^2)$ is the skin-friction coefficient, then $U_\tau = \sqrt{\tau_w/\rho}$ is the friction velocity, and we define $u_\tau = U_\tau/U_e = \sqrt{\frac{1}{2}C_f}$. When the flow is separated, so that near the wall the mean velocity is in the negative x direction, we take $U_\tau = -\sqrt{\tau_w/\rho}$ and $u_\tau = -\sqrt{\frac{1}{2}|C_f|}$.

*Formerly of Aerodynamics Department, Royal Aircraft Establishment, Farnborough where the work described in this paper was done.

F. T. Smith, S. N. Brown, (Eds.)
Boundary-Layer Separation
IUTAM Symposium London 1986
© Springer-Verlag Berlin, Heidelberg 1987

We choose as a basic family of velocity profiles a generalisation of the standard expression of Coles, due to Cross[6]:

$$u = \frac{u_\tau}{\kappa}\ (\ln y^+ + A) + C \sin^\chi (\tfrac{1}{2}\pi\eta) \qquad (1),$$

where $y^+ = y|U_\tau|/\nu$ is the standard non-dimensional wall coordinate; values usually taken for the constants κ and A are 0.41 and 2.13 respectively. Here, χ is an arbitrary parameter which, as Cross[6] has shown, can be related (for attached flows) to the departure from equilibrium of the non-dimensional velocity gradient parameter $\frac{\theta}{U_e}\frac{dU_e}{dx}$. His empirical curve for χ can be fitted by the function

$$\chi = 0.7 + 1.3 \exp(\ 250\Pi) \qquad (2),$$

where $\Pi = \frac{\theta}{U_e}\frac{dU_e}{dx} - \left(\frac{\theta}{U_e}\frac{dU_e}{dx}\right)_{EQ}$,

and the suffix EQ denotes equilibrium values (cf Refs 2,12).

The constant C has to be chosen so that u(1) = 1 ; hence

$$C = 1 - \frac{u_\tau}{\kappa}\left[\ln\ (\ |u_\tau|Re\delta) + A\right] \qquad (3),$$

where $Re = U_e\delta/\nu$ is the Reynolds number based on ζ. Eq.(1) can therefore be written in the alternative form

$$u = 1 + \frac{u_\tau}{\kappa}\ \ln\eta - C\,G(\eta), \qquad (1a),$$

where $G(\eta) = 1 - \sin^\chi (\tfrac{1}{2}\pi\eta)$.

From Eq. (1a), the displacement and momentum thicknesses of the boundary layer can be derived. Defining $\Delta^* = \delta^*/\delta$ and $\Theta = \theta/\delta$, we have

$$\Delta^* = u_\tau/\kappa + C\ f(\chi) \qquad (4),$$

where $f(\chi) = 1 - \int_0^1 \sin^\chi (\tfrac{1}{2}\pi\eta)d\eta = 1-\pi^{-\frac{1}{2}}(\tfrac{1}{2}\chi - \tfrac{1}{2})!/(\tfrac{1}{2}\chi)!$ (5);

and $\Theta = \Delta^* - C^2\ \{2f(\chi) - f(2\chi)\} - 2Cu_\tau g(\chi)/\kappa - 2(u_\tau/\kappa)^2$ (6),

where $g(\chi) = 1 + \int_0^1 \sin^\chi (\tfrac{1}{2}\pi\eta)\ln\eta\ d\eta$. (7)

Convenient approximations to $f(\chi)$ and $g(\chi)$ are given by Cross[6] ;

$$f(\chi) = \chi^{\frac{1}{2}}\ (0.7973 + 0.0304\chi) - 0.2146 - 0.2495\chi \qquad (5a)$$

and $g(\chi) = f/(0.434 + 0.288\ f + 0.135\ f^2 + 0.142\ f^3)$. (7a)

Eliminating C from Eqs. (3) and (5), we obtain

$$\Theta = \Delta^* - a\Delta^{*2} - b\ u_\tau\Delta^*/\kappa - c\ (u_\tau/\kappa)^2 \qquad (8),$$

where $a = \{2f(\chi) - f(2\chi)\}/\{f(\chi)\}^2$, $b=2(g/f-a)$ and $c= 2-a-b$.
An immediate deduction from Eq.(8) is that in the limit of infinite Reynolds number, when u_τ is negligibly small (but C remains arbitrary),

$$\textcircled{H} = \Delta^* - a\,\Delta^{*2} \qquad (9);$$

so that the shape factors H and H_1 are given by

$$H = \Delta^*/\textcircled{H} = 1/(1 - a\Delta^*)$$

and $$H_1 = (1 - \Delta^*)/\textcircled{H} = H(\Delta^{*-1} - 1).$$

They are therefore related by

$$H_1 = a\,H^2/(H - 1) - H \quad . \qquad (10)$$

For finite Reynolds numbers it is not possible to determine an explicit expression for H_1 , although Eq.(10) turns out to be a useful guide, as will be shown later.

With this family of profiles, both the shape of the velocity profile at separation and the corresponding values* of H depend on the value of χ:

$$u_{sep} = \sin^{\chi}(\tfrac{1}{2}\pi\eta),$$

$$H_{sep} = f(\chi)/\{f(2\chi) - f(\chi)\} \quad . \qquad (11)$$

Values of H_{sep} are plotted against χ in fig. (1), and the corresponding forms of the separation profiles are shown in Fig.(2). Also shown in Fig.(2) is the separation profile used by Le Balleur[7], namely $u_{sep} = 1 - (1 - \eta^{3/2})^2$; this is closely similar to Cross's profile for $\chi = 1.5$, with $H_{sep} = 3.5$, and therefore corresponds to a flow separating under a moderately strong adverse pressure gradient.

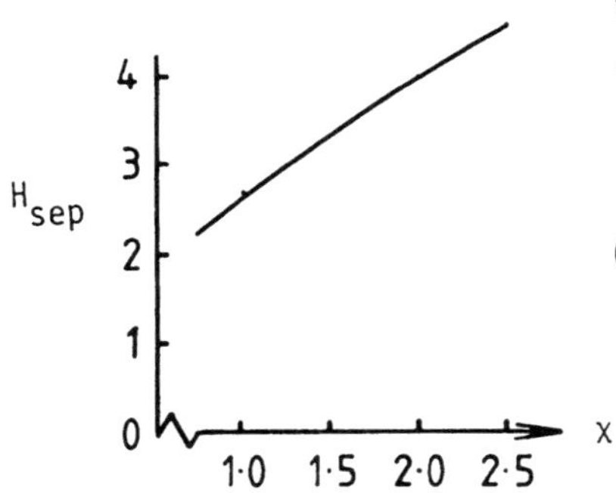

Fig.1.

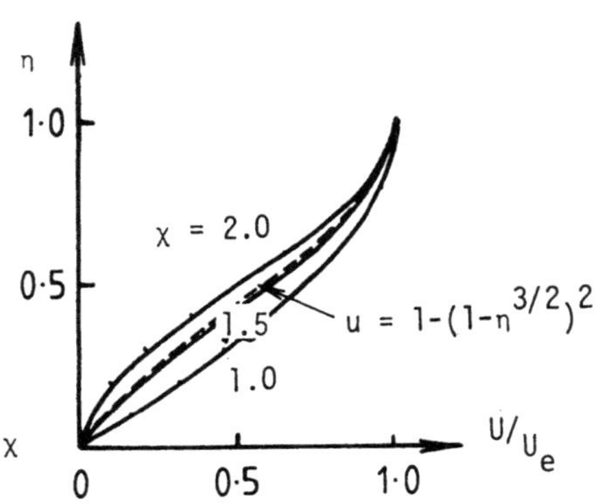

Fig.2.

*For compressible flows, the relevant shape parameter is $\bar{H}$, related to H by $\bar{H} = (H + 1)/\{1 + \tfrac{1}{2}(-1)rM_e^2\} - 1$.

From Fig.1 we see that the value of H (or $\bar{H}$) at separation, for a flow close to equilibrium ($\chi \simeq 2$), is about 4; similar values are found experimentally when separation occurs as the result of a long steady adverse pressure gradient, as in the rear stall of an aerofoil at low speeds. But if the separation is caused by a sudden rapid increase of pressure, produced for example by a forward-facing step or by a shock wave, then the value of H_{sep} (or $\bar{H}_{sep}$) is much smaller, between 2 and 3 according to circumstances. All this is in line with general experience in two dimensions[3], and represents the main reason for adopting Cross's family of profiles in the present context.

Even for attached flows, there are however two obvious - though minor - defects in the simple form of profile given by Eq.(1). These are :-

(a) the logarithmic form of the wall law,

$$u/u_\tau = \frac{1}{\kappa}(\ln y^+ + A),$$

is not valid for values of y^+ less than about 20. As a result, as separation is approached ($u_\tau \to 0$), a weak logarithmic singularity occurs in the expressions for both Δ^* and Ⓗ (and hence in the relation between H_1 and H), due to the fact that (see Eq. (3))

$$C \sim 1 - \frac{u_\tau}{\kappa} \ln(|u_\tau| Re_\delta) + O(u_\tau) \text{ as } u_\tau \to 0;$$

(b) at the outer edge of the boundary layer, the profile does not blend smoothly into the external inviscid flow, since (from Eq. (1a))

$$du/d\eta = u_\tau/\kappa \quad \text{when } \eta = 1.$$

Both these defects can be remedied by combining the modified wall law used by East[14] with a simple smoothing function at the outer edge, to give

$$u = \frac{u_\tau}{\kappa}\left[\tfrac{1}{2}\ln(1+\kappa y^{+2}/D) + \kappa D(1-e^{-y^+/D})\right] + E\eta^2(1-\eta) + C\sin^\chi(\tfrac{1}{2}\pi\eta) \tag{12}$$

where $D = 8.96$ (so that $\tfrac{1}{2}\ln(\kappa/D) + \kappa D = A$),

$$C = 1 - \frac{u_\tau}{\kappa}\left[\tfrac{1}{2}\ln(1+\kappa u_\tau^2 Re_\delta^2/D) + \kappa D(1-e^{-|u_\tau|Re_\delta/D})\right]$$

and

$$E = \frac{u_\tau}{\kappa}\left[\frac{\kappa u_\tau^2\, Re_\delta^2}{D + \kappa u_\tau^2 Re_\delta^2} + \kappa|u_\tau| Re_\delta\, e^{-|u_\tau|Re_\delta/D}\right].$$

Here, the first term (in square brackets) can easily be seen to satisfy the required conditions:

$$u/u_\tau \sim y^+ + O(y^{+3}) \quad \text{as } y+ \to 0$$
and $$u/u_\tau \sim (\ln y^+ + A)/\kappa \text{ when } y^+ \text{ is large;}$$
while the smoothing term $E\,\eta^2(1-\eta)$ ensures that $u'(1) = 0$.

Like the simpler equation (1), Eq.(12) defines u as a function of the three basic parameters u_τ $(=\pm\sqrt{\tfrac{1}{2}|C_f|}\,)$, Re_δ and χ. From it, the value of Δ^* can be found analytically, but Ⓗ has to be obtained by numerical integration. This has been done for a range of values of the basic parameters and the corresponding values of H and H_1 have been calculated. Following a suggestion of Gaudet[18], who found that for $\chi = 2$, H_1 could be well approximated by the expression

$$H_1 = \frac{(4-\beta)}{(2+\beta)}\left(\frac{3}{2}\,\frac{H^2}{H-\beta} - H\right) \tag{13}$$

where $\beta = 1 + 6.2/(\log_{10}Re_\theta)^{3.3}$,

we can generalise this for arbitrary values of χ to

$$H_1 = \left(\frac{1-f}{f}\right)\left(\frac{H_{sep}-\beta}{(a-1)H_{sep}+\beta}\right)\left(\frac{aH^2}{H-\beta} - H\right) \tag{14}$$

where $a = \{2f(\chi) - f(2\chi)\} / f^2(\chi)$ (see Eqs. (8) and (11))

and $H_{sep} = f(\chi)/\{f(2\chi) - f(\chi)\}$.

When $Re_\theta \to \infty$, $\beta \to 1$ and Eq. (14) reduces to the limiting form (10); when $\chi = 2$, $H_{sep} = 4$, $f = \tfrac{1}{2}$ and $a = \tfrac{3}{2}$ so that Eq. (13) is recovered. The accuracy of Eq. (14) has been checked for a range of values of χ between 1 and 3.

For fully separated flows, when u_τ is negative, the use of mean velocity profiles like Eqs. (1) or (12) is open to the objection that the standard log-law of the wall, $\frac{u}{u_\tau} \sim (\ln y^+ + A)/\kappa$, is no longer strictly valid in a reversed-flow region. This fact has been demonstrated in a number of recent experiments [4,5,8,9], in which measurements have been made very close to the wall using both pulsed hot-wires[4,8] and laser anemometry[5,9]. Simpson[10] found that his data[5] correlated reasonably well according to the relation

$$\frac{U}{U_{min}} = B\left[\frac{y}{N} - \ln(y/N) - 1\right] - 1 \quad (\text{for } y<N) \tag{15}$$

(with B = 0.3),

where U_{min} is the maximum negative velocity and N is the value of y at which it occurs. Dianat and Castro[4] found that their measurements (in a more rapidly separating flow than that of Simpson) correlated better with B = 0.24; while Devenport[8] suggests that B should be allowed to vary so as to provide a smooth join, at $y = y_1$ say, with the limiting form $U/U_\tau \sim y^+$ as $y \to 0$. This leads to the equations

$$B(e^{1/B} - 1) = \tfrac{1}{2} C_f Re_\delta (N/\delta)\ (U_e/U_{min}) \qquad (16)$$

and $$y_1/N = e^{-1/B} \qquad (17)$$

Thus, from any experiment in which all the quantities on the right hand side of Eq. (16) have been measured, B (and y_1) can be calculated; Devenport's values of B lay between 0.24 and 0.37.

Alternatively, if we assume a suitable mean value for B, say 0.3, then Eq.(16) provides a convenient way of estimating C_f in a separated flow. If we define y_0 to be the value of y at which U returns to zero (see Sketch), so that

$$y_0/N - \ln(y_0/N) = 1 + 1/B,$$

giving $y_0/N = 6.1$ when $B = 0.3$, then we find that

$$C_f \simeq \frac{100}{Re_\delta} \frac{u_{min}}{\eta_0} \qquad (18)$$

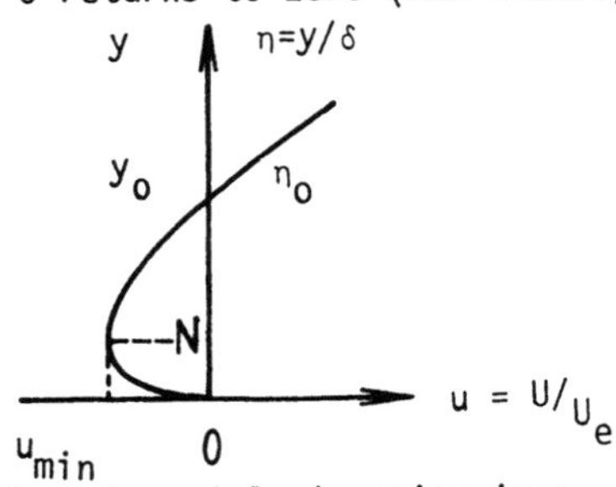

It appears therefore that, although there is a 'quasi-log' region in a reversed flow, it does not fit with the standard form for attached flows. As a result, the use of the standard log-law leads to reversed flow profiles in which, for a given value of U_{min}, the mean (negative) velocities very close to the wall are too low, as also is the value of $|C_f|$. However, as we shall show in Section 3, it is nevertheless still possible to obtain reasonable agreement with experiment as regards the overall shape of the velocity profiles; and this should be sufficient for the purpose of deriving the relation between the shape factors H_1 and H , which is the main reason for the present work.

3. Comparison with experimental measurements.

The crucial stages of the present exercise are: first, to see how well, by appropriate choice of the parameter χ, profiles given by Eqs.(1) or (12) can be fitted to experimentally measured profiles for separated flows; and second, to see whether the resulting values of χ follow a consistent pattern which can be used to derive an empirical relation for it. For this purpose data from four recent experiments have been used: three in low-speed flows and one in which the separation is shock-induced. In all of them, laser anemometry was used to measure the velocities in the reversed flow region.

The first two can conveniently be taken together. Hastings[9], using the boundary-layer tunnel at RAE Bedford, set up an 'equilibrium' separated flow on the flat roof of the tunnel by appropriate shaping of the flexible floor. Simpson et al[5], in a similar wind-tunnel, produced a pressure distribution which simulates that on an aerofoil at high incidence; this is shown in Fig.3, together with the resulting values of H. Separation takes place at X = 3.5, after which H increases rapidly, reaching a maximum value of 9.5. There are four stations in the separated flow region at which mean velocities have been measured, labelled here for convenience A,B,C and D. The velocity profiles are shown in Fig.4, together with the best fits to them

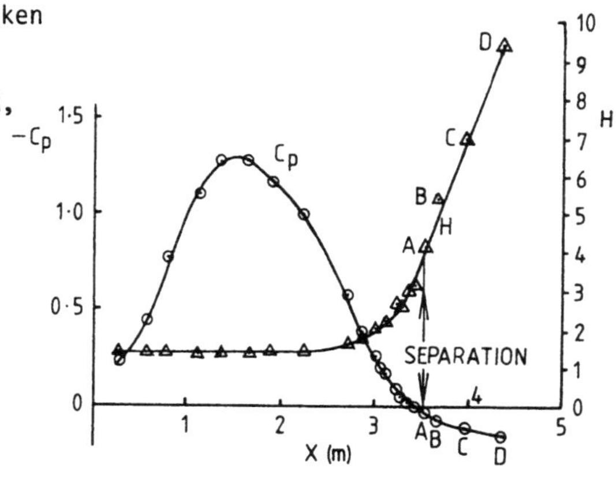

Fig.3 Experiment of Simpson et al[5]

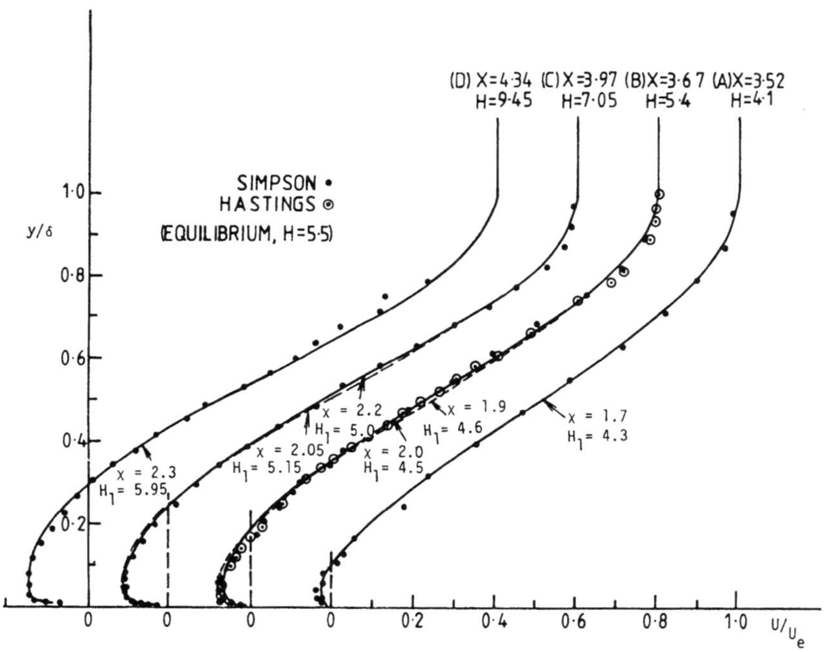

Fig.4 Measurements of Simpson et al[5] and Hastings[9] (equilibrium)

which could be obtained by the present method; in each case, the measured values of H and Re_θ have been used. In the case of Station (B), at which H = 5.4, the values of u measured in Hastings's equilibrium separated flow[9] are also included; it will be seen that these are closely similar to those obtained (under very different conditions) by Simpson. In all cases, the chosen theoretical profiles show good agreement with experiment; and in particular the discrepancy which does exist very close to the wall (see Section 2) is hardly visible on this 'global' scale.

The third experiment is also to Hastings[11], who measured the separated flow developing on the uppersurface of a NACA4412 aerofoil at an angle of incidence of 12.5^0, close to the ultimate stalling condition. The tests were made on a model of 1m chord spanning the larger dimension of the 13 x 9ft (3.96 x 2.47m) low-speed wind tunnel at RAE Bedford; the chord Reynolds number was 4.2 x 10^6. The pressure distribution is shown in Fig.5, which also includes the results of calculations by Williams's method[20] (which mades use of the (H_1,H) relation derived in the present paper). The velocity profiles on the upper surface measured between 75% chord and the trailing edge are shown in Fig.6. In this case, it appears that the value of χ at separation is higher than 2 (about 2.4), so that the value of H_{sep} is about 4.2. As the separation develops, it becomes increasingly more difficult to obtain as good a fit with the present profile family as was possible with the two previous experiments. In particular, a distortion occurs in the measured shapes near y/δ = 0.6 which implies that a theoretical profile which is a good fit near the wall is poor in the outer region, and vice-versa. The values of χ that are shown here were chosen as a compromise to give a reasonable overall fit to the data.

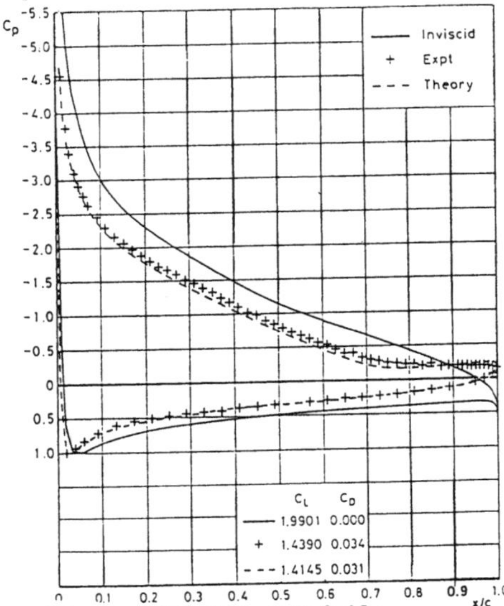

Fig.5 NACA4412 aerofoil: Pressure distribution,α=12.5^o

The last case we consider is that of the experiment of Delery[12] on a separated flow induced by a shock wave, produced by a circular-arc bump on the floor

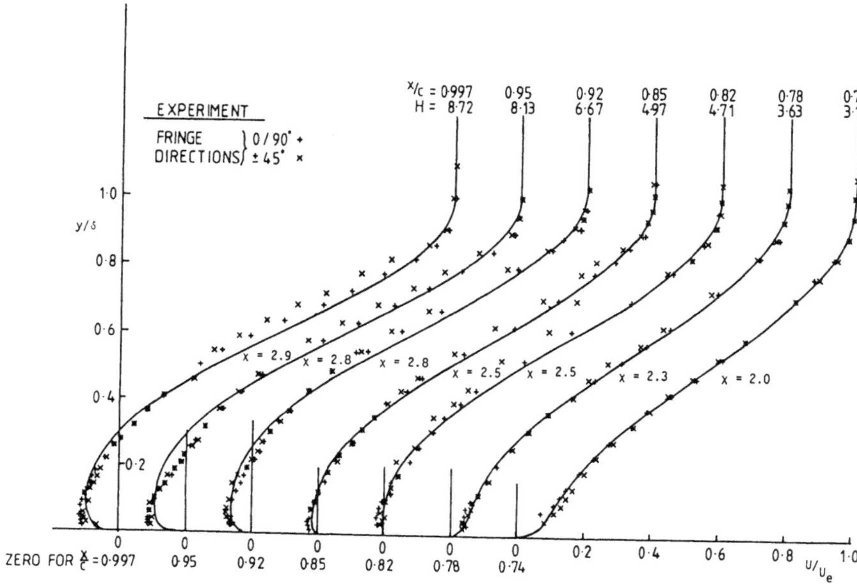

Fig.6. Hastings' measurements on NACA4412 aerofoil: $\alpha = 12.5^o$

of a transonic wind tunnel. Measured velocity profiles (taken from the published figures of Ref. 12) are shown in Fig.7. Here, the Mach number upstream

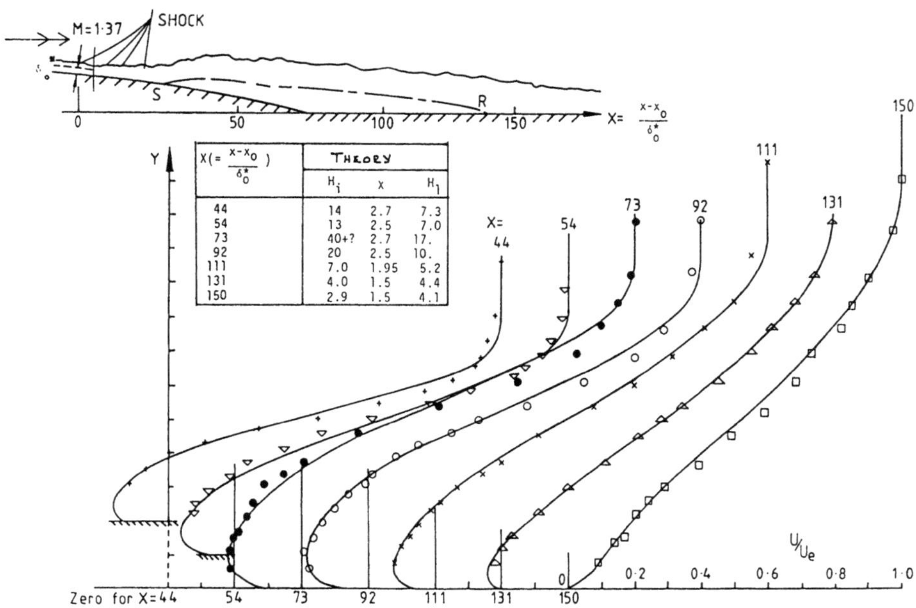

X(= (x-x0)/δ0*)	Theory H_i	χ	H_1
44	14	2.7	7.3
54	13	2.5	7.0
73	40+?	2.7	17.
92	20	2.5	10.
111	7.0	1.95	5.2
131	4.0	1.5	4.4
150	2.9	1.5	4.1

Fig.7. Experiment of Delery[12]

of the shock wave is 1.37, and this leads to a rapidly devloping separation bubble downstream. The 'incompressible' shape factor H_i increases to a maximum of over 40 at the trailing edge of the bump, and then decreases gradually until reattachment. The profiles shown here are all (except the last) in the separation region. Again, by a suitable choice of χ it has been possible to obtain a reasonable fit to all the profiles - perhaps only 'fair' in the most severely separated cases, but excellent downstream as the flow approaches reattachment. Note that, at reattachment, the value of χ, about 1.5, is well below the 'standard' value of 2, and this suggests that the flow has retained the characteristic which, according to Eq. (2), it must have had at separation at the foot of the shock wave.

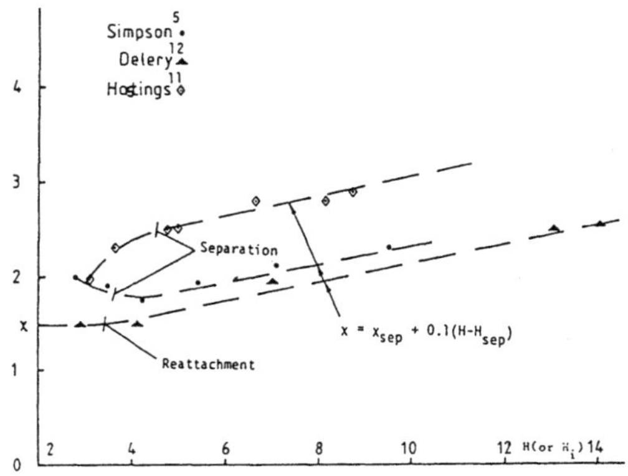

Fig.8. Variation of χ with H

The values of χ that have been obtained in these comparisons with experiment are shown in Fig.8; from which it will be seen that, in all cases, the 'best fit' values increase steadily after separation*, roughly according to the linear relation between H (or H_i)

$$\chi = \chi_{sep} + 0.1(H - H_{sep}) \qquad . \qquad (19)$$

This applies for values of H up to about 15; the very limited evidence suggests that for higher values of H the rate of increase of χ needs to be moderated slightly.

If we make this assumption, then the corresponding curves of the vital shape factor H_1 against H (or $\bar{H}$) can be derived, using the approximate equation (14). Five of these are shown in Fig.9, for values of χ up to separation (χ_0) of 1.0(0.5)3.0. The value 5 x 10^4 has been chosen for Re_θ , which is typical of most of the experiments; though for separated flows the dependence of H_1 on Reynolds number is almost negligible. A large number of experimental points is shown; in addition to the cases already mentioned, results from several experiments in attached flows (approaching separation) are also included, covering a range of situations in which the pressure gradient varies

*in the case of Delery's experiment, this remark applies as H_1 decreases towards reattachment.

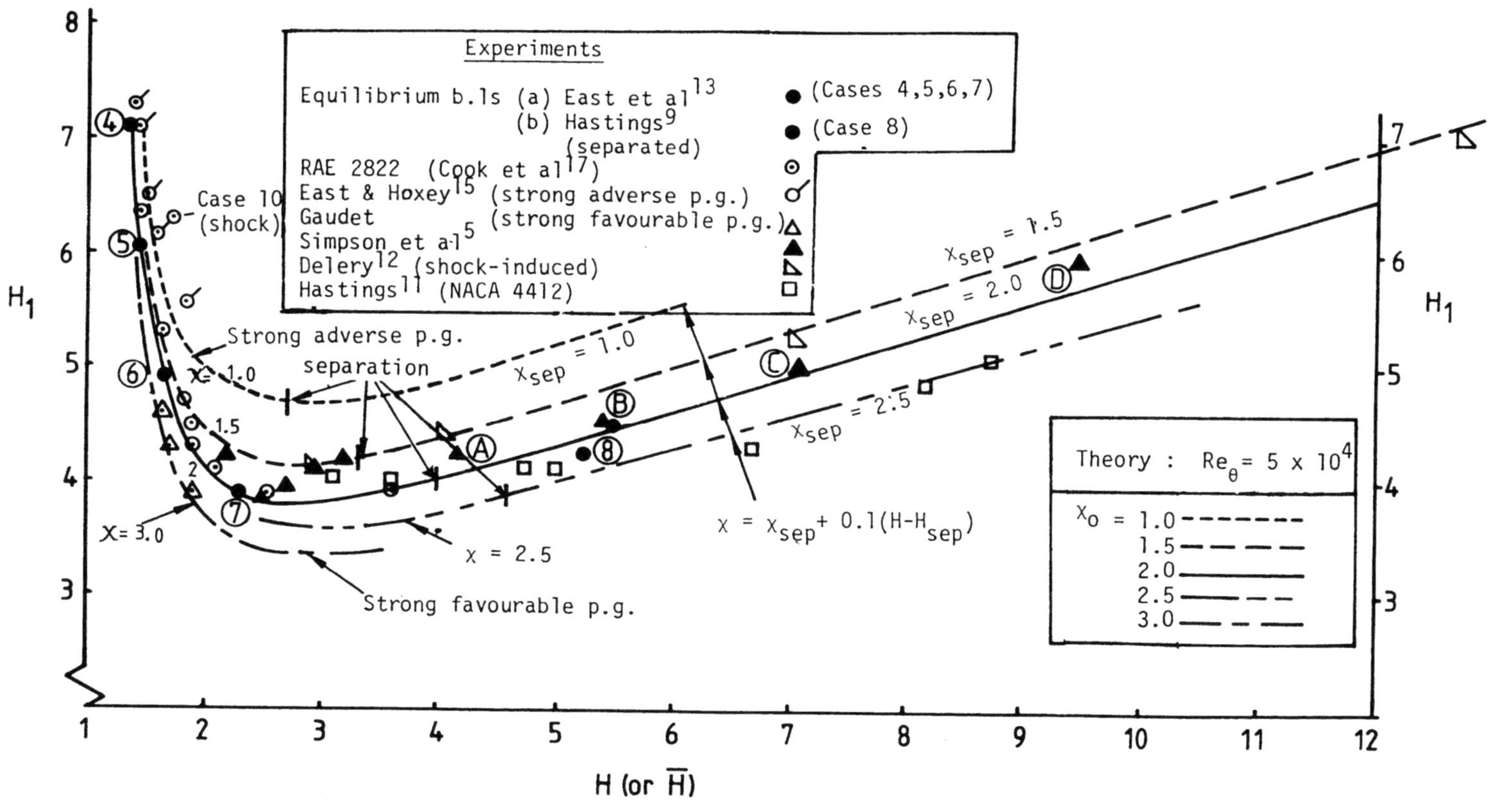

Fig.9. Variation of H_1 with H.

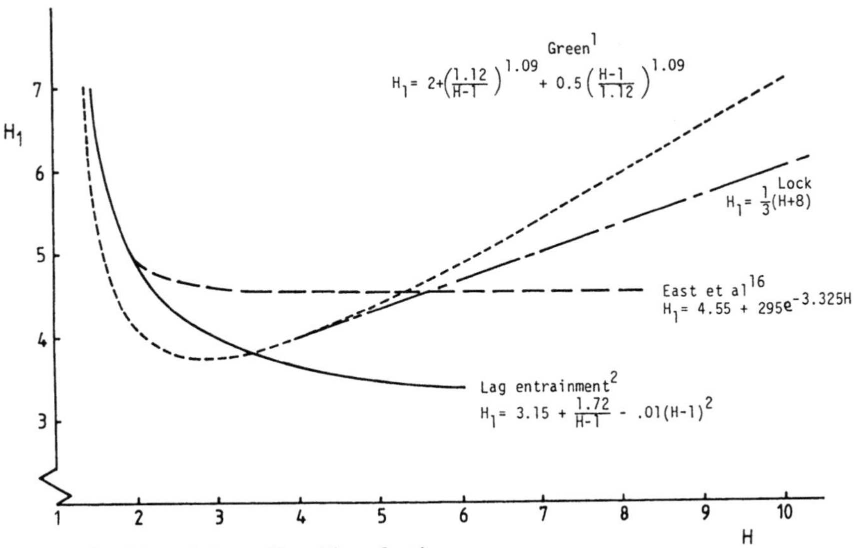

Fig 10. Other (H_1, H) relations.

between strongly adverse to strongly favourable. In all cases, it will be seen that the present method does successfully follow the general trend of the relevant experimental data.

Finally, it is worth noting that the curve for $x_0 = 2$, which represents a convenient mean path through all the data, can be closely approximated by the expression used by Williams[20] in his separated flow calculations:

$$H_1 = 2+1.5\{1.12/(H-1)\}^{1.09} + 0.5\{(H-1)/1.12\}^{1.09} \quad \text{(for } H<4) \qquad (20)$$

$$= (H + 8)/3 \quad \text{(for } H>4) \quad .$$

This is shown in Fig.10, together with other expressions for H_1 which have been suggested. These include the equation originally given by Green[1], which forms the first part of Eq.(20); the formula

$$H_1 = 3.15 + 1.72/(H-1) - 0.01(H-1)^2 \qquad (21)$$

used in the standard lag-entrainment method[2]; and the modification to it suggested by East et al[16]:

$$H_1 = 4.55 + 295e^{-3.325 \cdot H} \quad \text{(for } H>1.8) \quad . \qquad (22)$$

For flows involving shock-induced separation this may well be more satisfactory than Eq.(20); and indeed it is clear that a unique relation between H_1 and H (or $\bar{H}$) simply does not exist.

References

1. J.E.Green. Application of Head's method to the prediction of turbulent boundary layers and wakes in compressible flow. RAE Tech.Rep.72079 (1972).

2. J.E. Green; D.G. Weeks; J.W.F. Brooman. Prediction of Turbulent boundary layers and wakes by a lag-entrainment method. ARC R&M 3791 (1973).

3. S.J. Kline; J.G. Bardina; R.C. Straun. Correlations of the detachment of two-dimensional turbulent boundary layers. AIAA J., 21, No.1 (1983).

4. M. Dianat; I.P. Castro. Measurements in separating boundary layers. 15th ICAS Congress, London (1986).

5. R.L. Simpson; Y.T. Chew; B.G. Shivaprasad. The structure of a separating turbulent boundary layer (Part I). J. Fluid Mech., 157, p305 (1985).

6. A.G.T. Cross. Caclulation of compressible three-dimensional turbulent boundary layers with particular reference to wings and bodies. British Aerospace (Brough) Rep. YAD3379 (1979).

7. J.C. Le Balleur. Strong matching method for computing transonic flows including wakes and separations. La Recherche Aerospatiale 1981-3 (1981).

8. W.J. Devenport. Separation bubbles at high Reynolds number: measurement and computation. Ph.D. Thesis, Cambridge University (1985).

9. R.C. Hastings; K.G. Moreton. An investigation of a separated equilibrium turbulent boundary layer. In 'Laser Anemometry in Fluid Mechanics', Lisbon (1982).

10. R.L. Simpson. A model for the back-flow mean-velocity profile. AIAA J., 21, p142 (1983).

11. R.C. Hastings; B.R. Williams. Studies of the flow field near a NACA4412 aerofoil at nearly maximum lift. To appear in Aero. J. of Roy. Aero. Soc. (1986).

12. J. Delery. Experimental investigation of turbulence properties in shock/boundary layer interaction. AIAA J., 18, No.7 (1980).

13. L.F. East; W.G. Sawyer; C.R. Nash. An investigation of the structure of equilibrium turbulent boundary layers. RAE Tech. Rep. 79040 (1979).

14. L.F. East. A prediction of the law of the wall in compressible three-dimensional turbulent boundary layers. RAE Tech. Rep. 72178 (1972).

15. L.F. East; R.P. Hoxey. Low-speed turbulent boundary layer data. ARC R&M 3653 (1969).

16. L.F. East; P.D. Smith; P.J. Merryman. Prediction of the devlopment of separated turbulent boundary layers by thelag-entrainment method. RAE Tech. Rep. 77046 (1977).

17. P.H. Cook; M.A. McDonald; M.C.P. Firmin. Aerofoil RAE2822: Pressure distribution and boundary layer measurements. AGARD AR 138, A6 (1979).

18. L. Gaudet. Private communication.

19. L. Gaudet. Experimental investigation of the turbulent boundary layer at high Reynolds number and a Mach number of 0.8. Aero. J. of Roy. Aero. Soc., 90, No.893 (1986).

20. B.R. Williams. The prediction of separated flows using a viscous-inviscid interaction method. Aero. J. of Roy. Aero. Soc., 89, No.885 (1985).

Analysis of Three-dimensional Separated Flow Using Interacting Boundary-Layer Theory

DAVID E. EDWARDS
United Technologies Research Center
East Hartford, CT 06108

Summary

An Interacting Boundary-Layer Theory (IBLT) procedure which employs a quasi-simultaneous technique to couple a finite difference representation of the viscous flow to an integral representation of the inviscid flow is presented for the analysis of incompressible three-dimensional separated flow. The model problem studied in this investigation is laminar flow over a flat plate with a protuberance. Three series of computations are made to 1) assess the accuracy of the present method and 2) examine the effect of Reynolds number on the structure of three-dimensional laminar flow.

1. Introduction

Viscous effects can have a substantial impact on the aerodynamic performance of external and internal flow configurations. For a significant number of high Reynolds number flows of practical interest, the viscous effects even in the presence of boundary-layer separation are confined to a relatively small layer near the surface. In the past, many two-dimensional aerodynamic problems (see Ref. 1 for a review) have been represented by Interacting Boundary-Layer Theory (IBLT) whereby the flow is divided into regions of viscous and inviscid flows with the two regions coupled through the displacement thickness. However, since the majority of the fluid flows in nature are three dimensional and with the rapid advances in the development of supercomputers, it is timely to turn our attention to the analysis of strongly interacting three-dimensional flows. The focus of this present investigation is to determine if the IBLT methodology, which has been applied successfully in analyzing two-dimensional flows, can be extended to analyze three-dimensional strongly interacting flows.

During the past ten years, several research efforts have been directed toward the extension of the IBLT approach to three-dimensional flows. The principal focus of that work has been to develop inverse (pressure not prescribed) viscous techniques, either finite difference or integral, which can be applied in the analysis of 3D separated flow [2,3,4,5,6]. Research has only begun to develop a complete IBLT approach for three-dimensional flows. Wigton and Yoshihara [7] developed a 3D IBLT procedure which uses an integral inverse boundary-layer method and a 3D full potential analysis coupled using a semi-inverse technique. They applied their method to analyze a wing/fuselage

F. T. Smith, S. N. Brown, (Eds.)
Boundary-Layer Separation
IUTAM Symposium London 1986

configuration in transonic flow. However, Wai and Yoshihara [8] reported encountering numerical difficulties using Wigton and Yoshihara's IBLT procedure and have raised questions about using an IBLT approach. Chen and Wu [9] have developed a 3D IBLT procedure which uses an integral inverse boundary-layer method and an integral representation of incompressible inviscid flow coupled using the quasi-simultaneous methodology of Veldman [10]. Smith [11] has also developed a 3D IBLT procedure which closely resembles his earlier method for the triple-deck equations. Both Chen and Wu [9] and Smith [11] have used their procedures to analyze incompressible laminar flow over a flat plate with a protuberance but at this time, only limited results have been published.

The overall objective of the present study is to extend the developments discussed above to produce a three-dimensional IBLT procedure. The model problem studied in this investigation is incompressible laminar flow over a flat plate with a protuberance. Three series of computations are presented in this paper. The first series of cases evaluates the accuracy of the results of the present IBLT method in comparison with the results of a thin-layer Navier-Stokes procedure developed by Davis, et al. [12] for flow over a flat plate with a dent. Two additional series of calculations are made to evaluate the effect of Reynolds number on flow over a flat plate with either a dent or a bump. Several conclusions are drawn from this study; the principal conclusion is that the Interacting Boundary-Layer Theory (IBLT) methodology can be extended to analyze three-dimensional separated flow. In addition, the results of the present analysis for the class of flows investigated in this study conform to a topological law which states that for a smooth surface the number of nodal singular points must exceed the number of saddle singular points exactly by two. This topological law has been proposed by mathematicians (see Kaplan [13]) and has been shown by Lighthill [14] and others to be satisfied in three-dimensional separated flow over a closed body.

2. Analysis

The procedure used in this study of three-dimensional separated flow is Interacting Boundary-Layer Theory (IBLT) whereby the entire flow field is represented as regions of viscous and inviscid flow with the two regions coupled through the viscous displacement thickness. The inviscid solution, written in Cartesian coordinates, is obtained from a three-dimensional small disturbance representation [15] of an incompressible, irrotational flow field where the effect of the viscous flow is imposed by analyzing the inviscid flow over a displacement body instead of the actual geometrical surface. The components of surface velocity, u_e and w_e, are given by the following expressions:

$$u_e = u_\infty + \frac{1}{2\pi}\int_{-\infty}^{\infty}\int_{-\infty}^{\infty} A\frac{x-\xi}{r^3}d\xi d\varsigma \ , \quad w_e = w_\infty + \frac{1}{2\pi}\int_{-\infty}^{\infty}\int_{-\infty}^{\infty} A\frac{z-\varsigma}{r^3}d\xi d\varsigma \qquad (1)$$

where

$$r = [(x - \xi)^2 + (z - \varsigma)^2]^{1/2} \tag{2}$$

and

$$A = \frac{\partial}{\partial \xi}[u_\infty y_w + u_e \delta^*] + \frac{\partial}{\partial \varsigma}[w_\infty y_w + w_e \delta^*] \tag{3}$$

In the above equations, u_∞ and w_∞ are the free stream velocity components, y_w is the geometrical surface and δ^* is the displacement thickness. In two-dimensional flows the displacement thickness has a closed form integral definition. However, in three-dimensional flows, the displacement thickness does not have a closed form definition and it is desirable to remove this flow quantity from the set of unknowns. Moore [16] deduced that the three-dimensional displacement thickness is governed by a partial differential equation which for incompressible flow is written as

$$\frac{\partial}{\partial x}(u_e \delta^* - u_e \delta_x) + \frac{\partial}{\partial z}(w_e \delta^* - w_e \delta_z) = 0 \tag{4}$$

where Moore defined the integral thicknesses δ_x and δ_z as

$$\delta_x = \int_o^\infty (1 - \frac{u}{u_e})dy, \quad \delta_z = \int_o^\infty (1 - \frac{w}{w_e})dy \tag{5}$$

By introducing the mass flow parameters $m_x = u_e \delta_x$ and $m_z = w_e \delta_z$, Eq. (3) can then be rewritten as

$$A = \frac{\partial}{\partial \xi}[u_\infty y_w + m_x] + \frac{\partial}{\partial \varsigma}[w_\infty y_w + m_z] \tag{6}$$

For this study, the inviscid flow is represented using Eqs. (1), (2) and (6). The numerical technique used to solve the double surface integrals in Eq. (1) is a three-dimensional extension of a method developed by Davis and Werle [17] to solve the Cauchy-Hilbert integral.

The flow in the viscous region is assumed to be governed by Prandtl's boundary-layer equations, expressed in body-surface coordinates as shown in Fig. 1. The nondimensional incompressible laminar boundary-layer equations are represented in the present study by two "stream-like" function definitions and two momentum equations which are written as:

$$\frac{\partial \psi}{\partial \eta} = hu \tag{7}$$

$$\frac{\partial \phi}{\partial \eta} = hw \tag{8}$$

$$h^2 u \frac{\partial u}{\partial x} - h(\frac{\partial \psi}{\partial x} + \frac{\partial \phi}{\partial z})\frac{\partial u}{\partial \eta} + h^2 w \frac{\partial u}{\partial z} = h^2(u_e \frac{\partial u_e}{\partial x} + w_e \frac{\partial u_e}{\partial z}) + \frac{\partial^2 u}{\partial \eta^2} \tag{9}$$

$$h^2 u \frac{\partial w}{\partial x} - h(\frac{\partial \psi}{\partial x} + \frac{\partial \phi}{\partial z})\frac{\partial w}{\partial \eta} + h^2 w \frac{\partial w}{\partial z} = h^2(u_e \frac{\partial w_e}{\partial x} + w_e \frac{\partial w_e}{\partial z}) + \frac{\partial^2 w}{\partial \eta^2} \tag{10}$$

where y, v, ψ, and ϕ have been scaled in the usual manner by $\sqrt{Re}$, where Re is the Reynolds number. The continuity equation is automatically satisfied through the use of ψ and ϕ which are related to the normal velocity component, v, by

$$v = -(\frac{\partial \psi}{\partial x} + \frac{\partial \phi}{\partial z}) \tag{11}$$

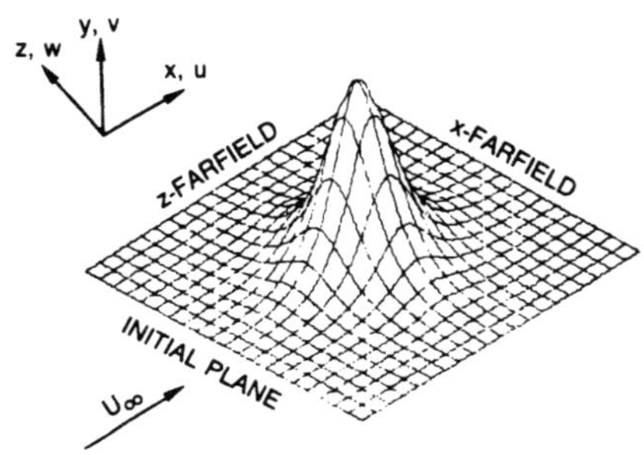

Fig. 1 Schematic of flow over a flat plate with a protuberance.

The normal coordinate $\eta = y/h$ where $h=h(x,z)= \delta_x$ has been introduced to reduce the boundary-layer growth in the η -coordinate. Boundary conditions for the foregoing viscous equations are given by:

at the surface

$$\eta = 0, \quad u = w = \psi = \phi = 0 \tag{12}$$

and at the boundary-layer edge

$$\eta \to \infty, \quad u \to u_e, \quad w \to w_e, \quad \psi \to u_e y - m_x, \quad \phi \to w_e y - m_z \tag{13}$$

where m_x and m_z are the mass flow parameters described above.

A system of viscous equations with u, w, ψ, ϕ, m_x and m_z as unknowns is formed from Eqs. (7) through (10) subject to the boundary conditions given in Eqs. (12) and (13).

Two additional conditions must be specified in order to close this system of equations. The conditions specified depend upon the mode (direct, inverse, or quasi-simultaneous) which is used to solve the equations. In this study, a three-dimensional extension of Veldman's quasi-simultaneous methodology [10] is used and the two additional conditions come from localized approximations of Eq. (1) resulting in relations of the following form:

$$u_e = u_e(m_x, m_z), \quad w_e = w_e(m_x, m_z) \tag{14}$$

The viscous equations with the two local inviscid relations are solved with an implicit finite difference approach in which derivatives in the x-direction are discretized using first-order upwind approximations and derivatives in the η- and z-directions are discretized using second order central differencing. Newton iteration is used to solve the viscous equations at each x and z location, as the solution is marched from a plane of initial data and a plane of symmetry. Multiple global sweeps are made through the computational flow field until the maximum change per sweep of both m_x and m_z is less that .0001. Once the solution is converged, the pressure coefficient is obtain from Bernoulli's equation

$$C_p = 1 - (u_e^2 + w_e^2) \tag{15}$$

3. Results and Discussion

Results are presented for the analysis of three-dimensional incompressible separated flow using the present IBLT method. The model problem used in this study is laminar flow over a flat plate with either a dent or bump located downstream of the leading edge. The flat plate with the protuberance, as shown schematically in Fig. 1, is formed by

$$y_w = t\ sech[4x - 10]sech[4z] \tag{16}$$

where the sign of t determines if the flat plate has either a dent or bump and the leading edge of the plate is assumed at x = 0. The oncoming flow for this model problem is aligned with the x coordinate and henceforth u_∞=1 and w_∞=0 is assumed. Along the vertical plane of symmetry (z=0) the surface geometry is identical to the planar trough shape studied by Carter and Wornom [18] for two-dimensional separated flow.

Three series of computations have been made in this study. The first series of cases involves assessing the accuracy of the present IBLT method with a thin-layer Navier-Stokes (TLNS) procedure developed by Davis, et al. [12] for flow over a flat plate with a dent (t=-.06) at Re= 8×10^3. A second series of calculations is made to evaluate the effect of Reynolds number on flow over a flat plate with a dent (t=-.03). The final series of calculations is made to evaluate the effect of Reynolds number on flow over a flat plate with a bump (t=.03).

In each of the cases presented the present method uses a uniform grid in the x- and z-directions where the x-z computational domain is $1<x<4$ and $0<z<1.5$. A non-uniform mesh of 21 grid points is applied in the normal grid where h= δ_x is used to control the growth of the boundary layer in the computational space. Symmetry conditions are assumed along z=0 and the Blasius flat plate solution is assumed at x=1. For all the cases presented, the IBLT procedure required 30 to 70 global iterations to obtain a converged solution.

3.1 Comparison with thin-layer Navier-Stokes solutions

Several comparisons [12,19,20] have been made between the present IBLT method and the approach of Davis, et al. [12] which uses a vorticity formulation of the thin-layer Navier-Stokes (TLNS) equations. One flow case is laminar flow over a flat plate with a dent (t=-.06) at Re = 8×10^3. Both analyses used the same uniform grid consisting of 81 points in the x-direction and 21 points in the z-direction. The IBLT method used 21 points in the normal direction across the boundary layer while the TLNS method used 81 points in the normal direction with at least 32 points in the viscous region. Both procedures have been tested with finer normal grids with little change in the computed results.

A global comparison of the results of the two analyses is shown in Fig. 2 where contours from both analyses are shown for the integral thickness, δ_x, axial component of skin friction, $C_{f_x} = 2\frac{\partial u}{\partial y}_w$, and spanwise component of skin friction, $C_{f_z} = 2\frac{\partial w}{\partial y}_w$. From this figure, it can be seen that the results of the two analyses show nearly similar trends for the three flow quantities. It is observed in Fig. 2(c) that both analyses have spikes in the spanwise skin friction which dampen out quickly. These numerical disturbances are due to the assumption of a Blasius flow at the initial data plane indicating that either the protuberance in the flat plate should be moved further downstream or different initial conditions should be applied. At several spanwise locations, detailed comparisons between the two analyses in terms of δ_x, C_{f_x} and C_{f_z} are shown in Figs. 3(a), 3(b) and 3(c), respectively. From Fig. 3, it is observed that the agreement is good between the solutions for this flow case where separation occurs since the main stream velocity component (see Fig. 3(b)) undergoes flow reversal. This favorable comparison between the present IBLT method and the TLNS method of Davis, et al. demonstrates that the Interacting Boundary-Layer Theory methodology can be applied to the analysis of strongly interacting three-dimensional flows. Further, this comparison shows numerically that the displacement body concept is valid, at least for the three-dimensional strongly interacting flows examined in this investigation.

3.2 Flow over a flat plate with a dent

A second series of flows were analyzed for the laminar flow over a flat plate with a dent (t=-.03). The purpose of this series of calculations was to evaluate the effect of Reynolds number on this class of flows. Three cases were examined in this investigation where the Reynolds number was $8x10^3$, $4x10^4$ and $8x10^4$ respectively. For each case, the IBLT analysis used a uniform surface grid consisting of 121 points in the x-direction and 31 points in the z-direction. The effect of the Reynolds number on this flow can be seen in Fig. 4, which shows the limiting streamlines near the vicinity of the dent for each of the calculations. These limiting streamlines are the paths taken by particles located an infinitesimal distance above the surface and are determined by the equation

$$\frac{dz}{dx} = \frac{C_{f_z}}{C_{f_x}} \tag{17}$$

using a technique developed by Duck and Burggraf [21]. It is noted in Fig. 4 that the flow is moving from left to right. At Re= $8x10^3$, the flow is attached and tends to follow the surface of the flat plate (Fig. 4(a)). At Re= $4x10^4$ and $8x10^4$, the flow is separated (flow reversal) and a different flow structure forms. Figures 4(b) and 4(c) show the formation along the symmetry plane of two singular points on the surface (saddle point of separation and nodal point of reattachment) as the Reynolds number is increased. Lighthill [14] has shown that the number of nodal points must exceed the number of saddle points exactly by two for a closed body. Since the model problems in this study can be regarded as a local region on such a body, then any additional nodes and saddle points must cancel in number. It is seen from Fig. 4 that this condition is satisfied. Also in Fig. 4, note that for both Re= $4x10^4$ and $8x10^4$, there is a dividing limiting streamline (sometimes referred to as a separation line) emanating from the saddle point of separation which results in forming a region that is not accessible from the upstream and that the region grows as the Reynolds number is increased. Similar types of flow structure have been observed in the work of Duck and Burggraf [21] , where laminar flow over a flat plate with a protuberance at infinite Reynolds number was examined using triple-deck theory.

Further information about the effect of Reynolds number on this flow is shown in Fig. 5. Figure 5(a) shows the pressure coefficient C_p at three spanwise stations for each of the Reynolds number flows as well as the inviscid flow. The effect of the viscous flow is to reduce the pressure gradients that were observed in the inviscid flow. It is observed in Fig. 5(a) that the lower the Reynolds number the more the pressure gradients have been reduced. Figures 5(b) and 5(c) show the axial and spanwise components of skin friction at various spanwise stations. From these figures, one can conclude that the effect of increasing the Reynolds number is to increase the interaction between the viscous and inviscid flow resulting in a reverse flow region. Hence, as in planar laminar flows, the extent of separated viscous flow increases with Reynolds number for flow over a flat plate with a dent.

3.3. Flow over a flat plate with a bump

A third series of calculations were made to evaluate the effect of Reynolds number on laminar flow over a flat plate with a bump (t=.03). Three cases were examined where the Reynolds number was $8x10^3$, $4x10^4$ and $6x10^4$ respectively. For each case, the IBLT analysis used a uniform surface grid consisting of 121 points in the x-direction and 31 points in the z-direction. The effect of the Reynolds number on this flow can be seen in Fig. 6, which shows the limiting streamlines near the vicinity of the bump for each of the calculations. At Re= $8x10^3$, the flow is attached and tends to follow the surface of the flat plate (Fig. 6(a)). At Re= $4x10^4$ and $6x10^4$, the flow is separated (flow reversal) behind the bump and a different flow structure is observed. Figures 6(b) and 6(c) show the formation along the symmetry plane of two singular points on the surface (nodal point of separation and saddle point of reattachment) as the Reynolds number is increased. As in the previous model flow problem, the condition that any additional nodes and saddle points must cancel in number is satisfied. Notice in Fig. 6(b) that for Re= $4x10^4$, the flow is close to incipient separation along the symmetry plane. Also in Fig. 6, note that for both Re= $4x10^4$ and $6x10^4$, there is a dividing limiting streamline approaching the saddle point of reattachment which results in forming a region that is not accessible to the downstream and that the region grows as the Reynolds number is increased.

Additional information about the effect of Reynolds number on this flow is shown in Fig. 7. Figure 7(a) shows the pressure coefficient C_p at three spanwise stations for each of the Reynolds number flows as well as the inviscid flow. It is observed, as in the previous model problem, that the lower the Reynolds number the more the pressure gradients have been reduced. Figures 7(b) and 7(c) show the axial and spanwise components of skin friction at various spanwise stations. From the above figures, it is apparent that the extent of 3D separated flow increases with Reynolds number for laminar flow over a flat plate with a bump.

4. Concluding Remarks

An Interacting Boundary-Layer Theory (IBLT) procedure which employs a quasi-simultaneous technique to couple a finite difference representation of the viscous flow to an integral representation of the inviscid flow has been presented for the analysis of incompressible three-dimensional separated flow. The model problem used in this investigation is laminar flow over a flat plate with a protuberance. Three series of computations were made to 1) assess the accuracy of the present method and 2) examine the effect of Reynolds number on the structure of three-dimensional laminar flow.

Several conclusions have been made from this investigation. The principal conclusion is that the IBLT methodology, which has been used very successfully for two-dimensional flows, has been extended to analyze three-dimensional strongly interacting flow. Favorable comparison between the results of the IBLT analysis with a TLNS analysis of Davis, et al. [12] demonstrate numerically that the 3D displacement surface concept is valid, at least for the class of three-dimensional flow studied in this investigation. It was shown in this investigation that for three dimensional flows, as in planar flows, the extent of laminar separated flow increases with Reynolds number. In addition, for the class of flows investigated in this study (finite Reynolds number laminar flow over a flat plate with a protuberance), the results of the present analysis conform to the topological law discussed by Lighthill [14], specifically that the number of nodal points must exceed the number of saddle points exactly by two for a closed body.

Acknowledgements

The work reported in this paper was supported by United Technologies Research Center under the corporate research program and the Office of Naval Research under Contract N00014-81-C-0381. The author expresses his gratitude to Dr. R. L. Davis of UTRC for providing the results of his numerical procedure which was used for comparison with the present method. In addition, the author acknowledges the interest and helpful suggestions provided by Dr. J. E. Carter of UTRC.

References

1. McDonald, H. and Briley, W. R., "A Survey of Recent Work on Interacted Boundary-Layer Theory for Flow with Separation," Numerical and Physical Aspects of Aerodynamic Flows II, Springer-Verlag, 1984.
2. Cousteix, J. and Houdeville, R., "Singularities in Three Dimensional Turbulent Boundary Layer Calculations and Separated Phenomena," AGARD CP No. 291, 1980.
3. Smith, P. D., "The Numerical Computation of Three Dimensional Turbulent Boundary Layers," Royal Aircraft Establishment Technical Memorandum Aero 1945, 1982.
4. Delery, J. M. and Formery, M. J., "A Finite Difference Method for Inverse Solutions of 3-D Turbulent Boundary Layer Flow," AIAA paper 83-0301, 1983.
5. Edwards, D. E. and Carter, J. E., "Analysis of Three-Dimensional Separated Flow with the Boundary-Layer Equations," AIAA Paper CP 85-1499, 1985.
6. Van Dalsem, W. R. and Steger, J. L., "The Efficient Simulation of Separated Three-Dimensional Viscous Flows Using The Boundary-Layer Equations," AIAA paper 85-4064, 1985.
7. Wigton, L. and Yoshihara, H., "Viscous-Inviscid Interactions with a Three-Dimensional Inverse Boundary Layer Code," Boeing Report D6-51715, 1982.
8. Wai, J. and Yoshihara, H. , "Transonic Turbulent Separation on Swept Wings - A Return to the Direct Method," AIAA paper 84-1582, 1984.
9. Chen, Z. L. and Wu, J. M., "Approximate Viscous/Inviscid Interacting Method for Laminar and Turbulent Flows with Separation," AIAA paper 84-0267, 1984.
10. Veldman, A. E. P., "A Numerical View on Strong Viscous-Inviscid Interaction," NLR MP 83048 U, 1983.
11. Smith, F. T., "Steady and Unsteady Boundary-Layer Separation," Annual Review of Fluid Mechanics, Vol. 18, Annual Reviews Inc., 1986.

12. Davis, R. L., Carter, J. E. and Hafez, M., "Three-Dimensional Viscous Flow Solutions with a Vorticity- Stream Function Formulation," AFOSR report prepared under contract F49620-84-C-0032, May, 1986.
13. Kaplan, W., Ordinary Differential Equations , Addison-Wesley, 1958.
14. Lighthill, M. J., "Attachment and Separation in Three-Dimensional Flow," Laminar Boundary Layers, Oxford University Press, 1963.
15. Ashley, H. and Landahl, H., Aerodynamics of Wings and Bodies , Addison-Wesley, 1965.
16. Moore, F. K., "Displacement Effect on a Three Dimensional Boundary Layer," NACA Report 1124, 1953.
17. Davis, R. T. and Werle, M. J., "Progress on Interacting Boundary-Layer Computations at High Reynolds Number," Numerical and Physical Aspects of Aerodynamic Flows , Springer-Verlag, 1982.
18. Carter, J. E. and Wornom, S. F., "Solutions for Incompressible Separated Boundary Layers Including Viscous-Inviscid Interaction," NASA SP-347, 1975.
19. Carter, J. E., Davis, R. L., Edwards, D. E. and Hafez, M. M., "Three-Dimensional Separated Viscous Flow Analyses," Presented at the Tenth International Conference on Numerical Methods in Fluid Dynamics, Beijing, China, June, 1986.
20. Davis, R. L., Carter, J. E. and Hafez, M. M., "Three-Dimensional Viscous Flow Solutions with a Vorticity Stream Function Formulation," AIAA Paper 87-0601, 1987.
21. Duck, P. W. and Burggraf, O. R., "Spectral Solutions for Three-Dimensional Triple Deck Flow over Surface Topography," J. Fluid Mech., Vol. 162, 1986.

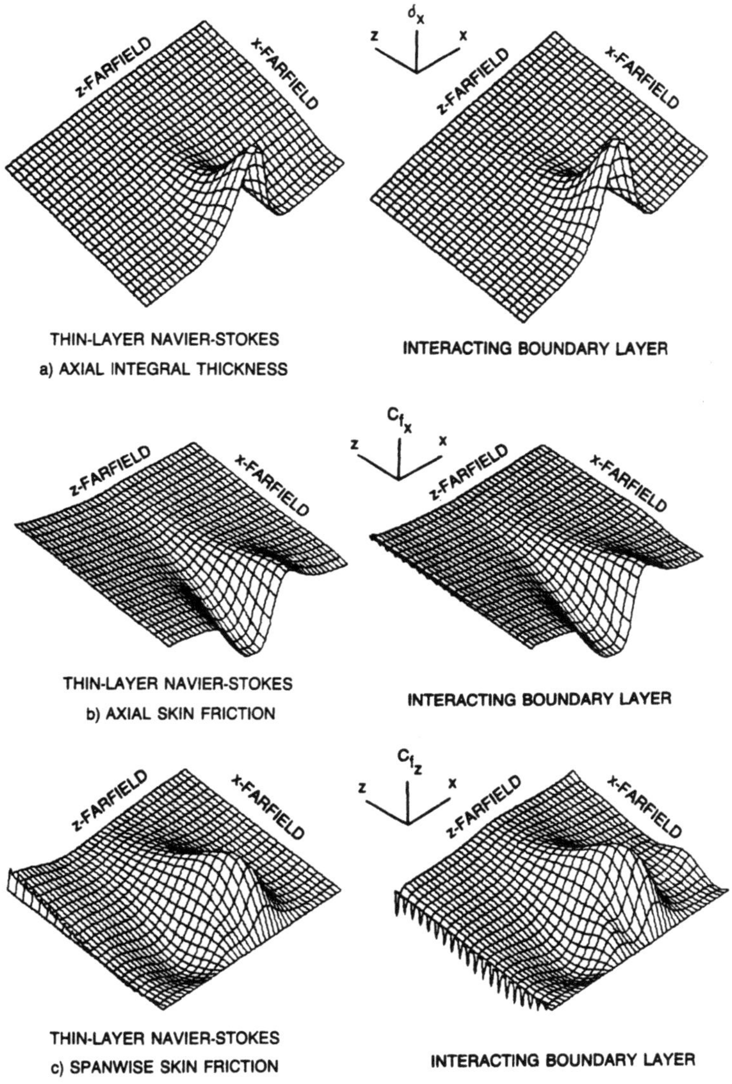

Fig. 2 **Flow over a flat plate with a dent (t=-.06), Re=8000. Global comparison of results of IBLT and TLNS.**

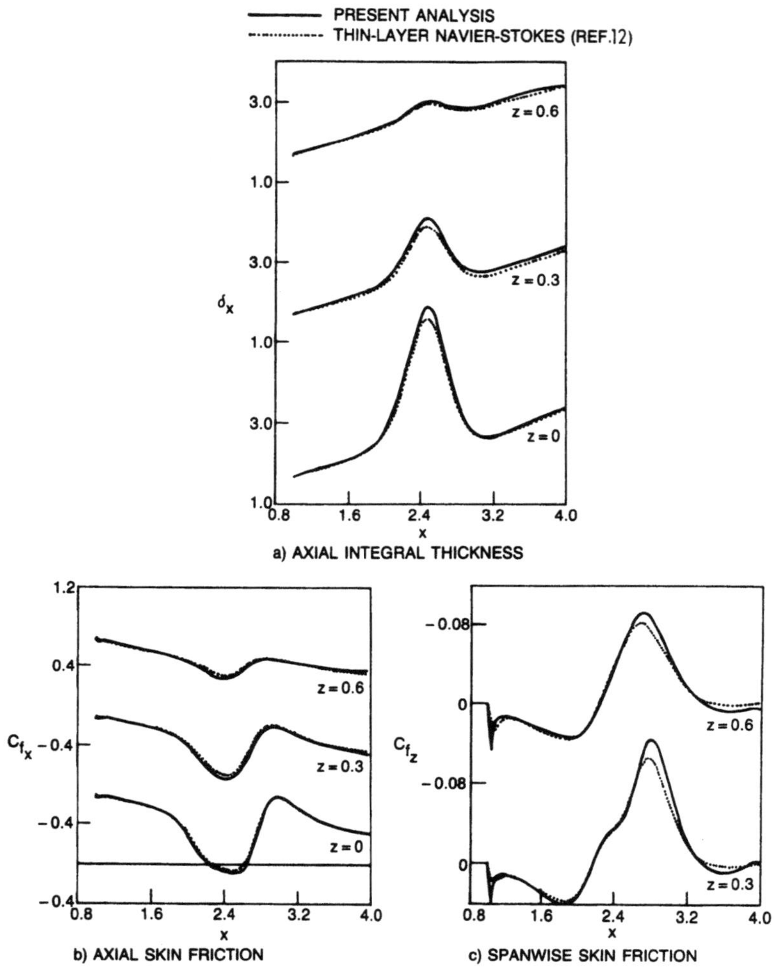

Fig. 3 Flow over a flat plate with a dent (t=-.06), Re=8000. Detailed comparison of results of IBLT and TLNS.

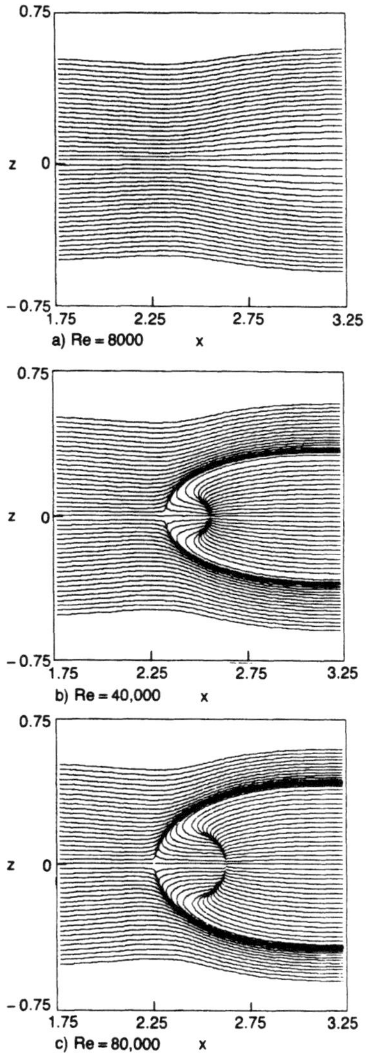

Separation at x = 2.286

Reattachment at x = 2.581

Separation at x = 2.248

Reattachment at x = 2.657

Fig. 4 Effect of Reynolds number on limiting streamlines for flow over a flat plate with a dent (t=-.03).

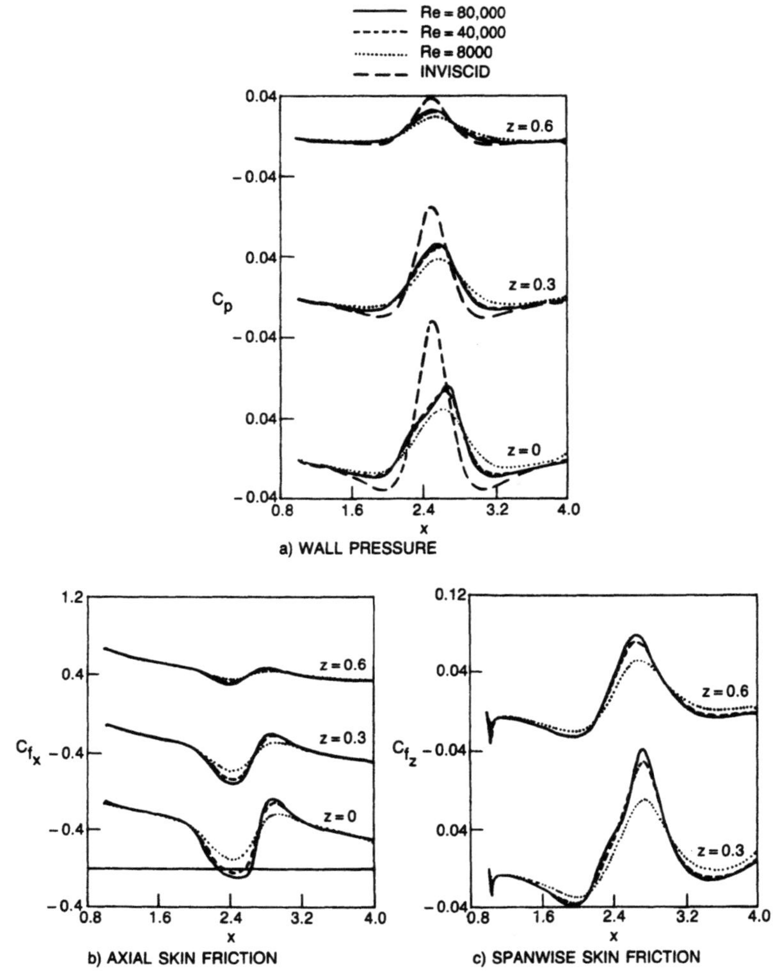

Fig. 5 **Effect of Reynolds number on flow over a flat plate with a dent (t=-.03).**

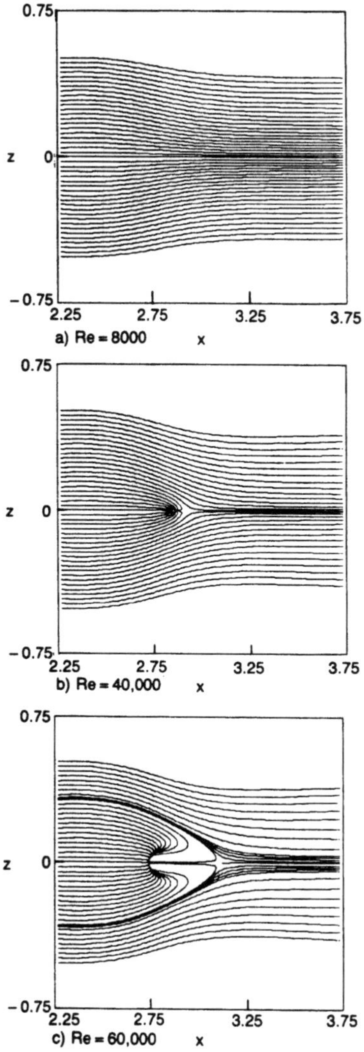

Separation at x = 2.766

Reattachment at x = 3.002

Separation at x = 2.709

Reattachment at x = 3.117

Fig. 6 **Effect of Reynolds number on limiting streamlines for flow over a flat plate with a bump (t=.03).**

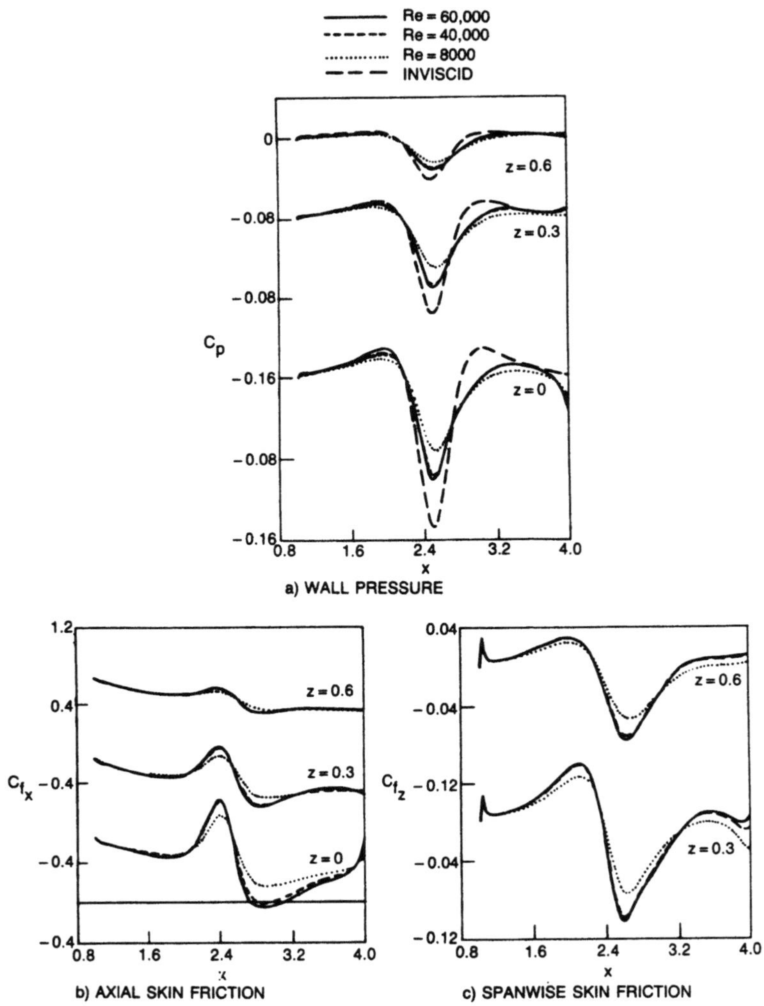

Fig. 7 Effect of Reynolds number on flow over a flat plate with a bump (t=.03).

A Review of Two-dimensional Turbulent Separated Flow Calculation Methods

ROGER L. SIMPSON

Department of Aerospace and Ocean Engineering
Virginia Polytechnic Institute and State University
Blacksburg, VA 24061 USA

In recent years, when reliable experimental data have become available for nominally two-dimensional in the time-mean separating and reattaching turbulent flows, much effort has been devoted to developing calculation methods. In 1985 Simpson [1] reviewed many experiments and calculation methods for such flows. One purpose of this brief paper is to include some more recent calculation method results.

Another purpose is to summarize from ref. [1] some of the physical flow phenomena as revealed from experiments that are important. Unless we properly model the physical phenomena, we cannot develop robust calculation methods that require only a few adjustable empirical numerical parameters.

The emphasis here is on the results of different turbulence models. Only investigations where comparisons were made with reliable experimental data will be discussed. No particular discussion of numerical techniques is given. Inviscid simulation methods that use vortex dynamics are briefly reviewed. Integral and differential time-mean methods for separating flows are given; only differential methods for reattaching flows are discussed. Interactive inviscid/shear layer calculation procedures are required because of the influence of the flow in these two regions on each other.

Some of the earlier calculation method review by Simpson [1] is included here. References given by a date in parentheses are listed in ref. [1] in order to reduce the length of this paper.

1. The Nature of a Separating Turbulent Boundary Layer

As a turbulent boundary layer undergoes an adverse pressure gradient, the flow near the wall decelerates until some backflow first occurs at incipient detachment (ID), defined as the near wall location where the fraction of time that the flow moves downstream γ_{pu} is 0.99. Large eddies, which bring outer region momentum toward the wall, supply some downstream flow. These large eddies grow rapidly in all directions and agglomerate with

F. T. Smith, S. N. Brown, (Eds.)
Boundary-Layer Separation
IUTAM Symposium London 1986

one another to decrease the average frequency of passage as detachment ($\bar{\tau}_w$ = 0; γ_{pu} = 0.5) is approached. Substantial pressure gradient relief begins near intermittent transitory detachment (ITD) defined as γ_{pu} = 0.8 near the wall.

These large-scale structures supply the turbulence energy to the near wall detaching flow. The velocity fluctuations in the backflow region are greater than or at least comparable to the mean backflow velocities. Since the free-stream flow is observed to be rather steady, the near wall fluctuations are not mainly due to a flapping of the entire shear layer, but due to turbulence within the detached flow. Even though the outer region mean velocity profiles look like those for a free mixing layer, the inner third is substantially different. The traditional attached flow "law-of-the-wall" type of mean velocity profile based on a wall shearing stress is not valid for the backflow.

The intermittently forward flow (γ_{pu} > 0) in the mean backflow region can be due to only two effects. Either high momentum forward flow moves occasionally toward the wall, or high momentum turbulent motions away from the wall set-up instantaneous streamwise pressure gradients that are impressed onto the low momentum wall region to produce instants of forward flow. Both of these effects contribute to the large eddy turbulent diffusion of energy from the outer region where it is produced to the backflow. Turbulent diffusion and dissipation are the main terms in the backflow turbulence energy balance (Simpson et al., 1981a). The small mean backflow does not come from far downstream. The small Reynolds shearing stresses in this wall region must be modeled by relating them to the turbulence structure and not to local mean velocity gradients. The mean velocity profiles in the backflow are a result of time-averaging the large turbulent fluctuations and are not related to the cause of the turbulence. Of course, this mechanism for supplying the backflow may be dominant only when the thickness of the backflow region is small as compared with the turbulent shear layer thickness, as in the Simpson et al. (1981a) flow. Experiments (Fox and Kline, 1962) on separation in wide-angle diffusers indicate that the mean backflow can come from far downstream when the thickness of the backflow region is comparable to the thickness of the forward flow. However, the measurements of Ashjaee et al. (1980) indicate that for large transitory stall, γ_{pu} is never zero! Thus, some forward flow is supplied to the near wall region by the large eddies.

2. The Nature of Separation from Two-Dimensional Sharp-Edged Bluff Bodies and Reattachment

The main obstacle to understanding this class of flows comes from the fact that the major detachment occurs near the sharp edges of the body, with accompanying large variations in velocity and pressure around the detachment location. Here intermittent detachment is located very near detachment. The flow downstream of detachment is strongly dependent upon the upstream velocity distribution and the local geometry of the surface around the detachment location. Downstream of detachment the thickness of the energetic outer region flow is comparable in size to that of the backflow region. The zone of recirculating fluid is a substantial portion of the entire detached shear layer. At some distance downstream the turbulent shear layer reattaches to the surface or mixes with an adjacent stream in the near wake.

Although the backward-facing step is the simplest reattaching flow, the flow field is still very complex. The upstream boundary layer detaches at the sharp corner forming a free-shear layer. If the boundary layer is laminar, transition begins soon after detachment, unless the Reynolds number is very low.

The separated shear layer appears to be much like an ordinary plane-mixing layer through the first half of the separated flow region, but differs from the plane-mixing layer in one important aspect: the flow on the low-speed side of the shear layer is highly turbulent, as opposed to the low-turbulence-level stream in a typical plane-mixing layer experiment. The separated shear layer curves sharply downwards in the reattachment zone and impinges on the wall. Part of the shear-layer fluid is deflected upstream into the recirculating flow by a strong adverse pressure gradient. The recirculating flow region below the shear layer has a maximum measured backflow velocity that is usually over 20% of the free stream velocity. The shear layer is subjected to the effects of adverse pressure gradient and strong interaction with the wall in the reattachment zone. A rapid decay of Reynolds normal and shear stresses occurs within the reattachment zone and downstream. The outer part of the reattached shear layer still has most of the characteristics of a free-shear layer as much as 50 step heights downstream of reattachment. This observation demonstrates the persistence of the large-scale eddies which are developed in the separated free-shear layer.

This flow is very unsteady. Very large turbulent structures with length scales at least as large as the step height pass through the reattachment region. The length of the separation region fluctuates so that the instantaneous impingement location of the shear layer moves up and downstream by as much a ± 2 step heights. The non-dimensional frequency of this motion is $fX_R/U_o \approx 0.6$ to 0.8. Here X_R is the distance from the step to the location of reattachment and U_o is the inviscid flow velocity upstream of the step. The period between flow reversals in the reattachment region appears to be random with no apparent correlation between the near wall flow upstream and downstream of reattachment. Although the time T between flow reversals $X_R/TU_\infty \approx 0.09$, these flow reversals are short-lived during time t with $X_R/tU_\infty \approx 0.6$ (Driver et al., 1983).

The turbulence intensity level of the detached flow is 5 - 10% higher than for the plane-mixing layers, which is believed to be due to a very low frequency ($fX_R/U_o < 0.1$) vertical "flapping" motion of the reattaching shear layer less than 20% of the shear layer thickness. The flapping correlates with strong flow reversals in the vicinity of reattachment. There is a reduction in the reverse flow rate with abnormally short instantaneous reattachment lengths. The shear stress in the flow increases dramatically with longer instantaneous reattachment lengths. Driver et al. suggest that the flapping is produced when a particularly high momentum structure moves far downstream before reattaching. This would create a somewhat greater pressure gradient that would cause greater backflow at some time later. The flapping motion produces negligible contributions to the Reynolds shearing stress.

The largest structure in the flow originates from the roll-up and multiple pairing of spanwise vortices (Pronchick, 1983). This roll-up is similar to the vortex roll-up and pairing process seen in the plane free shear layer. The convective speed of these structures is about 0.6 U_o and the spanwise coherence or organization of these vortical structures starts to breakdown 3 step-heights downstream of detachment. The turbulence structure becomes fully three-dimensional upstream of reattachment.

From their flow visualization studies, Müller and Gyr (1982) explained this breakdown behavior conceptually. As soon as a vortex tube is bent or distorted, it starts to induce velocities on itself (Biot-Savart law). A vortex loop bent in the downstream direction moves away from the adjacent wall while a vortex loop bent in the upstream direction moves closer toward the wall. The velocity induced on the near wall vortex loop by its image vortex can cause it to move upstream relative to the surrounding flow, thus supplying some backflow and forming a very large streamwise length scale and a long time scale. The vortex tubes that are stretched in the streamwise direction lead to more intense mixing, greater Reynolds shearing stresses and turbulence intensities. Pronchick did not observe large-scale structures in the backflow zone. He observed that the backflow consists of small-scale turbulent fluid created by eddy impingements on the wall and directed upstream by the adverse pressure gradient.

3. Inviscid Simulation Methods

Several inviscid models have been used to calculate the approximate time-mean behavior of some detached flows. Simpson [1] discussed several inviscid models that describe massive detached flow zones around airfoils. These use pressure gradient relaxation and a dead flow zone to obtain lift and drag values that are reasonably good. Taulbee and Robertson (1972) and Perry and Fairlie (1975) used a frozen vorticity analysis to describe the mean flow over a forward-facing step and detaching and reattaching boundary layer, respectively. Good agreement was obtained with experimental mean streamlines. An advantage of such analyses is that the irrotational free-stream and the rotational flow can be computed simultaneously. However, since the entrainment of non-turbulent free-stream fluid and the diffusion of vorticity are neglected in these analyses, some errors are present.

Several simulations of the instantaneous behavior of high Reynolds number two-dimensional separated flows have been performed using discrete-vortex models. Ashurst et al. (1980) performed simulations for the backward-facing step, while Spalart and Leonard (1981) studies flow over a square body and an oscillating airfoil and Kiya et al. (1982) examined a blunt flat plate. A number of discrete vortices are born at the body surfaces and are tracked through the flow in a Lagrangian description using essentially inviscid equations of motion. Viscous effects spread away from these vortices at a distance approximately $(\nu t)^{1/2}$ where t is the age of the vorticity since its birth. A viscous analysis next to the surface supplies vorticity to the outer region flow.

The simulated flow patterns look very much like those observed in flow visualization experiments mentioned above. When detachment occurs from a sharp edge, the flow remains fairly two-dimensional until near reattachment. Mean velocity and turbulence profiles for two-dimensional vortices look qualitatively like experimental data, but there are still large differences. Kiya et al. also obtained reasonable calculations of the time-mean and rms values of the velocity and the surface-pressure fluctuations. When three-dimensional effects are included there is a significant improvement, especially near reattachment. The experimental results of section 2 show that three-dimensional effects must be modeled.

The work of Graham [2] and Jaroch and Graham [3] also shows the necessity of including three-dimensional vortices substantially downstream of a two-dimensional detachment from a sharp edge. Graham obtained good two-dimensional simulations for higher frequency oscillatory freestream flows, where the unsteadiness keeps the shed vortices more two-dimensional. In the steady and low frequency oscillatory cases, where more spanwise bending of vortex tubes occurs by self-induced velocities, the three-dimensional effects are essential features that must be modeled; two-dimensional models produce Reynolds shear stresses that are at least 40% too low.

4. Incompressible Flow Adverse-Pressure-Gradient-Induced Separation Models

All of the known models make use of the established ideas for attached turbulent boundary layers. Eddy viscosity, mixing length, turbulence kinetic energy, and entrainment models are used as well as empirical correlations. Here we will discuss integral and differential methods and present some of their results and deficiencies. Integral methods are possible, even though there are large regions of backflow because the displacement of streamlines from the surface, the backflow, and the pressure distribution can be computed simultaneously or iteratively with a given velocity profile model. On physical grounds [1], the local large-eddy structure strongly influences the local backflow so a downstream marching calculation method for the velocity field is possible. Unfortunately, many of the calculation methods have drawn too heavily from attached flow concepts without making use of available experimental knowledge about these flows. For example, most methods do not incorporate the correct physics for the backflow region: Namely, that turbulence diffusion and dissipation control the backflow behavior and that the backflow mean velocity profile is determined by the large-scale fluctuations which scale on the maximum shear stress [1]. Hopefully, future calculation methods will build on this experimental information.

4.A. Integral Methods

Cousteix (Kline et al., Vol. II, 1982) presented a review of integral method techniques. The set of global equations consists of an integral momentum equation, e.g., the von Karman equation with normal stresses [1], and a secondary equation, either an entrainment equation, a moment of momentum or a mean kinetic energy integral equation. For each set of two integral ordinary differential equations, there are more unknowns than equations, so closure relationships are needed. Three types of relationships are usually considered: (1) relationships among integral thicknesses or shape parameters; (2) a skin-friction relation; and (3) entrainment equations, dissipation integrals or some other shear-stress integral. In general, the large-eddy-dominated outer region flow away from the wall lags the behavior of an equilibrium flow under the same boundary conditions, so a model ordinary differential equation for the lag effect is used for calculating the entrainment coefficient or dissipation integral.

It should be pointed out that the practical problem is one with inviscid/turbulent shear flow interaction where the displacement thickness effect is not negligible and the potential flow and shear layer calculations are not independent. The normal approach for attached flows, i.e., specifying P(x) or $U_e(x)$ for calculations using the boundary layer equation (1), prevents convergence of the solution at detachment. This is due to the fact that the differential boundary layer equation has a weak singularity when $\tau_w = 0$. An inverse boundary layer technique can be used to eliminate this computational problem [1].

Most integral methods use Coles' law-of-the-wall and law-of-the-wake velocity profile (Coles and Hirst, 1969) to determine the relationships among integral thicknesses, the skin friction relation, or at least the forms of the relationships. This profile does not fit closely the mean velocity profiles of detached turbulent boundary layers [1], even when the sign of U_τ is changed to account for mean backflows. Thus, earlier researchers who used this profile with attached flow constants did not obtain very good results. As more experimental data have become available, better fits of data to the forms of the relationships among parameters have been obtained. A number of other less than satisfactory methods were reviewed by Simpson [1].

Recent improvements to some of these methods and ideas have been made, largely as a result of the availability of additional experimental data. Lyrio et al. (1981) used the form of the law-of-the-wall and the law-of-the-wake velocity profile to obtain the relationships among H, δ^*/δ, Re_{δ^*}, and $C_f/2$. Numerical values in these relations were determined by fitting experimental data. Different entrainment correlations were used for different conditions. Good estimates of the detaching flow behavior can be made since the correlations make use of data from many different experiments; far downstream of detachment poorer estimates are obtained. Bardina et al. (1982) and Strawn et al. (1983) applied this method in the inviscid/viscid calculation and design of diffusers.

Thomas (1984) examined several of these different ingredients to integral methods. Green (1966) and LeBalleur's (1981) [4] modified law-of-the-wall/wake model also was examined. LeBalleur shifted the wake function away from the wall to permit a constant region of reversed flow next to the wall. The East et al. (1977) lag-entrainment method was also examined. In non-iterative calculations with the freestream, Thomas obtained fair agreement with experimental results from which the correlations were derived.

Melnik and Brook [5] followed LeBalleur's work using his velocity profile, except they used Green's nonequilibrium lag equation for the entrainment instead of the integrated turbulent kinetic energy and dissipation equations employed by LeBalleur. They used some free shear layer concepts to develop modifications to the lag equation for separated flow and used some of Thomas' (1984) ideas. Constants were adjusted to produce agreement with the experimental data of Simpson et al. (1981) and Hastings and Moreton (1982). While they obtained fairly good calculations of C_p for transonic flow over a RAE 2822 airfoil, they pointed out that the mean velocity profile model did not agree with data and planned to improve this method.

Williams [6] and Hastings and Williams [7] used almost the same ideas in their method to calculate flow over a NACA 4412 airfoil, for which experimental data were obtained by Hastings and Moreton (1982). While very good C_L calculations were made, boundary layer integral parameters for an angle of attack of 13.5° agreed with data for an angle of attack of 12.23°; the difference may have been partly due to three-dimensional effects induced in the flow.

Several integral methods do a fair job of calculating important parameters of detached compressible flows. Most of their difficulties stem from weaknesses in their correlations that are present for low speed flows too. The results of East et al. (1977), LeBalleur (1981), and Whitfield et al. (1981) and the results of Thomas (1984) using the models of these methods show fairly good agreement with experimental results for flows with relatively slow variations. Some type of lag model must be used in these methods across shock waves and in highly non-equilibrium conditions.

Childs et al. (1981) extended the incompressible methods of Ghose and Kline (1976), Bardina et al. (1981), and Lyrio et al. (1981) to subsonic compressible flows. The major new element was a compressible shear flow method, which extended the incompressible correlations. Good pressure recovery calculations were obtained for the Ashjaee et al. (1980) 10° diffuser and the Sajben et al. (1977) planar diffuser with an inlet Mach number of 0.65.

For unsteady flow Lyrio and Ferziger (1983) used their steady free-stream flow lag entrainment model. Although their set of equations is also hyperbolic, they chose to compute the conditions for all times at each x location before moving to the next location. Computed results for a Houdeville and Cousteix adverse pressure gradient flow and a conical diffuser unsteady flow were in good agreement with experimental results as far downstream as the time-averaged detachment location. The computed results for these low amplitude oscillatory flows support the observation that the mean flow parameters of attached boundary layers are almost unaffected by oscillations. Farther downstream δ^* and $\bar{H}$ are overpredicted. As suggested by Lyrio and Ferziger, the shortcomings of the method in the presence of flow reversal may be that the Coles velocity profile and/or the entrainment correlation may not hold in the unsteady reversed-flow region. Simpson <u>et al</u>. (1983) and Simpson and Shivaprasad (1983) observed from their data that the ensemble-averaged or phase-averaged velocity profiles look like steady free-stream profiles for the same near wall γ_{pu} when $\partial\gamma_{pu}/\partial t < 0$. For $\partial\gamma_{pu}/\partial t > 0$, the profiles differed significantly from the steady free-stream results, but looked much like the Coles velocity profile. Since the Coles velocity profile model does not fit the steady free-stream velocity profile data well nor model the velocity profile hysteresis, the profile model used in this method needs to be modified.

Strickland, Oler, and Im (Francis and Luttges, 1984) used the Lyrio and Ferziger method to compute some unsteady airfoil flows. Qualitative agreement with data was obtained. Good quantitative agreement was observed for small angles of attack.

One idea that needs to be pursued in an integral method is the use of τ_{max} as the descriptive shearing stress. As discussed in section 1, this is the scaling shear stress for the shear layer upstream and downstream of detachment. The wall shear stress is very small in detaching flows and influences only the near wall region. Schofield (1981) presented a method that incorporates the maximum shear stress that is based on the Perry and Schofield (1973) correlation for strong-adverse-pressure-gradient equilibrium attached boundary layer. This method has not yet been extended to detached flows.

4.B. Differential Methods

A number of earlier differential calculation methods were reviewed by Simpson [1]. Here we will review only the more successful and more recent results. As for integral methods, inverse computational schemes must be used for differential methods to eliminate the weak singularity at $\tau_w = 0$.

The effective viscosity model of Cebeci and Smith (1974) has proved to be successful in representing attached boundary layer flows. In the inner region

$$\mu_t = \rho \ell^2 \frac{\partial U}{\partial y} \qquad (1)$$

with

$$\ell = 0.4y[1 - \exp(-y^+/A^+)] \qquad (2)$$

$$A^+ = \frac{26}{N}, \quad N = (1-11.8p^+)^{\frac{1}{2}}, \quad p^+ = \frac{\nu U_e}{U_\tau^3} \frac{dU_e}{dx}$$

and, in the outer region,

$$\mu_t = 0.0168\, U_e \delta^* \qquad (3)$$

As we can see in the figures, this model does not do well for detached flows [8-14]. Lorber and Covert [15] as well as other researchers reviewed by Simpson [1] also found that this model does not well represent unsteady separating flows.

Johnson and King (1984) proposed a hybrid Reynolds stress/eddy viscosity model for separating flows to account for experimentally-observed lower eddy viscosities than given by the Cebeci and Smith model. They proposed that the Cebeci-Smith model is adequate when turbulence energy production and dissipation rates are in balance of "equilibrium", but needs to be modified to account for strong non-equilibrium cases. Taking a cue from the experimentally-observed fact that the maximum turbulent shearing stress $-\rho\overline{uv}_{max}$ is the proper shearing stress scale for strong adverse pressure gradient and separating flows, they proposed an ordinary differential equation for $-\overline{uv}_{max} = \overline{g}^{-2}$ along the path of maximum turbulence kinetic energy that was derived from the turbulence energy equation:

$$\frac{dg}{dx} = \frac{a_1}{2U_M L_M}\left[1 - \left(\frac{g}{g_{eq}}\right)\right] + \frac{C_{dif}}{2\delta U_M[0.7-y/\delta/_M]}\left[1 - \left(\frac{\nu_{to}}{\nu_{to,\,eq}}\right)^{\frac{1}{2}}\right] \qquad (4)$$

The left side of this equation represents the convective change of shear stress; the first term on the right side is the imbalance between production and dissipation; the last term is due to turbulent diffusion. Johnson and King [8] show that the diffusion term represents 'bulk' or large-eddy diffusion. The large-eddy diffusion velocity scales on $(-\overline{uv}_{max})^{\frac{1}{2}}$ and is proportional to the physical entrainment velocity of non-turbulent fluid into the shear layer. It is important to include this effect because we know from section 1 that the turbulent diffusion process is an important feature of separated turbulent flows. Here, equilibrium values are given by

$$\nu_{t,\,eq} = \nu_{t,o\,eq}[1-\exp(-\nu_{ti,\,eq}/\nu_{to,eq})] \qquad (5)$$

$$\nu_{ti,\,eq} = \ell^2(-\overline{uv}_{max})^{\frac{1}{2}}/0.4y \qquad (6)$$

with ℓ from equation (4) and $A^+ = 15$,

$$\nu_{to,\,eq} = 0.0168\, U_e \delta^*/(1+5.5(y/\delta)^6) \qquad (7)$$

The dissipation length L_M is given by

$$L_M = 0.4 y_M \text{ for } y_M/\delta \leq 0.225 \qquad (8)$$

and

$$L_M = 0.09\delta \text{ for } y_M/\delta > 0.225.$$

C_{dif} is a modeling constant and U_M is the mean velocity at the maximum shearing stress location. These equations reduce to the form of the Cebeci-Smith model for the equilibrium case.

Because the eddy viscosity is lower in the detached flow zone, the Johnson and King model produces outer region mean velocity profiles, surface pressure and skin friction distributions, shape factor H distributions, and maximum shearing stresses that are in much better agreement with experimental values than the unmodified Cebeci-Smith model (Figures 1-5).

Figures 1 and 2 show very good calculations of the Simpson et al. (1981) diffuser flow. Good C_p, C_f, H, and δ^* results were obtained. For the Bachalo and Johnson (1979) axisymmetric bump flow, good C_p, H, δ^*, and velocity and shear stress profiles were obtained (Figures 3 and 4). Very good results were presented for a NACA 64A010 airfoil (Figure 5) and a DSMA 671 supercritical M = 0.72 airfoil. Coakley [12] found that the Johnson-King model gave superior results for surface pressure, skin friction, δ^*, lift and drag for 3 cases of transonic flow over a RAE 2822 airfoil as compared to the k-ε, Baldwin-Lomax (1978), Cebeci-Smith, and his own q-ω model. Lee and Pletcher [16] reported the same δ^* results as Johnson and King for the axisymmetric bump flow. Good results for C_p for a boat tailflow was also obtained.

Barnett and Carter [14] eliminated the damping factor in equation (3) when the flow detached and used a streamwise length exponential decay model to force a reduction in the eddy viscosity and model the history effect. The relaxation length was 30 δ_o^*, with δ_o^* calculated at detachment. Good qualitative calculations for flow over a curved backward-facing step were obtained. While this model is different than the Johnson and King model, it independently illustrates that lower eddy viscosities and relaxation effects are required to capture the physics of detached flows than given by the Cebeci-Smith model.

The k-ε model is not as good as the Johnson and King model, as discussed by Simpson [1]. Rodi and Scheuerer (1983) showed that the k-ε model gives consistently too high C_p values in attached adverse-pressure-gradient flows. This is due to small ε values relative to the production rate of k. An increase in ε production is necessary to correct this feature. Horstman and Johnson [13] found that the $k-\omega^2$ model gave results little different from the $k-\varepsilon$ model for the bump flow shown in Figure 3, which were poorer than the Johnson and King model. The curvature correction for the $k-\varepsilon$ model had little effect on results.

Rodi [17] pointed out that Visbal and Knight [18] found that a variation of the Baldwin-Lomax (1978) model does not calculate compressible separated flows well. Délery and Marvin [19] present a review of calculations of shock-induced separation using many of these methods, except those using the recent Johnson and King model.

5. Calculation Methods for Bluff Body Reattachment Flows

The $k-\varepsilon$ model has been widely tested by a number of researchers for two-dimensional bluff body flows, e.g., cavity, rib, and sudden expansion flows, where detachment ocurs at a well-defined location. A number of slightly different versions were used by Mansour, Pollard, Rodi et al., Launder et al,, and Ha Minh et al. (Kline et al., Vol. III, 1982) at the 1981 Stanford Conference to compute the Kim et al. (1978) backward-facing step flow.

Rodi (Kline et al., 1982) pointed out that all of the standard $k-\varepsilon$ models produce results for recirculating flows that are in much poorer agreement with experiment than results for simple shear layers (Vasilic-Melling, 1976; Gosman et al., 1977). All versions of the standard $k-\varepsilon$ model tested at the Stanford Conference produced reattachment lengths that were at least 20% too short for the Kim et al. flow. Since we know from section 2 that the flow behavior in the streamwise direction scales on the reattachment length X_R, it is no surprise that the computed C_p rises too rapidly and mean velocity profiles are in much error. Reynolds shearing stresses remain too high until about one X_R downstream of reattachment.

Launder et al. discuss the use of a Reynolds-stress transport model (RSM) or an algebraic stress model (ASM) instead of the Boussinesq viscosity model used in the standard $k-\varepsilon$ model. Launder and Leschziner (private communication, 1985) conclude that the ASM does best in the recirculation region; the $k-\varepsilon$ is best in the reattaching flow region. Rodi suggested that the RSM and ASM are not significantly better for recirculating flows than the $k-\varepsilon$ model. The ASM produces a greater X_R, in closer agreement with data, because the dissipation ε is increased and $-\overline{uv}$ is reduced in the detached flow zone.

Just as Rodi and Scheuerer (1983) observed for strong adverse-pressure-gradient flows many other investigators have concluded that the dissipation ε

must be higher than computed by a standard k-ε model to get good results. Marvin (1983) noted this for the results of Sindir (1982), some of which are shown in Figure 6 for the experiments of Driver and Seegmiller (1982). Clearly, the best results occur for computations carried out with the ASM and the ε equation modified to increase dissipation. Calculations of U and $-\overline{uv}$ profiles are in fair agreement with experiment. None of these models agree well with the turbulence kinetic energy balances shown by Driver and Seegmiller for detached flow. Mansour et al. (1983) used a rotation of turbulence term in the dissipation equation that reduced the dissipation and produced good X_R and C_p results, although k and $-\overline{uv}$ values were 30% and 10% low, respectively.

Others have used a correction similar to those employed for longitudinally curved flows, reasoning that the mean streamlines are curved. The algebraic stress model with curvature correction did fairly well at the Stanford Conference. Leschziner and Rodi (1981) suggest the use of the curvature modification to improve recirculating flow results. Gooray et al. (1983) used a functional C_μ similar to Leschziner and Rodi that accounts for streamline curvature. They obtained fairly good X_R, U, and C_p results for the Eaton and Johnston (1980) and Moss and Baker (1980) backward-facing step flows. Benodekar et al. (1983) used the curvature corrections of Launder et al. (1977) and obtained fairly good comparisons of U/U_e profiles for the Crabb et al. (1981) flow over a square rib.

Amano and Goel [20] also concluded that the ASM of Sindir (1982, 1983) is the best of the three models (k-ε, RSM, ASM) for the recirculation region of the Driver and Seegmiller flow. They suggested that the RSM does best for the redeveloping flow region. All these models underpredict the reattachment length. Amano and Goel [21] found that the triple-product turbulent correlations for the outer region of the Chandrsuda and Bradshaw (1981) backward-facing step flow were grossly underpredicted. This is not surprising because the large-eddy turbulence energy diffusion, which accounts for the magnitude of the triple-product turbulent correlations, is not modeled by the k-ε, RSM, and ASM models. Kodali and Amano [22] report poor agreement between k-ε model calculations and experimental results for axisymmetric flow normal to a circular disk.

Rodi [17, 23] summarized the situation with the k-ε model in the calculation of certain separated flows. The model performs well for confined axisymmetric flows with separation, and the same applies to confined plane flows in which the influence of an opposing wall is felt strongly, i.e. for large blockage ratios (Celik, Rodi, and Stamou [24]). When the opposing wall is moved away, the model tends to underpredict the size of the recirulation zone; the same is true also for unconfined separated flows (Majumdar and Rodi [25]) except for those situations where vortex shedding occurs and the steady-model approach is not suitable. Various modifications to the k-ε model have been suggested to improve its performance for separated flows, but none of them proved entirely satisfactory.

Ilegbusi and Spalding (1983) used a $k-\overline{\omega^2}$ model to compute the Kim et al. flow, with apparently better computations than the standard k-ε model. The reattachment length and maximum shear stress were well computed for one Tropea backward-facing step flow. The authors state that this method also works well for simple turbulent flows. More examination of this method for detached flows is needed.

6. Conclusions

From experimental information it is clear that large eddy transport or diffusion of turbulence energy is very important in separating and reattaching turbulent flows. The entrainment of non-turbulent free stream fluid and the lateral bulk large-eddy transport velocity scale on the maximum shear stress.

This author is clearly impressed with the performance of the Johnson and King turbulence model in the comparisons with experimental data. This model approximately accounts for the large eddy diffusion effect on the "non-equilibrium" flow condition by using an ordinary differential equation for the maximum shear stress. While not making the numerics substantially more complicated, this model is capable of calculating subsonic flows and transonic airfoil and diffuser flows with shock waves. More calculations using this model are needed to test and expand the range of applicability.

Vortex simulation methods provide insights about the instantaneous flow behavior, but fail in calculating good quantitative features of reattaching flows unless three-dimensional effects are included.

Integral methods can calculate the behavior of airfoil flows where the shape of the velocity profile does not change too much along the flow and can be described by a sufficiently universal velocity profile model. Differential methods are more able to account for rapid changes in velocity profiles.

The k-ε, ASM, and RSM models do not contain a large-eddy diffusion term. This partially explains why the k-ε model does not do well for large unconfined separated flows, but does reasonably well for high blockage confined flows where smaller scale motions have much more influence.

Future modeling improvements should more closely represent the experimentally-observed flow behavior. Much other discussion about turbulent separated flows is contained in [1] .

REFERENCES

1. Simpson, R.L., Two-Dimensional Separated Flow, AGARD-AG-287, 1, NATO, June 1985.

2. Graham, J.M.R., "Numerical Simulation of Steady and Unsteady Flow About Sharp-Edged Bodies," Proc. Int. Symp. Separated Flow Around Marine Structures, Norwegian Inst. Tech., Trondheim, pp. 347-373, June 26-28, 1985.

3. Jaroch, M.P.G. and Graham, J.M.R., "An Evaluation of the Discrete Vortex Method as a Model for the Flow Past a Flat Plate Normal to the Flow with a Long Wake Splitter Plate," submitted to J. Fluid Mechanics.

4. LeBalleur, J.C., "Strong Matching Method for Computing Transonic Viscous Flows, Including Wakes and Separations," La Recherché Aérospatiale no. 1981-3, 21-45, English Translation.

5. Melnik, R.E. and Brook, J.W., "Computation of Viscid/Inviscid Interaction on Airfoils with Separated Flow," Third Symp. Numerical and Physical Aspects of Aero. Flows, Long Beach, CA, pp. (1-21)-(1-37), Jan. 21-24, 1985.

6. Williams, B.R., "The Prediction of Separated Flow using a Viscous-Inviscid Interaction Method," Aero. Royal Aero. Soc., pp. 185-197, May 1985.

7. Hastings, R.C. and Williams, B.R., "Studies of the Flow Field Near an NACA 4412 Aerofoil at Nearly Maximum Lift," Third Symp. Numerical and Physical Aspects of Aero. Flows, Long Beach, CA, pp. (7-23)-(7-37), Jan. 21-24, 1985; to be published in Aero. J. of Royal Aero. Soc.

8. Johnson, D.A. and King, L.S., "A Mathematically Simple Turbulence Closure Model for Attached and Separated Turbulent Boundary Layers," AIAA J., 23, 11, pp. 1684-1692, November 1985.

9. Johnson, D.A., "Predictions of Transonic Separated Flow with an Eddy-Viscosity/Reynolds-Shear-Stress Closure Model," AIAA 85-1683, Cincinnati, OH, July 16-18, 1985.

10. Johnson, D.A. and King, L.S., "Transonic Separated Flow Predictions Based on a Mathematically Simple, Nonequilibrium Turbulence Closure Model," NASA TM 86826, October 1985.

11. King, L.S., "Separated Transonic Airfoil Flow Calculations with a Nonequilibrium Turbulence Model," NASA TM 86830, November 1985.

12. Coakley, T.J., "Impact of Turbulence Modeling on Numerical Accuracy and Efficiency of Compressible Flow Simulations," 10th Inter. Conf. on Numerical Methods in Fluid Dynamics, Beijing, China, June 23-27, 1986.

13. Horstman, C.C. and Johnson, D.A., "Prediction of Transonic Separated Flows," AIAA J., 22, 7, pp. 1001-1003, July 1984.

14. Barnett, M. and Carter, J.E., "Analysis of Turbulent Separated Flows Including Heat Transfer," AIAA 86-0113, Reno, NV, Jan. 1986.

15. Lorber, P.F. and Covert, E.E., "Unsteady Airfoil Boundary Layers - Experiment and Computation," Third Symp. Numerical and Physical Aspects of Aero. Flows, Long Beach, CA, pp. (5-39)-(5-47), Jan. 21-24, 1985.

16. Lee, D.S. and Pletcher, R.H., "Application of a Viscous-Inviscid Interaction Method to Predict Transonic Separated Flows," Third Symp. Numerical and Physical Aspects of Aero. Flows, Long Beach, CA, pp. (1-75)-(1-85), Jan. 21-24, 1985.

17. Rodi, W., "Turbulence Modelling," Proc. Int. Symp. Separated Flow around Marine Structures, Norwegian Inst. Tech., Trondheim, pp. 261-288, June 26-28, 1985.

18. Visbal, M. and Knight, D., "The Baldwin-Lomax Turbulence Model for Two-Dimensional Shock-Wave/Boundary-Layer Interactions," AIAA Journal, 22, pp. 921-928, 1984.

19. Délery, J. and Marvin, J.G., Shock-Wave Boundary Layer Interactions, AGARD-AG-280, NATO, February 1986.

20. Amano. R.S. and Goel, P., "Triple-Velocity Products in a Channel with a Backward-Facing Step," AIAA J., 24, 6, pp. 1040-1043, June 1986.

21. Amano, R.S. and Goel, P., "Computations of Turbulent Flow Beyond Backward-Facing Steps Using Reynolds-Stress Closure," AIAA J., 23, 9, pp. 1356-1361, 1985.

22. Kodali, V.S. and R.S. Amano, "Prediction of Turbulent Flows Behind Bluff Bodies by Using an Algebraic-Stress Model," AIAA 86-1061, Atlanta, GA, May 12-14, 1986.

23. Rodi, W., Lectures II and III from "Examples of Turbulence-Model Applications," Turbulence Models and Their Applications, Collection de la Direction des Etudes et Recherches Editions Eyrolles, Paris, 1984.

24. Celik, I., Rodi, W. and Stamou, A.I., "Prediction of Hydrodynamic Characteristics of Rectangular Settling Tanks," Proc. Inter. Symp. on Refined Flow Modelling and Turbulence Measurements, Iowa City, IA, Sept. 16-18, 1985.

25. Majumdar, S. and Rodi, W., "Numerical Calculations of Turbulent Flow Past Circular Cylinders," Third Symp. Numerical and Physical Aspects of Aero. Flows, Long Beach, CA, pp. (3-13)-(3-25), Jan. 21-14, 1985.

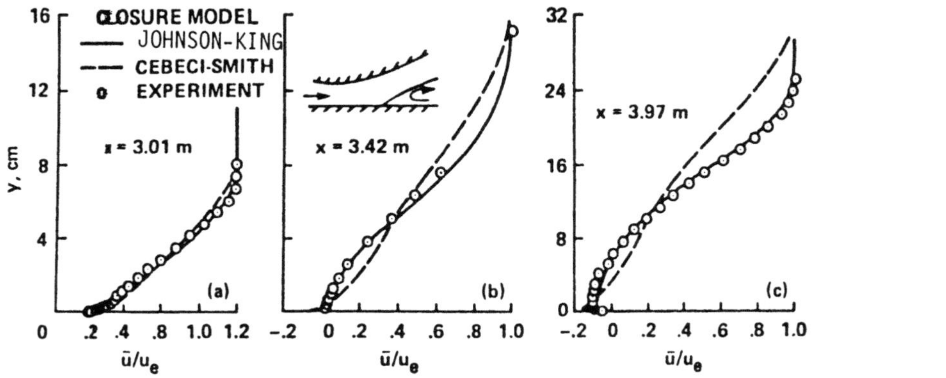

Fig. 1 Mean-velocity comparisons [8, 10] for the low-speed diffuser of Simpson et al. (1981) a) upstream of separation; b) at separation; c) downstream of separation.

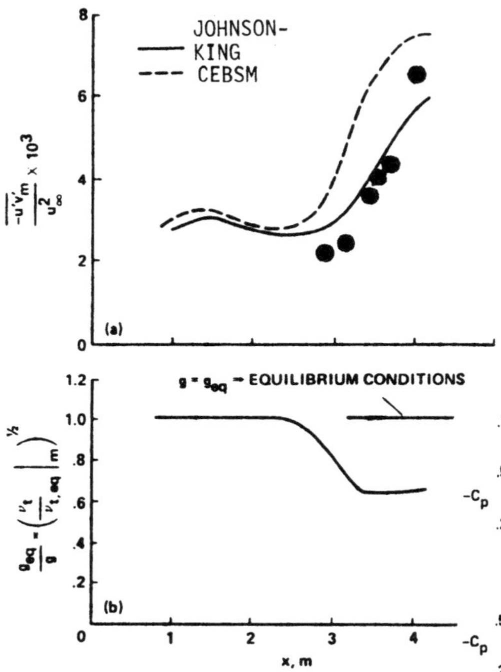

Fig. 2 Simpson et al. (1981) low-speed flow, "history" effects: a) maximum Reynolds shear stress development; b) nonequilibrium state of flow [8]. Symbols experimental data.

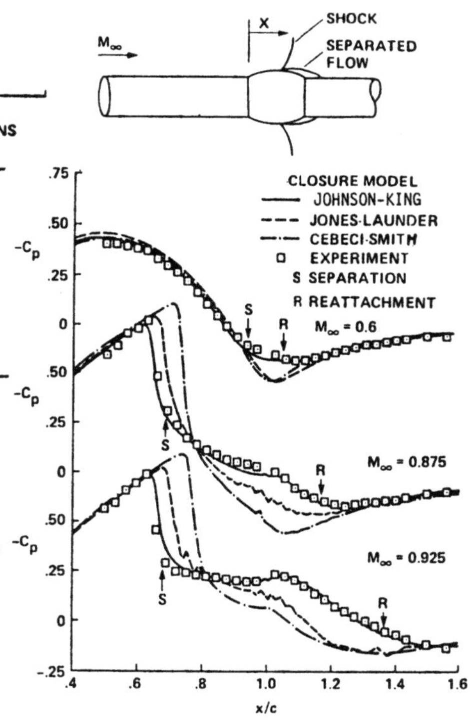

Fig. 3 Surface pressure comparisons [10] for the axisymmetric-bump flow of Bachalo and Johnson (1979).

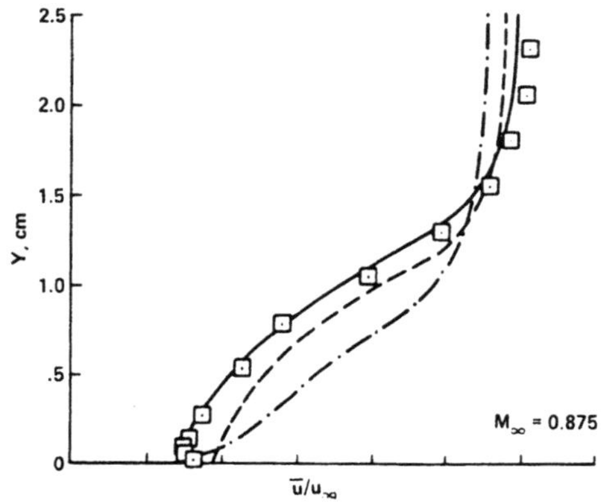

Fig. 4 Predicted, mean-velocity profiles [9] at model trailing edge, x/c = 1, axisymmetric bump flow.

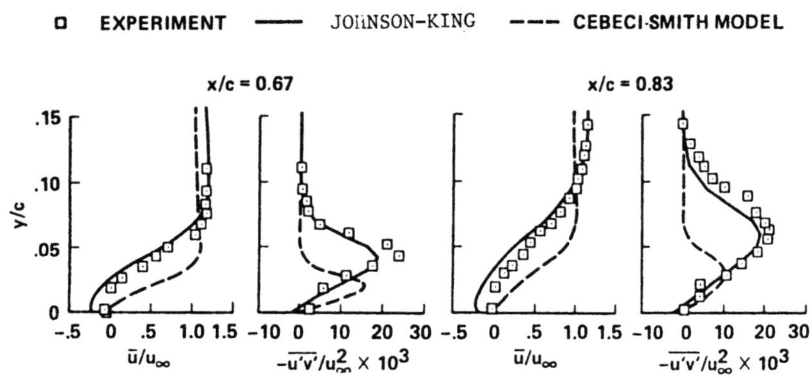

Fig. 5 Mean-velocity and Reynolds shear stress comparisons [10] at M_∞ = 0.8 and α = 6.2° for the NACA 64A010 airfoil section; data from Johnson and Bachalo (1980).

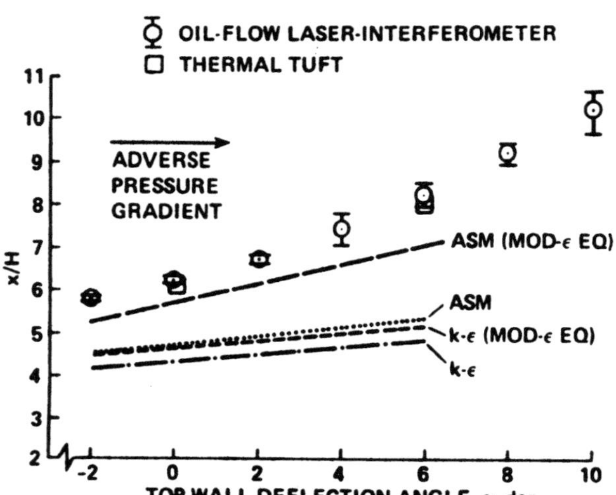

Fig. 6 Reattachment location versus top-wall deflection angle. Calculations of Sindir (1982). Data of Driver and Seegmiller (1982).

A new Interacting Boundary-Layer Formulation for Flows past Bluff Bodies

A. P. Rothmayer

Department of Aerospace Engineering
Iowa State University
Ames, Iowa 50011

Introduction

Many strong interaction phenomenon, including laminar separation, are described at high Reynolds numbers by the triple-deck theory of Neiland [1], Stewartson and Williams [2] and Messiter [3]. This asymptotic theory may be reproduced, in a composite sense, by the interacting boundary layer model. Interacting boundary-layer theory is formed by raising the second order boundary layer displacement-thickness effect to first order, allowing a strong interaction between the boundary-layer and the local outer inviscid flow. Numerous authors have tackled this problem numerically, producing a number of efficient algorithms for solving strong interaction problems. One of the most successful interacting boundary-layer algorithms is the quasi-simultaneous method of Veldman [4] and Davis and Werle [5]. This algorithm, like many others, represents the outer inviscid flow by a thin airfoil equation. In the Veldman-Davis method this thin airfoil equation is directly coupled into the outer edge boundary conditions in the boundary-layer. Numerous generalizations of the basic interacting boundary-layer method have been made to other classes of flows. However, in most cases some form of thin airfoil theory has been used to describe the outer inviscid flow. For example, in the massive separation problem of Rothmayer and Davis [6,7], which yielded solutions for large scale thin eddies over thin airfoils, a modified thin airfoil equation was used. Even in the problem solved by Carter and Vatsa [8], for the small leading-edge separation bubble on a thick airfoil, the airfoil surface was assumed to be a flat plate in the vicinity of the separation bubble and a thin airfoil equation was used. When thin airfoil theory is abandoned, researchers tend to return to a numerical solution of the full Euler equations. The goal of this study is to break away from the thin airfoil restriction while still retaining the formalism of a "thin-airfoil like" perturbation approach in treating the outer inviscid flow. The main reason for doing this is so that the efficient interacting boundary layer algorithms which have been based on thin airfoil theory, such as the Veldman-Davis method, may be directly generalized to flow past bluff bodies.

In the turbulent flow past an airfoil the fluid may remain attached, even for relatively thick airfoils. In addition, turbulent separation bubbles tend to be relatively short in length. Even though the methods presented in this study may eventually have applications to this more realistic problem of turbulent flow past an airfoil, this study will concentrate on the more unrealistic, but much simpler, problem of laminar separation at high Reynolds numbers. In general, laminar separation may only be avoided by using extremely thin airfoils closely aligned with the oncoming uniform stream. There are a number of exceptional cases, but it is generally believed that most bluff bodies possess extremely large separation regions. For two dimensional, incompressible flow

F. T. Smith, S. N. Brown, (Eds.)
Boundary-Layer Separation
IUTAM Symposium London 1986

past smooth bluff bodies, the high Reynolds number limit is given by Sychev [9] and Smith [10] This theory is based on a triple-deck embedded in a slightly singular Kirchhoff [11] free-streamline solution. The flow within the separation region near the body is forced by a slow entrainment through the shear layer. The only requirement of this structure is a slow reverse streaming entering the body scale separated region from the much larger downstream eddy. Smith [12] ha shown that a viscous eddy closure occurs for confined flows, such as a cascade c bluff bodies. The viscous closure prevents large return velocities from enterir the eddy and so preserves the body-scale Kirchhoff free-streamline solution. The situation is not so simple in external flows. Numerical calculations by Fornberg [15] point towards the asymptotic theory of Smith [14]. Smith's [14] theory for the flow past a single bluff body is based on a large O(Re) x O(Re) Sadovskii [15] constant vorticity eddy, which is essentially a very large Prandt Batchelor eddy. Smith's model allows for a re-entrant jet and so the separated region on the body scale must be solved as a non-constant vorticity eddy with velocity levels on the order of the free stream velocity. Fortunately, this severe departure from the Kirchhoff solution should still retain the basic form of the Sychev-Smith theory near the separation point. The main complication introduced by Smith's new model is a non-trivial dependence of the flow near separ ation on the eddy closure process occurring a very large distance downstream of the body.

The central problem considered in this study is to solve for flow past a circular cylinder. Such a flow should be described, in the high Reynolds number limit, by Smith's [14] model. However, since Smith's [14] model still remains to be verified, this study will employ the admittedly incorrect Kirchhoff [11] wake model. The reason for doing this are - i) the Kirchhoff model is the simplest eddy model and the easiest to calculate, ii) interacting boundary-layer solutions based on the Kirchhoff model will have the same structu near separation as those based on the more realistic Smith [14] model, iii) the interacting boundary-layer models presented in this study may easily be generalized to include more realistic eddy models as they become available.

It should be emphasized that the primary aim of this study is to provide a methodology for extending two dimensional, incompressible, thin-airfoil based interacting boundary-layer methods to models that are capable of solving for flows past bluff bodies. Keeping this in mind, the question of which wake model is appropriate should be viewed as a side issue.

The Boundary-Layer Equations in the Baseline Coordinate System

The governing equations within the boundary-layer are referenced to the baseline coordinate system shown in Fig. 1. This baseline coordinate curve lies within the boundary-layer and wake for a predominantly attached flow. For flows with large-scale separation, the baseline coordinate curve detaches from the vicinity of the body, near the separation point, and follows the shear-layer as it rides over the wake. The boundary-layer equations in this baseline coordinate system have been given by Davis and Werle [5]. In these equations, the streamwise and normal velocity components are non-dimensionalized by the free-stream velocity and scaled according to boundary-layer scaling laws. The arclength/normal baseline coordinates are treated in a similar fashion

$$s = s^*/L, \quad n = n^*/(L\,Re^{-1/2}) \quad , \qquad (1)$$

where the (*) denotes physical variables, L is a reference length, s is the arclength along the baseline coordinate curve and Re is the Reynolds number:

$$Re = U_\infty L/\nu. \tag{2}$$

In these variables, the no-slip boundary conditions are evaluated at $n = -t(s)$. To simplify the boundary conditions, the Prandtl transposition theorem is used, with

$$N = n + t(s) \tag{3}$$

If the the normal velocity is appropriately redefined then the boundary-layer equations retain their original form. However, the no slip boundary conditions are evaluated along $N = 0$. In order to accommodate the growth of the boundary layer in the s-direction, Davis and Werle [5] cast the equations in terms of Gortler variables referenced to the baseline coordinate curve

$$\xi = \int_o^s Ue_o(\bar{s})d\bar{s}, \qquad \eta = \frac{Ue_o(s)N}{2\xi}, \tag{4}$$

where Ue_o is the inviscid surface speed over the baseline coordinate curve. The final governing equations arrived at by Davis and Werle [5], after some manipulation, are

$$V_\eta + F + 2\xi F_\xi = 0 \tag{5}$$

and

$$F_{\eta\eta} - VF_\eta + FV_\eta + \beta_1 + (1 - \beta_o)F^2 = 0 \tag{6}$$

where F and V are the streamwise and normal velocity components in Gortler variables. There are two pressure gradient parameters present in these equations, the baseline pressure gradient

$$\beta_o = 2\xi Ue_{o_\xi}/Ue_o \tag{7}$$

and the actual pressure gradient

$$\beta_1 = 2\xi Ue\ Ue_\xi/Ue_o^2 \tag{8}$$

Following Davis and Werle [5], an additional equation is introduced for β_1:

$$\beta_{1_\eta} = 0 \tag{9}$$

This allows for a simultaneous solution, using a block tri-diagonal algorithm, of the continuity, streamwise momentum and β_1-equation, in keeping with the physics of strong-interaction problems. These equations are solved subject to the following boundary conditions - the F-matching condition

$$F \rightarrow Ue/Ue_o \quad \text{as } \eta \rightarrow \infty, \tag{10}$$

the v-matching condition

$$V - \eta V_\eta \to \sqrt{2\xi}(Ue\delta)_\xi \quad \text{as } \eta \to \infty \ , \tag{11}$$

and the no-slip condition

$$F(\xi,0) = V(\xi,0) = 0 \tag{12}$$

Additional information on this equation set, the baseline coordinate system and the numerical method presented in the next section may be found in Davis and Werle [5]. It should be noted at this point that these equations are valid for flow past bluff bodies or thin bodies so long as the baseline coordinate curve lies in, or near, the boundary-layer. In general, the outer inviscid flow must be solved subject to an injection condition from second order boundar layer theory:

$$W(\xi,0) = Re^{-\frac{1}{2}}\, Ue_o \frac{d}{d\xi}\,[Ue(\delta - t)] \ , \tag{13}$$

where δ is the boundary-layer displacement thickness and $t(\xi)$ is the distance from the baseline coordinate curve to the body surface. In a composite sense, this injection velocity allows the interacting boundary-layer model to capture a triple-deck strong interaction.

For the purpose of the numerical procedure, the solution to the boundary layer equation set is initiated in a region where the assumption of local self similarity is valid, say near a stagnation point on a blunt leading edge airfoil. This assumption amounts to setting the ξ-derivative to zero in Eqn. [5]. In addition it is assumed that the flow has a Falkner-Skan form with

$$Ue = U_o\, s^{\frac{\beta}{2-\beta}} \ , \quad \delta = \delta_o\, s^{\frac{1-\beta}{2-\beta}} \tag{14}$$

and with $\beta_\xi = 0$. Using these approximations, the boundary condition of Eqn. [11] may be written as

$$V - \eta V_\eta \to Ue\delta/\sqrt{2\xi} \tag{15}$$

for any self-similar flow. The equation set at the first point is solved in the same fashion as the interior of the flow. That is, β_1 is left as an unknown - to be determined by interaction with the outer inviscid flow.

At this point, the system of equations may be closed by determining the velocity response of the inviscid flow at the edge of the boundary-layer, Ue, to the injection velocity from the boundary-layer of Eqn. [13].

For flow past a bluff body, one might reasonably expect the response of the inviscid flow to be rather complicated. Understandably, previous researche have concentrated on the thin airfoil case where $Ue_o = 1$. This turns out to be unnecessary. The perturbed flow over the baseline coordinate curve may be foun from a relatively simple generalization of thin-airfoil theory. This fact allo bluff body problems to be solved with virtually the same code as is used for th airfoil problems. In the next section the "thin airfoil approach" for generati the outer inviscid flow will be reviewed. This review is meant to point out th strengths of thin airfoil theory when it is used as the outer inviscid flow solution in the recently developed quasi-simultaneous methods of Veldman [4] and Davis and Werle [5].

The Veldman-Davis Method

One of the most common methods of solving for the outer inviscid flow is to take the baseline coordinate curve to be on, or near, the x-axis. That is $Ue_o = 1$. The outer inviscid flow is found using thin airfoil theory. If the injection velocity of Eqn. [13] is prescribed everywhere along the x-axis, then a Cauchy-Hilbert form emerges for the thin airfoil integral

$$Ue = 1 + \frac{1}{\pi}\int_{-\infty}^{\infty} \frac{T'(\bar{\xi})d\bar{\xi}}{\xi - \bar{\xi}} , \tag{16}$$

where

$$T = Re^{-\frac{1}{2}}\, Ue(\delta - t). \tag{17}$$

This is the outer-inviscid flow equation employed by Davis and Werle [5] Veldman [4] and others. Alternatively, if the injection velocity of Eqn. [13] is prescribed from the leading-edge of an airfoil, $\xi = 0$, to some arbitrarily prescribed point ξ_1, and if a zero pressure region is prescribed from ξ_1 to, ∞ then the method of Cheng and Rott [16] may be used to generated a thin airfoil equation of the form

$$Ue = 1 + \frac{\sqrt{\xi_1 - \xi}}{\pi}\int_{o}^{\xi_1} \frac{T'(\bar{\xi})\, d\bar{\xi}}{(\xi - \bar{\xi})\sqrt{\xi_1 - \bar{\xi}}} \tag{18}$$

Rothmayer and Davis [6] used this particular equation to generate large-scale separation solutions which possessed extensive regions of constant pressure. The assumption of constant pressure from ξ_1 to ∞ is valid only if ξ_1 is placed within a true constant pressure region of the viscous solution. It was found by Rothmayer and Davis [6] that the initial pressure gradient singularity in Eqn. [18] would readily initiate a triple-deck free interaction near ξ_1. This local interaction near the separation point would then control the overall large-scale separation structure.

Element expressions on a uniform grid may be found for both Eqns. [16] and [18]. At the ξ_i grid point, the resulting quadrature formula takes on the form

$$Ue_i = g_i + \frac{1}{\pi\Delta\xi}\sum_{j=1}^{M}(T_j - P_j)D_{ij}. \tag{19}$$

For the Cauchy-Hilbert equation of Eqn. [16], Davis and Werle [5] used an integration by parts and a midpoint rule to generate Eqn. [19] with D_{ij} taking the form

$$D_{ij} = \frac{4}{1 - 4(i-j)^2} , \quad P_j = 0. \tag{20}$$

Rothmayer and Davis [6] used a similar procedure to produce the quadrature formula of Eqn. [19], but with

$$D_{ij} = \sqrt{\frac{\xi_o - \xi_i}{\Delta\xi}} \left[\frac{1}{(i - j + \frac{1}{2})\sqrt{M - j + \frac{1}{2}}} - \frac{1}{(i - j - \frac{1}{2})\sqrt{M - j - \frac{1}{2}}} \right],$$

$$P_{2_j} = T(\xi_o) + (\xi_j - \xi_o)\, T'(\xi_o). \tag{21}$$

Details of this derivation may be found in Davis and Werle [5] and Rothmayer and Davis [6]. Two important results follow from using the above thin-airfoil equations for the outer-inviscid flow. The first is that Eqn. [19] may be written in the form

$$Ue_i = \bar{C}_i + \bar{D}_i (Ue\delta)_i \quad , \tag{22}$$

which directly relates the inviscid surface speed Ue to the boundary-layer displacement thickness δ. The second important result is that Eqn. [22] may be used to provide a direct coupling between the outer inviscid flow and the boundary-layer - the Veldman-Davis quasi-simultaneous method. This method was first suggested by Veldman [4] and implemented in its final form by Davis and Werle [5]. The essence of the Veldman-Davis method is to use the thin-airfoil equation and the boundary-layer V-matching condition to eliminate Ue and δ from the $\eta \to \infty$ boundary condition. The boundary conditions, Eqns. [10] and [11], with $Ue_o = 1$, are finite differenced to give

$$F_{iN} = Ue_i \quad , \quad V_{iN} - \eta_N \left(\frac{V_{iN} - V_{iN-1}}{\Delta\eta} \right) = \sqrt{2\xi_i} \left[\frac{(Ue\delta)_i - (Ue\delta)_{i-1}}{\Delta\xi} \right] \tag{23}$$

and the thin airfoil equation, Eqn. [22], is used to eliminate $(Ue\delta)_i$ and Ue_i from Eqn. [23]. This results in a boundary condition written solely in terms of the Gortler variables within the boundary layer:

$$\left[- \frac{\sqrt{2\xi_i}}{\bar{D}_i \Delta\xi} \right] F_{iN} + \left[\frac{\eta_N}{\Delta\eta} \right] V_{iN-1} + \left[1 - \frac{\eta_N}{\Delta\eta} \right] V_{iN} =$$

$$- \frac{\sqrt{2\xi_i}}{\Delta\xi} \left[\frac{\bar{C}_i}{\bar{D}_i} + (Ue\delta)_{i-1} \right] \tag{24}$$

If a spatial quasi-linearization is used for the non-linearities then iteration is needed only to converge the boundary-layer displacement thickness. That is, if the equation set is being solved at the ith station then $j > i$, must take on guessed values in Eqn. [19]. However, if the ratio D_{ij}/D_{ii} is made small, then the effect of the numerical upstream influence in Eqn. [19] can be minimized and the method will converge faster. One important feature of the Veldman-Davis method is that the quadrature form used to generate Eqn. [19] may be used to control the form of D_{ij} and hence the convergence rate. This fact was used by Davis and Werle [5] in deriving Eqn. [20], and the dependence of the convergence rate of the interacting boundary-layer method on the ratio D_{ij}/D_{ii} was convincingly demonstrated for the cascade problem by Rothmayer and Davis [17].

The Veldman-Davis method provides a very efficient algorithm for solving the strong interaction problem. In addition the convergence rate can be manipulated, to some degree, by using different quadrature expressions for the inviscid thin airfoil equation. However, the strength of the Veldman-Davis method lies in the fact that thin airfoil theory can be used for the outer inviscid flow. The thin airfoil approach directly relates the inviscid surface speed to the boundary-layer displacement thickness through Eqn. [22]. In solving for flow past a bluff body it would be highly desirable to preserve this equation. Since all the boundary-layer equations and boundary conditions are trivially extendable to flows past bluff-bodies, all that is needed to implement the Veldman-Davis method in the bluff-body problem is a "thin-airfoil like" equation of the form of Eqns. [16] and [18]. The derivation of such an equation is the next topic to be addressed.

The Bluff-Body Problem

If the inviscid flow past a bluff body, or alternatively the baseline coordinate curve, is given then it is reasonable to expect that the boundary-layer will only generate a small perturbation to this flow. One of the first attempts to address the problem of flow past a bluff body was made by Carter and Vatsa [8], who considered small separation bubbles near the leading-edge of blunt leading-edge airfoil. They assumed that if the separation bubble were small enough then the body curvature could be neglected and the airfoil surface could be approximated by a flat plate. An inviscid surface speed Ue_o was generated using an inviscid solver over the entire airfoil. At this point Carter and Vatsa [8] assumed that the perturbation had the same form as the flow past a flat plate with the base flow, Ue_o, prescribed. For a general asymmetric flow, Carter and Vatsa [8] assumed that near the separation bubble

$$Ue = Ue_o + \frac{1}{\pi}\int_{-\infty}^{\infty} \frac{T'(\bar{\xi})\, d\bar{\xi}}{\xi - \bar{\xi}} \quad , \tag{25}$$

where Ue is the inviscid surface speed along the total displace surface. In a private communication Davis [18] suggested that a conformal mapping technique might be employed to transform the baseline curve to a plane in which thin airfoil theory could be applied directly. The interacting boundary-layer problem could then be solved in the transformed plane. Following this suggestion, Rothmayer [19] developed a "thin-airfoil like" theory using the conformal mapping approach. However, the resulting equation in the physical plane had a rather complicated form and was dependent on the conformal mapping. Special cases of Rothmayer's [19] problem and later work by Rothmayer and Davis [7], using a parabola as the baseline coordinate curve, suggested that a very simple "thin-airfoil like" equation could be developed for an arbitrary bluff baseline coordinate curve. In addition, it appeared that this equation could be made independent of the mapping and that it would take on a form very similar to Carter and Vatsa's [8], at least for the symmetric case. The simplest case to consider is a generalization of the Cauchy-Hilbert form of Eqn. [16]. In Eqn. [16], the baseline coordinate curve is the x-axis, since $Ue_o = 1$, and ξ is simply the x-coordinate. For a bluff baseline coordinate curve in the physical plane, see Fig. (1), the inviscid surface speed over the baseline curve is Ue_o and the injection condition from the boundary-layer into the outer inviscid flow is given by Eqn. [13]. The baseline coordinate curve is mapped to a stagnation flow plane using a conformal mapping of the form $\zeta = f(z)$, where the z-plane, $z = x + iy$, is the physical plane and the ζ-plane, $\zeta = \bar{\xi} + i\bar{\eta}$, is the transformed stagnation plane. The complex baseline velocity in the ζ-plane is assumed to take a general stagnation form

$$V_o(\zeta) = \tilde{u}_o - i\tilde{v}_o = M(\zeta - \bar{\xi}_o) \tag{26}$$

with an unknown stagnation point $\bar{\xi}_o$ and an unknown constant M. In a specific problem M and $\bar{\xi}_o$ would be determined from the mapping (see Davis [20]). Throughout this derivation the mapping function f(z) is taken to be completely arbitrary, as is the specific form of the stagnation flow. In addition the base flow V_o in the ζ-plane is assumed to be consistent with the base flow in the z-plane, that is

$$|\zeta_o'|\ |V_o|_{\bar{\eta}=o} = Ue_o \quad , \tag{27}$$

where, for convenience:

$$|\zeta_o'| = \left|\frac{d\zeta}{dz}\right|_{\bar{\eta}=o} \quad , \quad |z_o'| = \left|\frac{dz}{d\zeta}\right|_{\bar{\eta}=o} \tag{28}$$

The flow in the transformed plane then takes the form

$$V(\zeta) \sim V_o(\zeta) + Re^{-\frac{1}{2}}\, V_1(\zeta) + \ldots \tag{29}$$

with $V_i = \tilde{u}_i - i\tilde{u}_i$. The boundary condition on V_1 is given from the transformed boundary layer injection velocity of Eqn. [13] written in the variables of the ζ-plane

$$\tilde{v}_1(\bar{\xi},0) = |z_o'|\ Ue_o[Ue(\delta - t)]_{\xi} = [Ue(\delta - t)]_{\bar{\xi}} \tag{30}$$

In the ζ-plane a Cauchy-Hilbert thin airfoil problem results, with the solution

$$\tilde{u}(\bar{\xi},0) = \tilde{u}_o(\bar{\xi},0) + \frac{1}{\pi}\int_{-\infty}^{\infty} \frac{T'(\tilde{\xi})\, d\tilde{\xi}}{\bar{\xi} - \tilde{\xi}} \quad , \tag{31}$$

where $T = Re^{-\frac{1}{2}}Ue(\delta - t)$. This equation is transformed back to the physical plane simply by multiplying Eqn. [31] by $|\zeta_o'|$, giving

$$Ue = Ue_o + \frac{|\zeta_o'|}{\pi}\int_{-\infty}^{\infty} \frac{T'(\tilde{\xi})\, d\tilde{\xi}}{\bar{\xi} - \tilde{\xi}}. \tag{32}$$

However, the second term is still transformation dependent and written in the variables of the ζ-plane. Fortunately there is a natural connection between the variables of the ζ-plane and the Gortler variables defined with respect to the baseline coordinate curve. Using Eqn. [4]:

$$\xi = \int_0^s Ue_o(\bar{s})d\bar{s} = \int_{\xi_o}^{\xi} \tilde{u}_o(\bar{\xi},0)d\bar{\xi} = \frac{M}{2}(\bar{\xi} - \bar{\xi}_o)^2. \tag{33}$$

This implies that

$$\bar{\xi} = \bar{\xi}_o \pm \sqrt{\frac{2\xi}{M}} . \tag{34}$$

The plus/minus sign in the above equation indicates that a formal distinction must be made between the "upper" and "lower" surfaces of the body in the physical plane (see Fig. 1). That is, the Gortler variable, ξ, is positive on both sides of the stagnation point. Keeping this in distinction in mind and using the fact that

$$\sqrt{\frac{M}{2}}\,\frac{|\zeta_o'|}{\pi} = \pm\,\frac{Ue_o}{2\pi\sqrt{\xi}} \tag{35}$$

gives the final outer-inviscid flow equation over the upper surface:

$$Ue_+ = Ue_o\left\{1 + \frac{1}{2\pi\sqrt{\xi}}\int_0^{\infty} \frac{T_+'(u)\, du}{\sqrt{\xi} - \sqrt{u}} - \frac{1}{2\pi\sqrt{\xi}}\int_0^{\infty} \frac{T_-'(u)\, du}{\sqrt{\xi} + \sqrt{u}}\right\}. \tag{36}$$

Whereas for the lower surface:

$$Ue_- = Ue_o\left\{1 - \frac{1}{2\pi\sqrt{\xi}}\int_0^{\infty} \frac{T_-'(u)\, du}{\sqrt{\xi} - \sqrt{u}} + \frac{1}{2\pi\sqrt{\xi}}\int_0^{\infty} \frac{T_+'(u)\, du}{\sqrt{\xi} + \sqrt{u}}\right\}. \tag{37}$$

These equations may also be recast to bring out the symmetric and asymmetric contributions. However, because of a lack of space this will not be presented here. For the symmetric case $T_+ = - T_-$ and the above equations reduce to

$$Ue_{\pm} = Ue_{o}\left\{1 \pm \frac{1}{\pi}\int_{o}^{\infty} \frac{T'_{\pm}(u)\, du}{\xi - u}\right\}. \tag{38}$$

At this point there is really no distinction between the upper and lower surfaces, except for a trivial sign convention, and so the following equation may be used for the symmetric case:

$$Ue = Ue_{o}\left\{1 + \frac{1}{\pi}\int_{o}^{\infty} \frac{T'(u)\, du}{\xi - u}\right\} \tag{39}$$

This equation is in agreement with the proposals of Rothmayer amd Davis [7] for the possible generalized form for the symmetric case. In addition, it may be shown that if the airfoil surface is assumed to be locally flat, and a simple rotation/translation mapping is used, then this formulation reduces to the form given by Carter and Vatsa [8] (i.e. Eqn. [25]).

Several interesting features of Eqns. [36], [37] and [39] need to be examined. First off, these equations are dependent only on the baseline inviscid surface speed in the physical plane, Ue_o, which enters into the equations directly, and also indirectly through the baseline Gortler variable. More importantly, these equations are mapping independent. Since no special assumptions were made about the mapping f(z) and since the final perturbation equations in the physical plane should be unique, these equations are the general solution of the perturbed inviscid problem in the physical plane. The baseline flow Ue_o may be calculated by any method desired and so Eqns. [36], [37] and [39] are not restricted to a conformal mapping approach. Secondly, the boundary condition as $\eta \to \infty$ in the boundary-layer is $F = Ue/Ue_o$. At least for the symmetric case, the generalization of the Veldman-Davis method to the bluff body problem can be accomplished with trivial changes to a thin-airfoil based interacting boundary-layer code. The asymmetric case requires more work but the overall Veldman-Davis method is still preserved.

The more complicated Cheng-Rott case (Eqn. [18]) now needs to be considered. In the physical plane it is assumed that a change in boundary condition occurs at ξ_1 and ξ_2 (see Fig. 1). These two points correspond to $\bar{\xi}_1$ and $\bar{\xi}_2$ in the transformed plane. In the ζ-plane the boundary conditions now become

$$\tilde{v}_1(\bar{\xi},0) = (\text{Im } V_1)_{\bar{\eta}=o} = \frac{d}{d\bar{\xi}}\,[Ue(\delta - t)] \tag{40}$$

for $\bar{\xi}_1 < \bar{\xi} < \bar{\xi}_2$

and $\tilde{u}_1(\bar{\xi},0) = (\text{Re } V_1)_{\bar{\eta}=o} = 0$ (41)

for $\bar{\xi} < \bar{\xi}_1$ and $\bar{\xi} > \bar{\xi}_2$. Using the method of Cheng and Rott [16], this mixed boundary value problem is converted to a direct, i.e. Cauchy-Hilbert, problem for a new function $F(\zeta)$ where

$$F(\zeta) = V_1(\zeta)/H_o(\zeta) \tag{42}$$

To be consistent with the flow in the physical plane $H_o(\zeta)$ takes the form

$$H_o(\zeta) = i\sqrt{(\zeta - \bar{\xi}_1)(\zeta - \bar{\xi}_2)}. \tag{43}$$

The boundary conditions on $F(\zeta)$ are now

$$(\mathrm{Im}\ F)_{\bar{\eta}=o} = \begin{cases} \dfrac{[Ue(\delta - t)]_{\bar{\xi}}}{\sqrt{(\bar{\xi} - \bar{\xi}_1)(\bar{\xi}_2 - \bar{\xi})}} & \text{for } \bar{\xi}_1 < \bar{\xi} < \bar{\xi}_2 \\ 0 & \text{otherwise.} \end{cases} \tag{44}$$

A Cauchy-Hilbert problem is solved for the function $F(\zeta)$, yielding

$$F(\zeta) = \frac{-1}{\pi}\int_{-\infty}^{\infty} (\mathrm{Im}\ F)_{\bar{\eta}=o}\ \frac{d\tilde{\xi}}{\zeta - \tilde{\xi}} \tag{45}$$

This solution may be rearranged into the form

$$V_1(\zeta) = \frac{-i}{\pi}\sqrt{(\zeta - \bar{\xi}_1)(\zeta - \bar{\xi}_2)}\int_{\bar{\xi}_1}^{\bar{\xi}_2} \frac{T'(\tilde{\xi})\ d\tilde{\xi}}{(\zeta - \tilde{\xi})\sqrt{(\tilde{\xi} - \bar{\xi}_1)(\bar{\xi}_2 - \tilde{\xi})}} \tag{46}$$

From this equation the solution for the inviscid surface speed between $\bar{\xi}_1$ and $\bar{\xi}_2$ may be extracted:

$$\tilde{u}(\bar{\xi},0) = \tilde{u}_o(\bar{\xi},0) + \frac{Re^{-1/2}}{\pi}\sqrt{(\bar{\xi} - \bar{\xi}_1)(\bar{\xi}_2 - \bar{\xi})}\int_{\bar{\xi}_1}^{\bar{\xi}_2} \frac{T'(\tilde{\xi})\ d\tilde{\xi}}{(\bar{\xi} - \tilde{\xi})\sqrt{(\tilde{\xi} - \bar{\xi}_1)(\bar{\xi}_2 - \tilde{\xi})}}. \tag{47}$$

As in Rothmayer and Davis [6], a similar equation may be found for $T(\bar{\xi})$ outside of the region $(\bar{\xi}_1, \bar{\xi}_2)$. This aspect of the problem will not be considered in this study. Eqn. [47] is transformed to the physical z-plane through a multiplication by $|\zeta_o'|$. Rewriting the variables in terms of variables in the z-plane produces:

$$Ue_+ = Ue_o\left\{1 + \frac{\sqrt{(\sqrt{\xi} + \sqrt{\xi_1})(\sqrt{\xi_2} - \sqrt{\xi})}}{2\pi\sqrt{\xi}}\left[\int_o^{\xi_2} \frac{T_+'(u)\ du}{(\sqrt{\xi} - \sqrt{u})\sqrt{(\sqrt{\xi_1} + \sqrt{u})(\sqrt{\xi_2} - \sqrt{u})}} - \int_o^{\xi_1} \frac{T_-'(u)\ du}{(\sqrt{\xi} + \sqrt{u})\sqrt{(\sqrt{\xi_1} - \sqrt{u})(\sqrt{\xi_2} + \sqrt{u})}}\right]\right\} \tag{48}$$

over the upper surface, and

$$Ue_- = Ue_o \left\{ 1 + \frac{\sqrt{(\sqrt{\xi_1} - \sqrt{\xi})(\sqrt{\xi_2} + \sqrt{\xi})}}{2\pi\sqrt{\xi}} \left[\int_o^{\xi_2} \frac{T_+'(u)\,du}{(\sqrt{\xi} + \sqrt{u})\sqrt{(\sqrt{\xi_1} + \sqrt{u})(\sqrt{\xi_2} - \sqrt{u})}} - \int_o^{\xi_1} \frac{T_-'(u)\,du}{(\sqrt{\xi} - \sqrt{u})\sqrt{(\sqrt{\xi_1} - \sqrt{u})(\sqrt{\xi_2} + \sqrt{u})}} \right] \right\} \qquad (49)$$

over the lower surface. The details of transforming from the ζ-plane to the z-plane are essentially the same as for the Cauchy-Hilbert case. In the symmetric case $\xi_1 = \xi_2$, $T_+ = - T_-$ and the above equations effectively reduce to the following single equation:

$$Ue = Ue_o \left\{ 1 + \frac{\sqrt{\xi_1 - \xi}}{\pi} \int_o^{\xi_1} \frac{T'(u)\,du}{(\xi - u)\sqrt{\xi_1 - u}} \right\}. \qquad (50)$$

Again, this is the form predicted, by Rothmayer and Davis [7], using a simple analytic mapping. Since Eqn. [50] has virtually the same form as the thin airfoil equation [18], it is evident that the numerical method of Davis and Werle [5] and Rothmayer and Davis [6] may be carried over directly to the bluff body problem. The only important difference between the thin body problem and the bluff body problem is that in the thin body problem $Ue_o = 1$ whereas in the bluff body problem the baseflow Ue_o must be numerically calculated from the flow over a bluff baseline curve in space. Again, it should be stressed that Ue_o may be calculated using any number of methods, for example: conformal mappings, singularity distributions, or hodographs.

The Circular Cylinder Problem

The symmetric Cheng-Rott integral of Eqn. [50] is used in this section to calculate the two-dimensional, incompressible, laminar flow past a circular cylinder. It should be emphasized that the Cheng-Rott integrals, as given in Eqns. [48], [49] and [50], implicitly assume that the downstream wake has a Kirchhoff [11] form. This "Kirchhoff wake" problem should be viewed as a "model-separation" problem in which the wake effects must be corrected at a later date. The base flow is taken to be a singular Kirchhoff free-streamline solution with the inviscid "separation" point set just downsream of the point ξ_1. The baseline coordinate curve is then taken to be the combination of the body surface and the free-streamline. Note that the inviscid "separation" point is simply the point at which the baseline curve breaks away from the body surface. This free-streamline solution was calculated using the generalized Schwarz-Christoffel mapping code of Davis [20]. The interacting boundary-layer algorithm used to solve for the viscous flow is essentially the same as that used by Rothmayer and Davis [6,7] and will not be repeated here. The results for the wall shear stress

$$\tau_w = \frac{Ue_o^2}{\sqrt{2\xi}} \frac{\partial F}{\partial \eta}(\xi,0) = \frac{Ue_o^2}{\sqrt{2\xi_i}} \left(\frac{F_{i2}}{\Delta\eta} + \frac{1}{2}\beta_{1_i}\Delta\eta \right) + O(\Delta\eta^2) \qquad (51)$$

are shown in Fig. 2 for various Reynolds numbers, with θ being the angle in degrees from the upstream stagnation point. The Reynolds number used here is based on the free-stream velocity and the diameter of the cylinder (as is the case in Fornberg's [13] Navier-Stokes solutions). As Re increases, a very strong localized free-interaction developes near the separation point. In addition, since the physical surface vorticity is essentially Eqn. [51] multiplied by an $O(\sqrt{Re})$ factor, the results of Fig. 2 are in qualitative agreement with Fornberg's [13] surface vorticity results, both in sign and magnitude. The grid for the Re = 10^6 case was found to yield graphical accuracy for (n_ζ,n_η) = (121,151) distributed uniformly in (ζ,η). It was assumed that the lower Reynolds number cases could be accurately resolved by this same grid. The Re = 10^7, 10^8 case were both solved on a (n_ξ,n_η) = (201, 301) grid and while the Re = 10^7 case probably has adequate resolution, it may be seen that the Re = 10^8 case has reached the limits of the grid resolution on the (201,301) grid. It has been shown by Rothmayer and Davis [6] that this problem can easily be circumvented by using a simple stretching transformation. In addition, the Re = 10^9 case converged, but with even worse resolution near the separation point and this produced the blatantly incorrect result of the separation point moving away from the correct high Reynolds number limit solution (see later discussion). The Re = 10^{10} case rather quickly "diverged" from the massive separation solution and converged towards an attached solution. The upper limit, in Reynolds number, of the interacting boundary layer method appears to be set by the grid resolution near the separation point. Poor resolution initially produced convergent but incorrect results and eventually led to divergent solutions. The lowest Reynolds number case calculated in this study is Re = 200 (the Re = 100 case diverged). At this point it is not understood what caused the very low Reynolds number cases to diverge. In summary the interacting boundary-layer method converged from a Reynolds number of 200 to a Reynolds number of 10^9, with the upper limit in Reynolds number apparently being fixed by the grid resolution near the separation point.

The wall pressure coefficients for the various cases are shown in Fig. 3 with $C_p = 1 - Ue^2$. The localized adverse pressure gradient near separation is readily seen as Re increases. A comparison between the wall pressure at Re = 600 and Fornberg's [13] full Navier-Stokes results is shown in Fig. 4. In these calculations, the value of the pressure imposed in the eddy must be the free stream pressure. This means that the eddy pressure may be changed by simply changing the free-stream pressure. Such an effect is expected to occur since the overall large scale eddy would modify the flow before it reached the actual region of the body (see Smith [14]). In Fig. 4, another case was run with a modified wake pressure, which was fixed from Fornberg's [13] results for the Re = 600 case. Even though an incorrect wake model is being used, these results are in relatively good agreement with Fornberg's [13] full Navier-Stokes calculations. It is the authors opinion that any remaining discrepancy between these results may be removed by correcting the wake model and using higher order boundary-layer theory.

The results for the location of the separation points at various Reynolds numbers are shown in Fig. 5. Comparison is made here with the high Reynolds number limit of Kirchhoff [11] smooth separation and the low Reynolds number full Navier-Stokes results of Fornberg [13]. The slow approach to the expected high Reynolds number limit solution is readily observable in this figure. This slow approach is predicted by the asymptotic theory of Sychev [9] and Smith [10]. In their theory, the distance between the viscous separation point and the inviscid limit solution is $O(Re^{-1/16})$ and the viscous separation point must always be downstream of the Kirchhoff [11] smooth separation point.

The primary drawback of this interacting boundary-layer procedure is its inability to correctly predict the eddy pressure. This shortcoming will lead to gross mispredictions of both the lift and drag on an airfoil. However, it appears that there are a number of ways to remove the assumption of a constant pressure Kirchhoff wake while still retaining the formalism of the Cheng-Rott approach. This matter is currently under investigation.

Conclusion

In this study an equation has been developed for the perturbed flow past an arbitrary baseline coordinate curve with a small injection velocity. Even though this equation was derived using a conformal mapping approach, it was shown that the final equation could be cast in a form which was independent of the mapping. The fact that such an equation could be generated for the perturbed inviscid flow meant that the Veldman-Davis quasi-simultaneous method for thin body interacting boundary-layer problems could be directly generalized to the flow past bluff bodies. The generalization of two thin-airfoil forms were considered in this study: a Cauchy-Hilbert form and a simple Cheng-Rott form. The flow past a circular cylinder was found using the symmetric version of the simple Cheng-Rott form. For the circular cylinder problem, massive separation solutions could be easily calculated over an extensive Reynolds number range, from a Reynolds number of 200 to a Reynolds number of 10^9. The low Reynolds number cases were found to be in reasonable agreement with the full Navier-Stokes results of Fornberg [13]. It is the authors opinion that the technique presented in this study can be used to provide bluff body generalizations of many two-dimensional, incompressible, thin-airfoil based interacting boundary-layer methods.

Acknowledgement

The author wishes to express his appreciation to Dr. M. J. Werle, Dr. J. E. Carter and Dr. M. Barnett of the United Technologies Research Center for their encouragement and many helpful comments during the course of this research. This study was funded by the United Technologies Research Center (monitored by Dr. J. E. Carter) and by a National Science Foundation Presidential Young Investigator Award (monitored by Dr. S. C. Traugott). The author would also like to thank the late Professor R. T. Davis for providing the conformal mapping code used in this study, for suggesting the problem, and for his many helpful suggestions over the years that this study evolved.

References

1. Neiland, V. Ia., "Supersonic Flow of a Viscous Fluid Around a Separation Point," presented at the 3rd pan-Soviet meeting on Theoretical and Applied Mechanics, 1968.

2. Stewartson, K., and Williams, P. G., "Self-Induced Separation," Proc. Roy. Soc., Series A 312, 1969.

3. Messiter, A. F., and Enlow, R. L., "A Model for Laminar Boundary-Layer Flow Near a Separation Point," SIAM J. Appl. Math., Vol. 25, No. 4, 1973.

4. Veldman, A. E. P., "New Quasi-Simultaneous Method to Calculate Interacting Boundary-Layers," AIAA Journal, Vol. 19, 1981.

5. Davis, R. T., and Werle, M. J., "Progress on Interacting Boundary-Layer Computations at High Reynolds Numbers," Numerical and Physical Aspects of Aerodynamic Flows, Springer-Verlag, 1982.

6. Rothmayer, A. P., and Davis, R. T., "Massive Separation and Dynamic Stall on a Cusped Trailing-Edge Airfoil," Numerical and Physical Aspects of Aerodynamic Flows, 3rd ed., 1985.

7. Rothmayer, A. P., and Davis, R. T., "Progress on the Calculation of Large-Scale Separation at High Reynolds Numbers," presented at the ICASE Symposium on Vortex Dominated Flows, proceedings to be published by Springer-Verlag, 1985.

8. Carter, J. E., and Vatsa, V. N., "Analysis of Airfoil Leading-Edge Separation Bubbles," AIAA Journal, Vol. 22, No. 12, 1984.

9. Sychev, V. V., "On Laminar Separation," Mech. Zhid. i Gaza, Vol. 3, 1972.

10. Smith, F. T., "The Laminar Separation of an Incompressible Fluid Streaming Past a Smooth Surface," Proc. Roy. Soc., Series A 356, 1977.

11. Kirchhoff, G., "Zur Theorie freier Flussigkeitsstrahlen," J. reine angew Math, Vol. 70, 1869.

12. Smith, F. T., "Large Scale Separation and Wake Closure/Reattachment - The Cascade Problem," United Technologies Research Center Report, UTRC 83-13, to appear in J. Phys. Math., 1983.

13. Fornberg, B., "Steady Viscous Flow Past a Circular Cylinder up to a Reynolds Number of 600," J. Comp. Phys., (to appear), 1986.

14. Smith, F. T., "A Structure for Laminar Flow Past a Bluff Body at High Reynolds Numbers," to appear in J. Fluid Mech., Vol. 155, 1985.

15. Sadovskii, V. S., "Vortex Regions in a Potential Stream with a Jump of Bernoulli's Constant at the Boundary," Prikl. Math. Mech., Vol. 35, transl. Appl. Math. Mech., Vol. 35, 1971.

16. Cheng, H. K., and Rott, N., "Generalizations of the Inversion Formula of Thin Airfoil Theory," J. Rat. Mech. An., No. 3, 1954.

17. Rothmayer, A. P. and Davis, R. T., "The High Reynolds Number Finite Flat-Plate Cascade," AIAA Journal, Vol. 21, No. 5, 1985.

18. Davis, R. T., private communication, 1982.

19. Rothmayer, A. P., "The Development of a Comprehensive Two-Dimensional Linearized Airfoil Theory," AIAA 9th Annual Mini-Symposium on Air Science and Technology, Wright Patterson AFB, Dayton, OH, 1983.

20. Davis, R. T., "Numerical Methods for Coordinate Generation Based on Schwarz-Christoffel Transformations," AIAA 79-1463, presented at the 4th Computational Fluid Dynamics Conference, 1979.

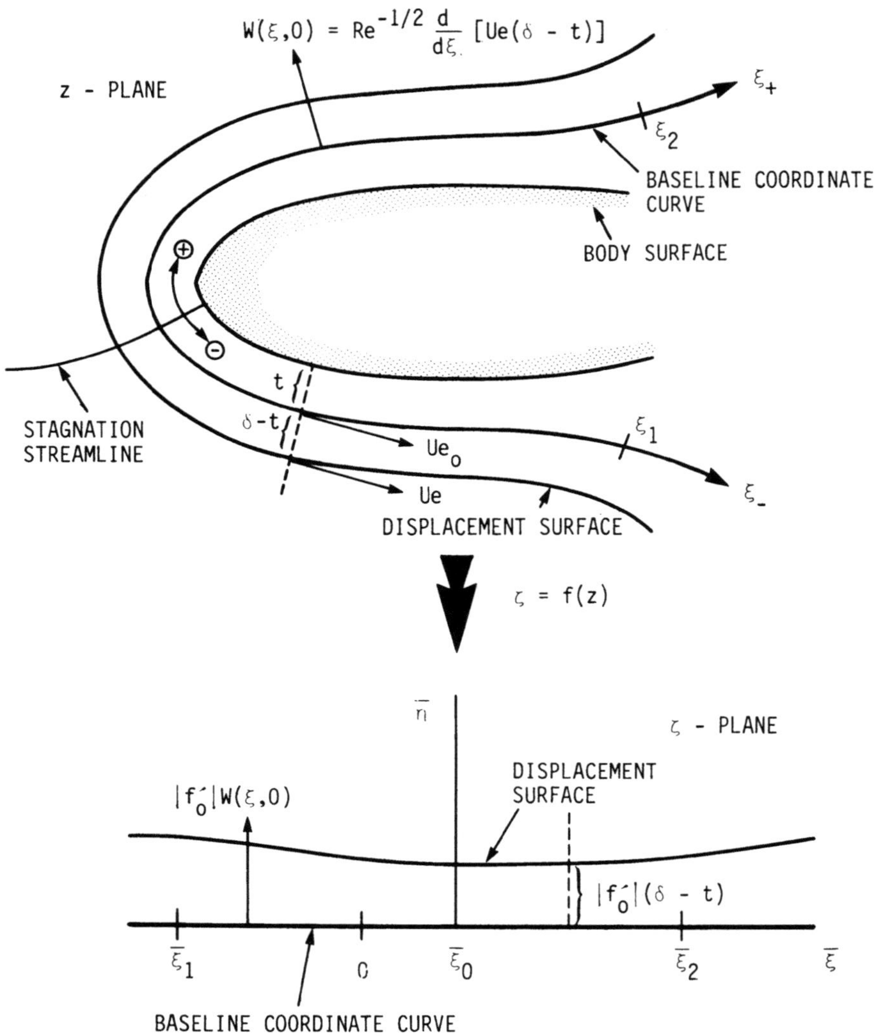

Fig. 1. The conformal mapping geometry.

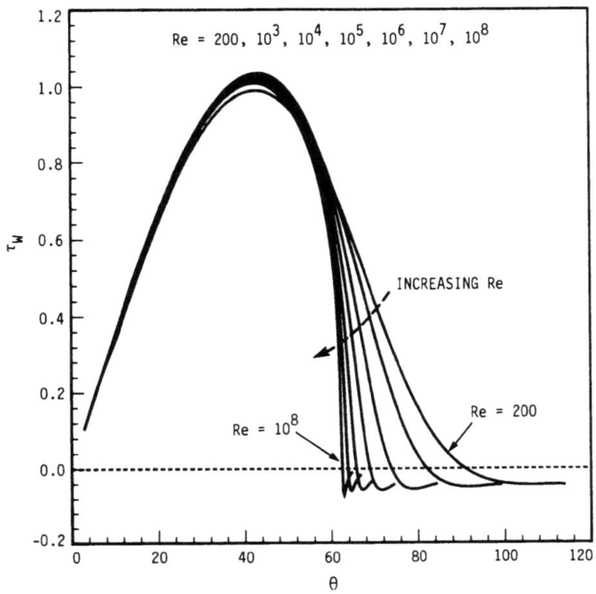

Figure 2. Wall shear stress for large-scale laminar separation on a circular cylinder.

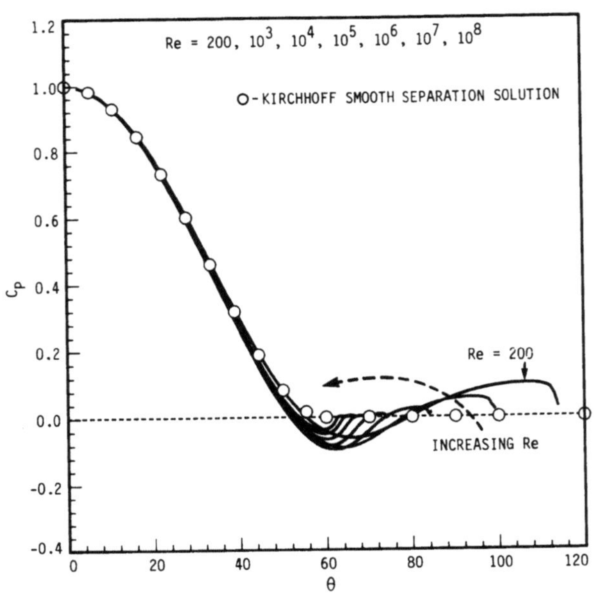

Figure 3. Surface pressure coefficient for flow past a circular cylinder.

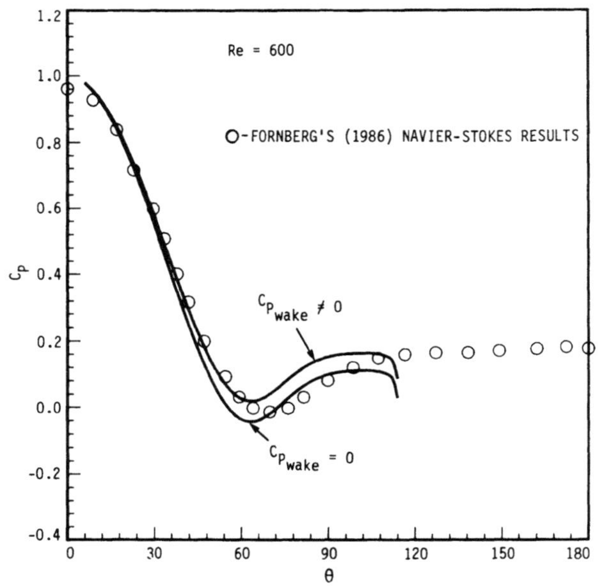

Figure 4. A comparison of surface pressure coefficients with the Navier-Stokes results of Fornberg (1986) at a Reynolds number, based on diameter of 600.

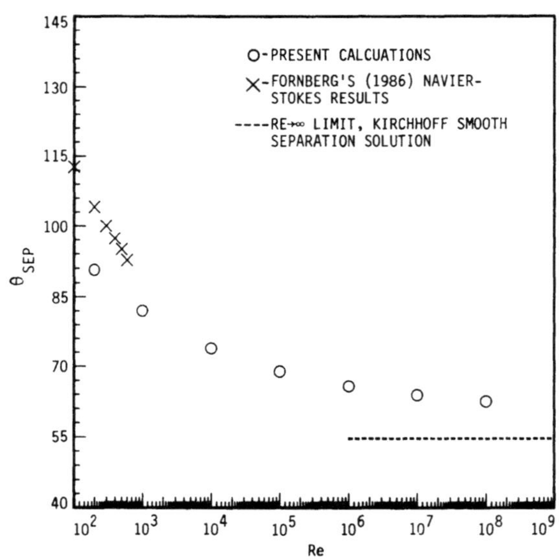

Figure 5. The location of the separation points for flow past a circular cylinder.

Session 4

A new Asymptotic Theory of Turbulent Boundary Layers and the Turbulent Goldstein Problem

R.E. MELNIK

Grumman Corporate Research Center
Bethpage, New York 11714-3580

Introduction

In this paper we present a new asymptotic theory for the description of fully developed turbulent boundary layers and we apply it to determine the limiting behavior of solutions to the boundary layer equations near the point of zero skin friction in a prescribed adverse pressure gradient. In his classic paper, [1] Goldstein addressed this problem for laminar flow, while here we consider the corresponding problem for turbulent flow.

Rational asymptotic methods have been applied previously by Yajnik [2], Mellor [3], and others to turbulent boundary layer flows in the limit of large Reynolds numbers. In these, the solutions were expanded in terms of a single small parameter that served as a measure of the Reynolds number effect. Because of the small defect nature of the approximation, the asymptotic solution was restricted to attached flow. In the present paper we present a new asymptotic formulation that is not subject to the deficiencies of the Yajnik-Mellor theory and can be applied to turbulent separated flow. It is based on a new two-parameter expansion of the turbulent boundary layer equations. In the work reported here, we employ a zero equation turbulence model and the two small parameters, $\varepsilon = 0(\ln Re)^{-1}$ and α where the latter is a small parameter appearing in the turbulent closure.

In the first part of the paper, we apply the two-parameter asymptotic approach to the problem of a turbulent boundary

F. T. Smith, S. N. Brown, (Eds.)
Boundary-Layer Separation
IUTAM Symposium London 1986

layer in a general prescribed pressure distribution. Numerical results are then presented for a family of self-preserving turbulent flows characterized by constant values of the Clauser parameter $\beta = (\delta^*/\tau_w)dp/dx$. Then in the second part of the paper, we examine the behavior of solutions of the asymptotic theory near the point of zero skin friction in a prescribed pressure distribution.

In the present paper, we provide only an abridged discription of the analysis and the main results. Further details are given in an expanded version of this paper in [4].

Asymptotic Theory of Turbulent Boundary Layers

We address the problem of incompressible two-dimensional turbulent flow over a smooth surface in a prescribed pressure distribution. We employ a Cebeci-Smith [5] (C-S) zero-equation model for the turbulent shear stresses and also the Edwards-Carter [6] modification of this model. We neglect the intermittency and the near wall damping terms in these models and instead match the solution to the classical logarithmic law of the wall. The C-S model involves a parameter, $\alpha = 0.0168$, which in the present paper, is treated as a basic small parameter in a formal asymptotic analysis of the boundary layer equations.

All lengths are referred to a characteristic dimension of the body; velocities to free stream velocity; pressure and shear stress to twice free stream dynamic pressure. In order to place the equations in a form appropriate for small α, we introduce stretched variables, $\bar{y} \equiv y/\alpha$, $\bar{v} \equiv v/\alpha$, $\bar{\tau} \equiv \tau/\alpha$, $\bar{\delta}^* \equiv \delta^*/\alpha$, and $\bar{\theta} \equiv \theta/\alpha$ where y and v denote the coordinate and velocity component normal to the surface, τ is the turbulent shear stress, and δ^* and θ are the displacement and the momentum thicknesses, respectively. With these scalings, the boundary layer equations are,

$$uu_x + \bar{v}u_{\bar{y}} = U_e U_{ex} + \bar{\tau}_{\bar{y}} , \qquad u_x + \bar{v}_{\bar{y}} = 0 \tag{1}$$

$$\bar{\tau}/u_{\bar{y}} = \bar{\nu}_t = \begin{cases} \bar{\delta}^* U_e H_r^{-j} & \text{for } \bar{y} > \bar{y}_e \\ \alpha^{-1}\kappa^2\bar{y}^2|u_{\bar{y}}| & \text{for } \bar{y} < \bar{y}_e \ (\kappa = 0.41) \end{cases} \tag{2}$$

where $\bar{y}_e$ is the boundary where the outer and inner forms of the eddy viscosity are equal. The quantity H_r is equal to the ratio of the shape factor to its value for a zero pressure gradient boundary layer, H_0. For the C-S model, j = 0 while, for the Carter Edwards model, j = 1 and $H_0 = 9/7$.

The matching of the outer solution to the law of the wall layer produces a solution for the skin friction and introduces a Reynolds number dependency into the overall solution through a parameter ε,

$$\varepsilon \equiv 0\ (\ell n Re)^{-1} \tag{3}$$

In the present approach we will seek a formal asymptotic solution in terms of a two-parameter expansion, $\alpha \to 0$, $\varepsilon \to 0$. From a matching of the outer and wall layer solutions, it will be evident that a distinguished limit arises in which a parameter, $\chi = \varepsilon/\alpha^{1/2}$ is fixed in the double limit, $\varepsilon, \alpha \to 0$. In this limit the solution in the outermost region develops a two-layer structure composed of an "outer" layer extending across most of the boundary layer and a thin sub-layer near the wall. We call the latter the "equilibrium" layer since it is virtually identical to the equilibrium layer of Townsend [7]. These two layers appear in addition to the usual wall layer. Thus, in the double limit α, $\varepsilon \to 0$, the solution develops in the three-layer structure sketched in Fig. 1.

The leading order solution in the outer region is governed by the full nonlinear boundary layer equations with $\mathcal{Y}_e = 0$ to $O(\alpha^{1/2})$. The outer layer solution thus depends on only the outer turbulent viscosity. In order to match the solution in the equilibrium layer, region 2, Fig. 1, the outer solution must, in general, have a nonzero slip velocity at the wall. This structure applies to attached flows not too near separation. For incipient and fully separated flow, a similar structure arises (see Fig. 1), but the solution in the two equilibrium layers, regions 2 and 5, differs, as will be discussed next.

The solution in the equilibrium layer can be represented as a perturbation about a slip velocity, $u_s(x)$, that is to be determined by a matching to the outer solution. The velocity components u, v and turbulent shear stress τ are represented by scaled variables, U,V and G defined by

$$u = u_s[1 + \alpha^{1/2} \gamma_s U (x,Y)] \tag{4}$$

$$v = \alpha^{3/2} u_s \gamma_s b[-(U_{ex}/U_e)Y + \alpha^{1/2}\gamma_s V (x,Y)] \tag{5}$$

$$\tau = \alpha u_s^2 \gamma_s^2 G (x,Y) \tag{6}$$

where Y is a stretched coordinate defined by,

$$Y \equiv y/\alpha^{3/2} \gamma_s(x)b(x) . \tag{7}$$

Here b(x) is an order one function that is introduced to transform the patch point of the turbulent viscosity to Y = 1. The function $\gamma_s(x)$ is the scaled friction velocity defined in terms of the scaled skin friction,. $\bar{\tau}_w \equiv \tau_w/\alpha$ by,

$$\gamma_s = \sqrt{\bar{\tau}_w/u_s^2} \tag{8}$$

For later reference, we note that the usual skin friction coefficient is given in terms of γ_s by the relation, $C_f = 2\alpha u_s^2 \gamma_s^2$. To lowest order, the momentum equation is,

$$G_Y = \alpha^{1/2} \Lambda(x) + O(\alpha) \quad , \Lambda = \frac{b}{\gamma_s u_s^2} [u_s u_{sx} - U_e U_{ex}] \tag{9}$$

The equilibrium layer (region 2) equations can be solved as an expansion in powers of $\alpha^{1/2}$. However, a singularity develops in these solutions as the slip velocity approaches zero at separation, and this leads to the requirement for a different type of solution in the equilibrium layer (region 5, Fig. 1).

Instead of developing separate solutions for the two equilibrium layer regions (regions 2 and 5, Fig. 1), we will construct a single solution which is uniformly valid in both regions. We do this by formally introducing a new parameter, $\lambda(x) \equiv \alpha^{1/2}\Lambda(x)$, which we treat as an O(1) function in carrying out the asymptotic expansions of Eq (9). In integrating Eq (9) we impose the boundary condition G(0,x) = 1. This condition arises from the matching of the shear stress to the wall layer solutions (where $\tau = \tau_w$). Thus, by introducing λ into Eq (9), integrating once and substituting the resulting expression for G into the expressions for the turbulent eddy viscosity, we arrive at the following equations for U:

$$
\begin{aligned}
(\kappa\, b_0/b)U_Y &= 1 + \lambda Y \qquad \text{for } Y > 1 \\
\kappa^2 y^2 |U_Y| U_Y &= 1 + \lambda Y \qquad \text{for } Y < 1
\end{aligned}
\tag{10}
$$

From the requirements that the turbulent viscosity and velocity gradient be continuous across the patch point, Y = 1, we arrive at the following expression for scale parameter, b:

$$
b = b_0/\sqrt{1+\lambda}\,, \qquad b_0 = \frac{U_e}{u_s}\,\frac{\bar{\delta}^*}{\kappa \gamma_s^2 H_r^j} \tag{11}
$$

Then, from Eq (11) and the definition of λ, we arrive at the following equation for λ:

$$
\lambda\sqrt{1+\lambda} = \lambda_0 \equiv \alpha^{1/2}\,\Lambda_0 \tag{12}
$$

$$
\Lambda_0 = [b_0/\gamma_s u_s^2][u_s u_{sx} - U_e U_{ex}] \tag{13}
$$

The squaring of Eq (12) leads to a cubic equation with three solutions for $\lambda(\lambda_0)$. The conditions that λ be real and have the same sign as λ_0 are sufficient to identify the single root of physical significance. Equation (10) can be integrated to give the following expression for U(Y):

$$U(Y) = \kappa^{-1} \begin{cases} (Y + \frac{1}{2}\lambda Y^2)/\sqrt{1+\lambda} & \text{for } Y > 1 \\ \ln(Y/\sqrt{1+\lambda}) + 2\,[\sqrt{1+\lambda Y} - 1] & \\ \quad - 2\ln[(\sqrt{1+\lambda Y}+1)/2] - D(\lambda) & \text{for } Y < 1 \end{cases} \qquad (14)$$

where

$$D(\lambda) = (1+1/2\ \lambda)/\sqrt{1+\lambda} + 2[\sqrt{1+\lambda}-1] - \ln[\tfrac{1}{4}(\sqrt{1+\lambda}+1)^2\sqrt{1+\lambda}] \ . \qquad (15)$$

The skin friction parameter, γ_s, is determined by matching the inner limit of the equilibrium layer solution to the logarithmic law of the wall. Details of the matching are given in [4].

Here we simply quote the final result for the friction parameter as,

$$\gamma_s = \chi/\{1 + [\alpha^{1/2}\chi/\kappa][\ln(U_e\bar{\delta}^* H_r^{-j}) + D(\lambda)]\} \qquad (16)$$

where $\chi \equiv \varepsilon/\alpha^{1/2}$ is the similarity parameter defined by

$$\chi \equiv [\kappa^{-1}\ln(\alpha^2 Re/\kappa) + B]^{-1}/\alpha^{1/2} \qquad (B = 4.9) \qquad (17)$$

This relation shows that the variable part of γ_s is $O(\alpha^{1/2})$ except in regions where $D(\lambda)$ is unbounded.

Typical values of the dimensionless parameters are, α = 0.0168, Re = 10^6, ε = 0.031 giving χ = 0.24. Since the definition of λ involves γ_s, Eq (16) is an implicit relation for the friction factor. The equilibrium layer solution above is uniformly valid in both the attached flow (region 2) and separated flow (region 5) regions, provided only that $D(\lambda)$ is bounded in the latter.

To complete the solution we require an inner boundary condition for the outer solution and a relation to determine the slip velocity, $u_s(x)$. These are determined by matching the outer limit of the equilibrium layer solution to the inner limit of the outer solution.

The formal matching analysis is given in [4]. It leads to the following expressions for the slip velocity and normal derivative of the outer solution at the wall,

$$u_s(x) = U_{out}(0,x), \quad \frac{\partial U_{out}}{\partial \bar{y}} = \frac{U_{out}^2 \gamma_s^2}{U_e \bar{\delta}^* H_r^{-j}} \quad \text{for } \bar{y} = 0 \qquad (18)$$

where γ_s is an implicit function of u_s determined from Eq (16). Equation (18) can be derived directly from the outer turbulent viscosity relation with $\bar{\tau} \to \bar{\tau}_w$ for $\bar{y} \to 0$. Equation (18) is the principal new result derived in this study. It provides the missing boundary condition needed to complete the outer solution.

Self-Preserving Flows

In order to evaluate the accuracy of the two-parameter theory, we apply it to the problem of self-preserving (or equilibrium) turbulent flows [7, 8] and compare the resulting solutions with the exact numerical solutions of Mellor-Gibson [8]. A particular self-preserving flow is characterized by the value of two parameters: a dimensionless pressure gradient $\beta \equiv \frac{\delta^*}{\tau_w} \frac{dp}{dx}$ (Clauser parameter), and a displacement thickness Reynolds number $R_{\delta^*} \equiv U_e \delta^*/\nu$, with the dependency on the latter being very weak.

The streamwise velocity of a self-preserving flow is represented in the following defect similarity form, [8]

$$U = U_e[1 - \gamma f'(\eta;\gamma)] \;, \quad \eta = y(\gamma/\delta^*), \quad \gamma = \sqrt{\tau_w}/U_e \tag{19}$$

where the defect stream function f satisfies a certain nonlinear ordinary differential equation in which the friction factor, γ, appears as a parameter. Mellor and Gibson [8] provide numerical solutions of a reduced form of the similarity equation that is accurate to $O(\gamma^2) = O(\alpha)$. They employed the same reduced form of the C-S turbulence model used in the present paper, and they matched their solution to the logarithmic law of the wall. The solutions for the friction parameter, γ, and the shape factor, H, are:

$$\gamma^{-1} = \kappa^{-1} \ln R_{\delta*} + B + A(\beta), \; A(\beta) \equiv \lim_{\eta \to 0} [f'(\eta) + \kappa^{-1} \ln \eta] \tag{20}$$

$$H = (1 - \gamma G)^{-1}, \qquad G \equiv \int_0^\infty f'^2(\eta)d\eta \tag{21}$$

In the present study we repeated the Mellor-Gibson calculations using a second-order box scheme and obtained results that showed three-place agreement with the original calculations in [8].

The asymptotic form of the similarity equations can be obtained by the following α scaling of Mellor and Gibson's equations,

$$\eta = \alpha^{1/2}\bar{\eta} \quad f = \bar{f} \quad \gamma = \alpha^{1/2}\bar{\gamma}\,, \quad G = \alpha^{-1/2}\bar{G} \tag{22}$$

The differential equation for $\bar{f}$ in the outer region can be found by transforming Mellor-Gibson's equation according to Eq (22) and then setting $\alpha = 0$ in the result. Details are given in [4]. Three boundary conditions are required, the first two of which are the same as in Mellor and Gibson: $\bar{f}(0) = 0$,

$\bar{f}'(\infty) = 0$. The third condition dealing with the limiting value of the velocity near the wall is changed to reflect the matching of the outer solution to the equilibrium layer rather than to the logarithmic law of the wall. From Eq (18) it follows that this boundary condition can be written in the form,

$$\bar{f}''(0) = -H_r^j \quad . \tag{23}$$

The asymptotic solution for the skin friction, given by Eq (16), can be written in the Mellor-Gibson form of Eq (20) with the slip constant "A" given by,

$$A = \bar{A}\,\alpha^{-1/2} + \kappa^{-1}[\ell n(\alpha/\kappa H_r^j) + D(\lambda)] \quad (\bar{A} \equiv \bar{f}'(0)) \; . \tag{24}$$

The parameter λ_0 appearing in the equation for λ is related to the similarity parameter by the following expression,

$$\lambda_0 \equiv \alpha^{1/2}\Lambda_0 \equiv \alpha^{1/2}\beta\left(\frac{1 - u_s^2/U_e^2}{\bar{\gamma}\,\kappa\,H_r^j}\right), \quad u_s = U_e[1 - \bar{\gamma}\,\bar{A}] \; . \tag{25}$$

We note that $\lambda_0 = O(\alpha^{1/2})$ for $\beta = O(1)$ and that the small λ_0 version of the asymptotic theory applies in this range. However, the order of magnitude of λ_0 increases for $\beta \to \infty$ and becomes $O(1)$ for $\beta/\bar{\gamma} = O(\alpha^{-1/2})$. As demonstrated in [8] the friction parameter, γ, becomes of $O(\beta^{-1/2})$ for $\beta \to \infty$. Together with Eq (25) this implies that $\Lambda_0 = O(\beta/\bar{\gamma}) = O(\beta^{3/2}) \to \infty$ for $\beta \to \infty$. Thus, after introducing scaled quantities $\bar{\beta}$, $\bar{\Gamma}$ we have,

$$\lambda_0 = \frac{\bar{\beta}^{3/2}(1 - u_s^2/U_e^2)}{\bar{\Gamma}\,\kappa\,H_r^j} = O(1) \text{ where } \bar{\beta} \equiv \beta\alpha^{1/3} \text{ and } \bar{\Gamma} \equiv \bar{\gamma}\,\beta^{1/2} \; . \tag{26}$$

It follows from these relations that the double limit $\beta \to \infty$, $\alpha \to 0$ is a distinguished limit in which $\bar{\beta}$ is a fixed $O(1)$ parameter.

A finite difference "box" scheme and Newton iteration was used to obtain numerical solutions of both the exact and asymptotic boundary value problems. The numerical computations were carried out for $j = 0$ (C-S model) and $R_{\delta*} = 10^5$ using the value, $\alpha = 0.0160$, employed in [8]. To the order of accuracy carried in each of the computations, the Mellor-Gibson solution can be considered an "exact" solution that is being approximated by the asymptotic solution for α, $\varepsilon \to 0$.

The comparison of the velocity profiles is given in Fig. 2 for $\beta = 0, 25$. Note that the solutions for $\beta = 25$ are given in the form $F'(\xi;\beta)$ of [8]. Included in the figure are the "exact" Mellor-Gibson solutions, the asymptotic solutions for the outer region and a "composite" solution (see [4]) that is uniformly valid in both the outer and equilibrium layers. We call attention to the smoothly varying behavior of the outer solution near the wall. The close correspondence between the exact and the composite solutions for the velocity profiles is indicative of the accuracy of the limiting asymptotic solution over a range of pressure gradients.

The asymptotic solutions for the variation, with β, of the shape parameter, G, and the friction factor, γ, are compared with the exact solutions in Fig. 3. The asymptotic solution for G includes an $O(\alpha)$ contribution from the equilibrium layer in addition to the $O(1)$ part from the outer layer solution. The comparisons indicate that the asymptotic theory is accurate to about 1% for the skin friction, 5% for the shape parameter.

The Turbulent Goldstein Problem

In [1] Goldstein addressed the problem of finding the detailed behavior of solutions of the laminar boundary layer equations in prescribed pressure distributions near the point of zero skin friction ($x=0$). The solution of the laminar problem developed in a characteristic two-layer structure with the inner solution depending on the limit process $y \to 0$ $x \to 0$,

$\eta = y/(-x)^{1/4}$ fixed. Goldstein showed that the skin friction approached zero with a square root behavior. His solution also explained why boundary layer solutions could not be continued past separation with a prescribed pressure distribution.

In this paper we address the turbulent flow counterpart of the Goldstein problem using the new two-parameter asymptotic theory. In this approach, the main task is to solve the outer layer equations for the variation of the slip velocity, $u_s(x)$, near the point of zero skin friction. The outer region problem consists of the full nonlinear boundary layer equations, Eq (1) and (2), the new matching condition, Eq (18), and the usual boundary conditions $v(0,x) = 0$, $u(\infty,x) = U_e(x)$. Because of the laminar-like character of the outer problem, we could closely follow Goldstein's approach. However, because of the change in boundary condition from a nonslip condition to Eq (18), the character of the singularity in turbulent flow differs greatly from the Goldstein solution.

Following Goldstein, [1], we introduce a new local length scale such that $p_x(0) = 1$ and we assume the pressure distribution can be expanded in the form

$$dp/dx\ (x) = (1 + p_1 x + \cdots) \tag{27}$$

where p_i are prescribed constants. The analysis is carried out using the Carter-Edwards modification of the Cebeci-Smith turbulence model.

For the turbulent flow problem, the solution in the outer region (region 4, Fig. 1) splits into the three-layer structure sketched in Fig. 4. The solution in the innermost layer, region 4', has a form similar to that of the inner layer of the Goldstein solution but differs in detail because of the absence of the no-slip condition in the turbulent flow problem. The solution in the outermost layer, region 4''', Fig. 4, is essentially inviscid and also similar to the corresponding Goldstein solution, except for an extra displacement of that layer due to the effect of the transition layer (absent in the laminar problem) region 4", Fig. 4. The solution in the transition layer is also essentially inviscid, but it is dominated by the full nonlinear inertia terms rather than by the linearized forms appropriate to region 4'''. The thickening effect of the transition layer leads to an unbounded logarithmic behavior of the displacement thickness and a corresponding unbounded vertical shift of the outer layer for $x \to 0$ (see [4] for details).

In the present paper we focus on the nature of the solution in the inner layer, region 4', Fig. 4. This analysis will be sufficient to establish the behavior of the slip velocity and skin friction at separation. It will also establish, $\Delta x = O(\alpha^{1/2})$, as the appropriate streamwise length scale near separation (Fig. 4).

Following Goldstein we seek a solution in the inner region in the form

$$u = 2U_e \xi^{m-1} f'(\xi,\eta) \ , \qquad \xi = (s)^{1/4} \ , \ \eta = t/2^{1/2}\xi$$

$$t \equiv -\bar{y}(U_{ex}/U_e)_{x=0} \ , \ s \equiv x(U_{ex}/U_e)_{x=0} \tag{28}$$

where f is a "reduced" streamfunction and primes denote differentiation with respect to η. If we require the inertia and viscous terms to be of the same order as the pressure gradient, then we find $m = 3$ and $n = 4$. These values are the same as those appearing in the laminar solution. As in the laminar case, $f(\xi,\eta)$ is expanded in integral powers of ξ,

$$f(\xi,\eta) = f_o(\eta) + \xi f_1(\eta) + \xi^2 f_2(\eta) + \dots \tag{29}$$

Logarithmic terms might be required at a given stage of the expansion, as in the laminar solution. The leading order solution is found to be governed by the following third order nonlinear boundary value problem,

$$f_o''' + 2f_o'^2 - 3 f_o f_o'' = 1 \tag{30}$$

$$f_o(0) = 0 \qquad f_o''(0) = 0 \tag{31}$$

The third boundary condition is that the solution not be exponentially large at infinity. The boundary value problem above differs from that of the laminar problem only through the replacement of the usual no-slip condition with the wake type boundary condition given in Eq (31). The boundary value problem has two solutions. The first, $f_o = (\eta^3)/6$, is identical to Goldstein's limiting singular solution for the laminar problem. For the turbulent problem, analysis of the higher order terms indicates that this solution is analytic at separation and generally not appropriate to problems for turbulent flows with prescribed pressure distributions.

The second solution, $f_o = \eta/\sqrt{2}$, leads to a singular solution with a slip velocity that has a square root behavior at separation. For the singular case, the first nontrivial contribution beyond f_o is the second order term, f_2 which turns out to be an eigenfunction with a corresponding eigenvalue, C_2. The first two terms lead to the following expression for the slip velocity,

$$u_s/U_e = \sqrt{-U_{ex}/U_e}\ \sqrt{-2x}\ \{1 + C_2\ \sqrt{-U_{ex}/U_e}\ \sqrt{-2x} + O(x)^{3/2}\} \tag{32}$$

where C_2 is the first eigenvalue. Thus, to leading order in x, the solution for the skin friction coefficient near separation, can be written in the form,

$$C_f = \alpha[2\ dU_e^2/dx]_{x=0}\ \gamma_s(x)^2 x \tag{33}$$

where γ_s is given by

$$\gamma_s^{-1} = \chi^{-1} + \alpha^{1/2}\kappa^{-1}[\ln(U_e H_o \overline{\theta})_{x=0} + D(\lambda)] \tag{34}$$

Substitution of the solution of the slip velocity into the equations defining λ_0 leads to the following expression for λ_0

$$\lambda_0 = \left(\frac{3C_2\overline{\theta}H_0}{2\kappa}\right)\gamma_s^{-3}|\hat{x}|^{-1} \quad (\hat{x} \equiv x/\alpha^{1/2}) \tag{35}$$

where $\hat{x}$ is a stretched variable defining the extent of the incipient separation region (Fig. 1). In this region $\hat{x} = O(1)$ and λ, λ_0, $D(\lambda)$ are all $O(1)$ so that the full form of the equilibrium layer solution is required. It follows that γ_s is constant and C_f is linear in x to $O(\alpha^{1/2})$. The singularity in $\lambda_0(\hat{x})$ indicates that the incipient separation region is not uniformly valid for small $\hat{x}$ and that at least one additional region is required to describe the approach to separation. The solution in this region has not yet been addressed.

As a preliminary assessment of the theory, we carried out a comparison of the asymptotic solution for the skin friction with an exact numerical solution of the full turbulent boundary layer equations. In this comparison, the numerical solution employed the Edwards-Carter version of the C-S turbulence model without intermittency but with the inner layer damping function. The computations extended directly to the wall where the nonslip condition was enforced. The pressure was taken to be constant over an initial distance followed by an abrupt switch to a region of constant adverse pressure gradient defined by $dU_e^2/dx = -0.25$. The calculations were started as a laminar flow and were switched to turbulent flow at a fixed point within the constant pressure region. The numerical solutions were carried out with a finite difference box scheme using successively refined grids to improve resolution near separation.

The overall numerical results for the skin friction are shown on a large scale in Fig. 5. The local behavior of the numerical solution near separation is compared with the asymptotic solution on a highly expanded scale in Fig. 6. The asymptotic result in this figure is based on the one-term expansion for the slip velocity (Eq (32) with $C_2 = 0$) and the relation for γ_s given by Eq (34) with $D(\lambda) = -1$. In order to fix the position of separation the asymptotic solution was required to pass through one point of the numerical solution. From Fig. 6 it appears that the one-term asymptotic solution has captured the essential character of the turbulent singularity.

The inviscid nature of the limit solution ($f_0 = \eta/\sqrt{2}$) explains why the present solution cannot be continued downstream of separation. In an inviscid flow, a positive pressure gradient cannot decelerate the reverse flow in the separated flow downstream of separation, as it must if the downstream flow is to be brought to rest at the separation point. This shows up in the analysis as an absence of a real continuation of the asymptotic solution across $x = 0$.

Concluding Remarks

In this paper we presented a new two parameter asymptotic theory for incompressible two-dimensional boundary layers in prescribed pressure distributions. The new theory eliminated the small defect approximation inherent in the earlier one-parameter theory and thus was able to deal with turbulent separated flows. The two-parameter theory leads to an outer region in which the velocity profiles are smoothly varying with a regular behavior at the wall. The asymptotic theory led to a new inner matching condition which replaced the standard nonslip or log matching conditions used in previous computations.

The application of the new asymptotic theory to the turbulent Goldstein problem established the nature of the separation point singularity in turbulent flow, at least, within the scope of a zero equation turbulence model. Results presented in the present paper established that the asymptotic solution was consistent with accurate numerical solutions of the full boundary layer equations near separation and that both solutions indicated a nearly linear variation of the skin friction near separation.

Acknowledgments

The author would like to thank Dr. R. Chow of the Grumman Corporate Research Center and Prof. G. Flynn of the New York State University at Farmingdale for carrying out the numerical solutions. He would also like to thank Dr. John Brook of the Grumman Corporate Research Center for many helpful discussions on self-preserving flows. This work was supported jointly by the U.S. Office of Naval Research Fluid Dynamics Program under Contract N00014-84-C-0337P0002 and by the Grumman Corporate IR&D program.

References

1. Goldstein, S.: Quart. J. Mech. Appl. Math, 1, 1948.

2. Yajnik, K.S.: J. Fluid Mech, 42, 1970.

3. Mellor, G.L.: Int. J. Engng Sci., 10, 1972.

4. Melnik, R.E.: Grumman Corp. Res. Cent. Report RE-722, 1986.

5. Cebeci, T.; Smith, A.M.O.: Analysis of Turbulent Boundary Layers, 1974.

6. Edwards, D.E.; Carter, J.E.: AIAA Paper 85-0371, 1985.

7. Townsend, A.A.: The Structure of Turbulent Shear Flows, second ed., 1976.

8. Mellor, G.L.; Gibson, D.M.: J. Fluid Mech., 24, 1966.

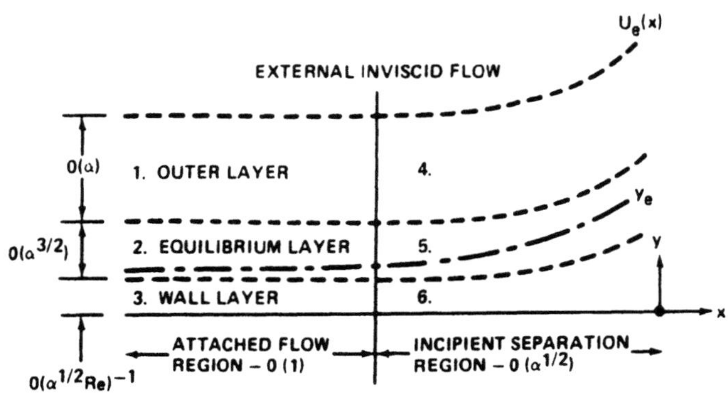

Fig. 1. Asymptotic three-layer structure for $\alpha \to 0, \epsilon\ (R_e) \to 0$

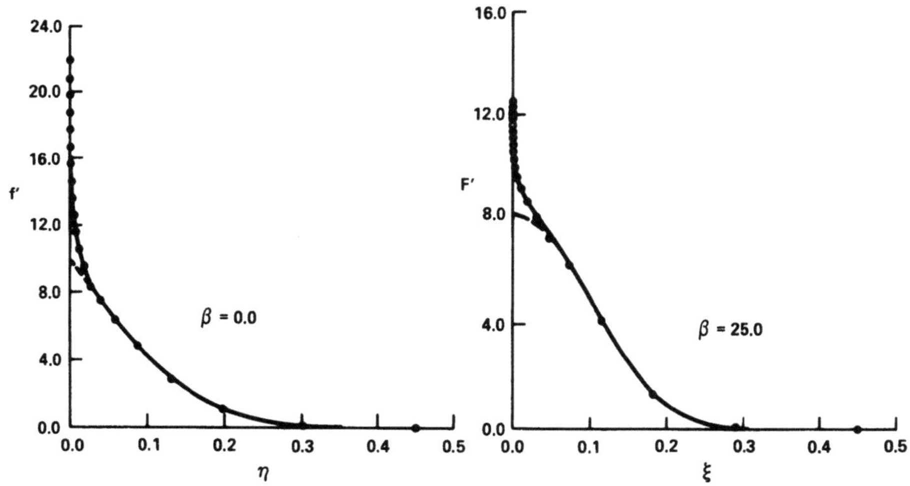

Fig. 2. Comparison of the "Exact" and "Asymptotic" solutions for the velocity profile. Exact ●●●, Asymptotic–Composite ——, and Asymptotic–Outer – – – –

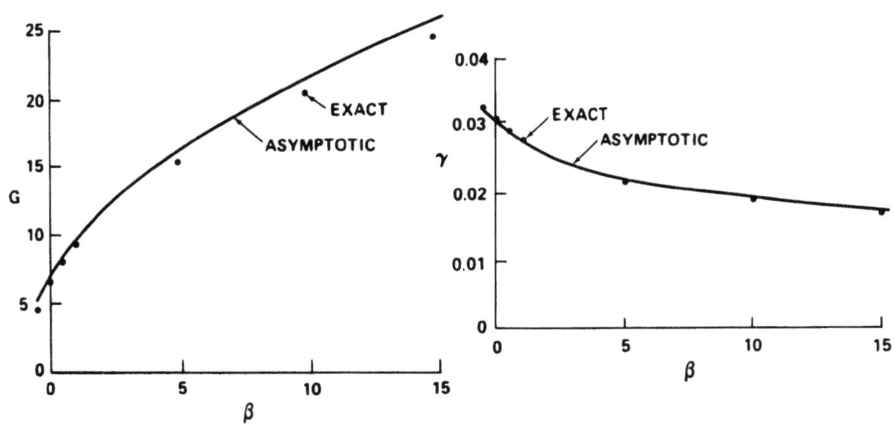

Fig. 3 Variation of the shape factor parameter "G" and friction parameter "γ" with pressure gradient

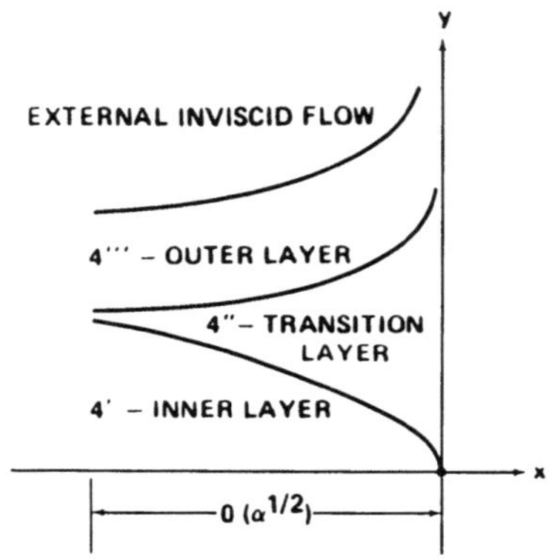

Fig. 4 Structure of the outer solution near separation (x = 0)

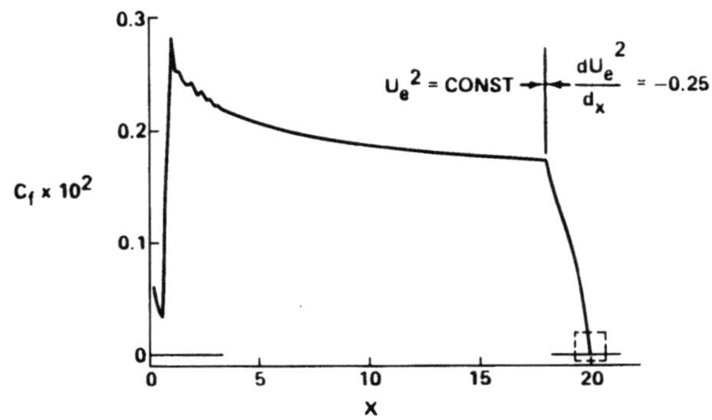

Fig. 5 Numerical solution of the full boundary layer equations for the skin friction

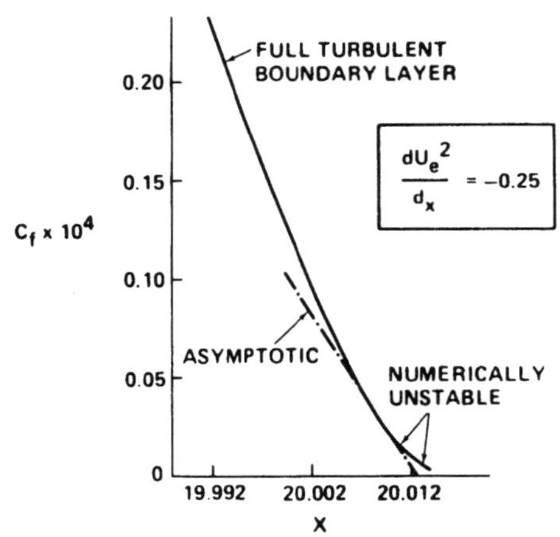

Fig. 6 Comparison of the asymptotic and numerical solutions of the full turbulent boundary layer solution near separation

Vortex Formation in Unsteady Flow near a Moving Wall

H.M. BADR and S.C.R. DENNIS

University of Petroleum and Minerals, Dhahran, Saudi Arabia
University of Western Ontario, London, Ontario, Canada

Summary

Results concerning the process of vortex formation in an example of unsteady two-dimensional flow near a moving wall are presented. The motion considered is that generated in a viscous fluid by a circular cylinder which suddenly starts from rest with a uniform translation perpendicular to and a uniform rotation about its axis. A study is made of the first vortex to form behind the cylinder over a range of Reynolds numbers. Then some features of the appearance of secondary vortices, the detachment of the primary vortex, and the ultimate coalescence of secondary vortices, are discussed. Comparisons made between results calculated respectively from unsteady boundary-layer theory and the Navier-Stokes equations for high Reynolds numbers will be presented in a later paper.

Introduction

This paper reports further numerical studies of a problem previously considered both theoretically and experimentally by Badr and Dennis [1], Coutanceau and Ménard [2] and Badr, Coutanceau, Dennis and Ménard [3, 4]. In this work the unsteady two-dimensional flow of a viscous incompressible fluid past a circular cylinder is considered. The cylinder is impulsively started from rest with uniform translational speed U perpendicular to its axis and uniform angular velocity ω_0 about this axis. The two basic parameters of the problem are the Reynolds number $Re = 2aU/\nu$ and the quantity $\alpha = a\omega_0/U$ where a is the radius of the cylinder and ν is the coefficient of kinematic viscosity of the incompressible fluid.

In [1]-[4] the motion has been studied for Re=200 and 500 and for the fairly low rotation rates $\alpha \leq 1$, in general. A number of interesting phenomena concerning vortex formation

F. T. Smith, S. N. Brown, (Eds.)
Boundary-Layer Separation
IUTAM Symposium London 1986

emerged from this work, some of which are considered in more detail in the present paper. For example, after an interval of time from the start of the motion a region of reversed flow, indicating the formation of a recirculating region of fluid, appears behind the cylinder. We have now investigated this vortex formation at the instant at which it starts over a Reynolds number range $100 \leq Re \leq 10{,}000$. The vortex forms at a point in the fluid whose radial distance from the wall of the cylinder decreases on a scale which appears to be approximately proportional to $Re^{-\frac{1}{2}}$ as Re increases, giving perhaps some indication of the nature of the corresponding phenomenon in boundary-layer flow.

Another feature studied is the transposition phenomenon between the closure point of a vortex which is about to detach from the cylinder and the closure point of a neighbouring, secondary, counter-rotating vortex. This phenomenon, which no doubt has a bearing on vortex detachment, was first reported experimentally at Re=200 in [2] and subsequently established numerically in [4], but in a case (Re=500, $\alpha=\frac{1}{2}$) where the experiments had been terminated prior to the occurrence of the phenomenon. Here we return to the case Re=200, $\alpha=\frac{1}{2}$ and establish that the phenomenon occurs in the calculations in excellent agreement with the observations. We also report in detail the phenomenon of the coalescence of two co-rotating vortices which merge to form a single vortex after a transposition of a closure point of one of them with a main vortex which then detaches into the wake. The calculations are presented for Re=500, $\alpha=\frac{1}{2}$ and are in excellent agreement with the schematic representation given in [4]. Several other features of the study will be presented later.

These include comparisons between the results of calculations based on the Navier-Stokes equations for high Re and the boundary-layer equations ($Re=\infty$) in which the effect of the pressure gradient normal to the cylinder wall is neglected. The principal defect of this neglect is that the resultant forces on the cylinder are incorrectly predicted.

Summary of the Basic Equations

For this illustrative case we consider a circular cylinder translating in the negative x direction of Cartesian coordinates with velocity U and rotating about its centre with angular velocity ω_0. With origin at the centre of the cylinder and axes translating with it but not rotating, the equations of motion are, in terms of a dimensionless vorticity ζ and stream function ψ given in [1]

$$2Re^{-1}(\partial^2\zeta/\partial\xi^2+\partial^2\zeta/\partial\theta^2)-(\partial\psi/\partial\theta)(\partial\zeta/\partial\xi)+(\partial\psi/\partial\xi)(\partial\zeta/\partial\theta) = e^{2\xi}\partial\zeta/\partial\tau \qquad (1)$$

$$\partial^2\psi/\partial\xi^2+\partial^2\psi/\partial\theta^2-e^{2\xi}\zeta = 0 \ . \qquad (2)$$

Here (r, θ) are polar coordinates and $\xi = \ell n(r/a)$, where a is the radius of the cylinder. The Reynolds number is $Re=2aU/\nu$ and $\tau=Ut/a$, where t is the actual time. The radial and transverse dimensionless components of velocity are given in terms of ψ by $u=e^{-\xi}\ \partial\psi/\partial\theta$, $v=-e^{-\xi}\ \partial\psi/\partial\xi$ respectively. The boundary conditions for (1) and (2) are

$$\psi=0,\ \partial\psi/\partial\xi=-\alpha \quad \text{when } \xi=0\ , \qquad (3a)$$

$$u\to\cos\theta,\ v\to-\sin\theta,\ \zeta\to0 \quad \text{as} \quad \zeta\to\infty\ . \qquad (3b)$$

Here $\alpha=a\omega_0/U$, which is a parameter measuring the surface rotational speed of the cylinder to its translational speed. The motion is started suddenly from rest both translationally and rotationally which corresponds to a good approximation to the experiments of [2], in which a cylinder is moved abruptly from rest in a tank of incompressible liquid. The mathematical structure of the initial flow is known from unsteady boundary-layer theory. The initial boundary-layer thickness at the surface of the cylinder is proportional to the parameter $\lambda = 2(2\tau/Re)^{\frac{1}{2}}$ and the appropriate boundary-layer coordinates and scale factors with which (1) and (2) are to be restructured to give the initial flow are given by

$$\xi=\lambda z,\ \psi=\lambda\Psi,\ \zeta=\omega/\lambda \ . \qquad (4)$$

The results presented in this paper have all been computed by transforming (1)-(3) into equations in ω, Ψ with independent variable z and a parameter λ and it is important to stress that no approximations are made anywhere in approximating the complete equations and boundary conditions (1)-(3). This is achieved by retaining all terms in λ, making the calculation procedure dependent on Re.

The treatment of (1)-(3) follows precisely, in fact, the exact treatment given by Collins and Dennis [5,6] to the corresponding case of symmetrical flow when $\alpha=0$. After the transformation of (1)-(3) using (4) the problem is solved numerically by assuming the expansions

$$\Psi(z,\theta,\tau)=\tfrac{1}{2}F_0(z,\tau)+\sum_{n=1}^{\infty}[F_n(z,\tau)\cos n\theta+f_n(z,\tau)\sin n\theta] \quad (5a)$$

$$\omega(z,\theta,\tau)=\tfrac{1}{2}G_0(z,\tau)+\sum_{n=1}^{\infty}[G_n(z,\tau)\cos n\theta+g_n(z,\tau)\sin n\theta] \quad (5b)$$

and deriving sets of time-dependent partial differential equations satisfied by the functional coefficients in (5). These equations and the numerical methods used to solve them are described fully in [1] but the most important feature of the method is the satisfaction of the boundary conditions as $z\to\infty$. It is shown in [1] that the conditions (3) in conjunction with the equation (2) can be used to deduce a set of conditions

$$\int_0^{\infty} e^{(2-n)\lambda z} g_n dz=2\delta_{n,1}; \quad \int_0^{\infty} e^{(2-n)\lambda z} G_n dz=0 \, (n=1,2....); \quad (6a)$$

$$\int_0^{\infty} e^{2\lambda z} G_0 dz=2\alpha , \quad (6b)$$

where $\delta_{n,1}$ is the Kronecker delta. These are used in conjunction with the conditions that $G_n(z,\tau)$, $g_n(z,\tau)$ must vanish as $z\to\infty$ to avoid specification of conditions for $F_n(z,\tau)$ and $f_n(z,\tau)$ as $z\to\infty$. The satisfaction of (6) ensures that the flow outside the boundary layer is correctly adjusted to the external conditions of the free stream. Moreover, satisfaction of (6b) ensures that the circulation round a circular contour of indefinitely large radius with centre at the centre of the cylinder remains zero for all time; it also ensures that the pressure remains single-valued throughout the flow field.

Vortex Formation Behind the Cylinder

The first calculated results to be presented concern the nature and position of the first vortex, or recirculating region of fluid, to be formed at the rear of the cylinder.

The illustrative results correspond to the case $\alpha = \frac{1}{2}$. For all Reynolds numbers, after the sudden start of the motion a stagnation point forms in the fluid where the oncoming stream meets the layer of fluid rotating with the cylinder wall. After a while its position stabilizes in the frontal region near the wall at about $\theta = 130^0$. Subsequently the first vortex forms in the fluid at the rear of the cylinder above the x axis, i.e. for fairly small $\theta > 0$, in a manner which varies somewhat with Re and α but of which figures 1a,b are typical. These figures show respectively the calculated

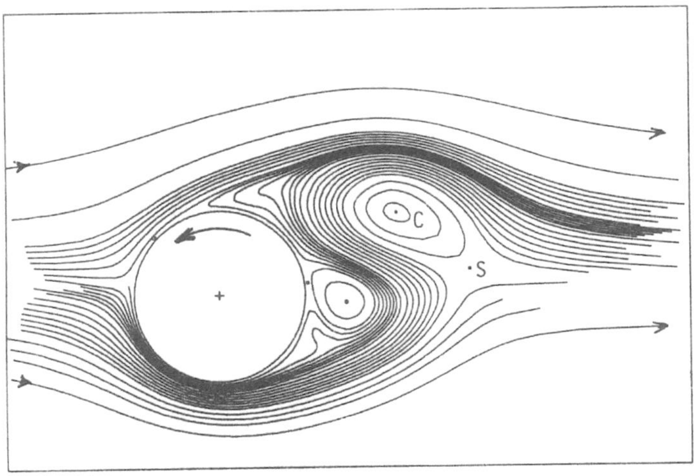

(a)

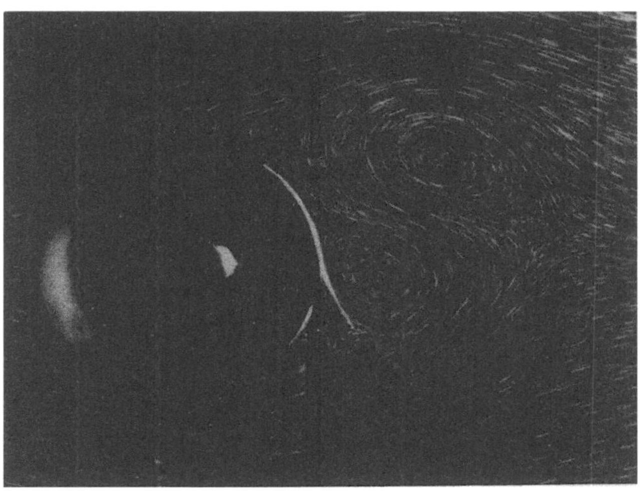

(b)

Figs. 1a calculated and 1b experimental streamlines for $Re=500$, $\alpha=\frac{1}{2}$, $\tau=6$.

and experimental structure of the instantaneous streamlines for Re=500, $\alpha = \frac{1}{2}$ at $\tau = 6$ some time after the first ($\theta > 0$) and second ($\theta < 0$) vortices have appeared. Each vortex has the basic structure noted in [7] consisting of a vortex centre C and a saddle, or closure, point S which are both stagnation points but of different kinds.

For the present we seek the instant of creation of the first vortex, which may be defined as the instant at which the stagnation points S and C first appear as a single coincident point. In fact we have to find the point at the rear of the cylinder for which u=v=0 at this one and only one point at some instant of time $\tau = \tau_{cs}$ (indicating the time at which S and C coincide). This determination is quite difficult, but is helped by the fact that the solutions are given in the form (5). Thus one can first find the angle θ_{cs} for which the radius passes through the required point and then, since the functions in the series (5) are given only at grid points in the ξ direction, find the required point by interpolation. It is then not difficult to find the variation of u and v along a radius through the point for which u=v=0. It is, however, more difficult to calculate the similar variation of u and v along a constant radial curve passing through this point; this has not yet been addressed.

Some typical calculations of the formation of the vortex were carried out for the case $\alpha = \frac{1}{2}$ and the range $10^2 \leq Re \leq 10^4$. The manner in which the initial centre of the vortex defined by the above procedure approaches the cylinder wall is shown in figure 2; it is approximately on a scale of $Re^{-\frac{1}{2}}$. The angle at which the vortex is formed varies very little over this range, from about $\theta = 12.2^0$ at Re=100 to $\theta = 14.5^0$ at R=10,000. The corresponding range of τ_{cs} is from $\tau_{cs} = 1.4$ at Re=100 to $\tau_{cs} = 0.69$ at Re=10,000. The instantaneous streamlines for the case Re=100 at the instant of birth of the vortex are shown in figure 3(a) and at a short time afterwards ($\tau = 1.5$) in figure 3b, at which time the vortex centre and closure point have clearly separated from each other. The initial vortex centre in figure 3a is identified by the streamline with the cusp-like point which coincides with C and S; there is a reversal of both radial and transverse components of velocity at this point. In figure 3b the vortex centre has hardly moved from the position at which it was created in figure 3a.

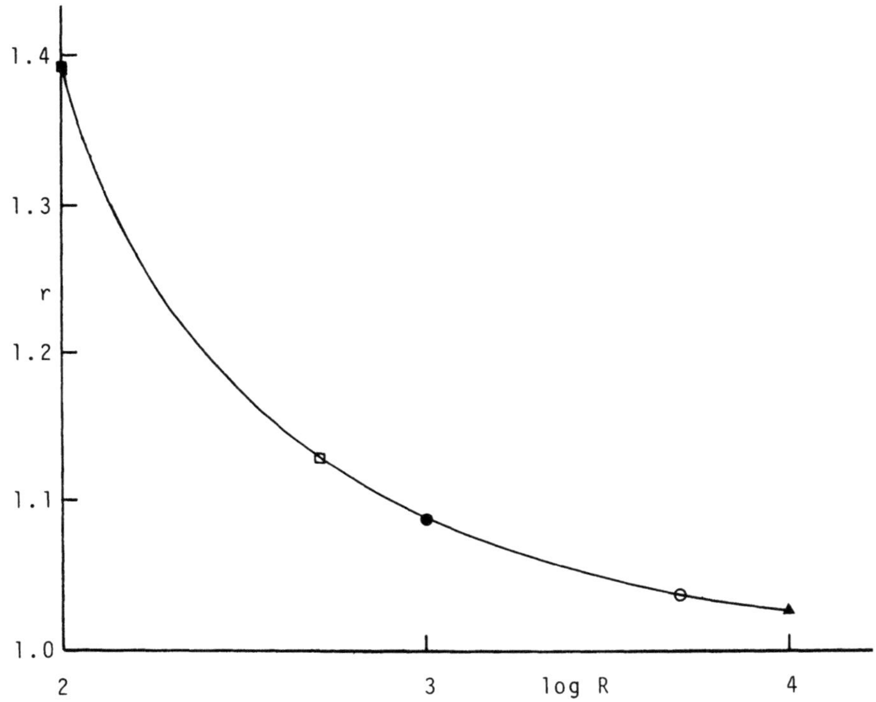

Fig. 2. Radial position of initial vortex centre against R.

Similar instantaneous streamlines at the birth of the vortex and at a short while afterwards are shown respectively for Re=500 in figures 4a, b and for Re=1000 in figures 5a, b. Results for $Re=10^4$ are too difficult to show clearly graphically, since the vortex formation is moving into the boundary layer and is close to the cylinder wall. However, the general trend of the results as $Re \to \infty$ and the formation occurs infinitesimally close to the wall seems to be that the conditions at the onset of the vortex formation are consistent with $u=v=\partial u/\partial\theta=0$ at the point where C and S coincide, i.e. when $\tau = \tau_{cs}$. In figure 6 we give the variation of u and v over the radius which passes through C=S at the instant of

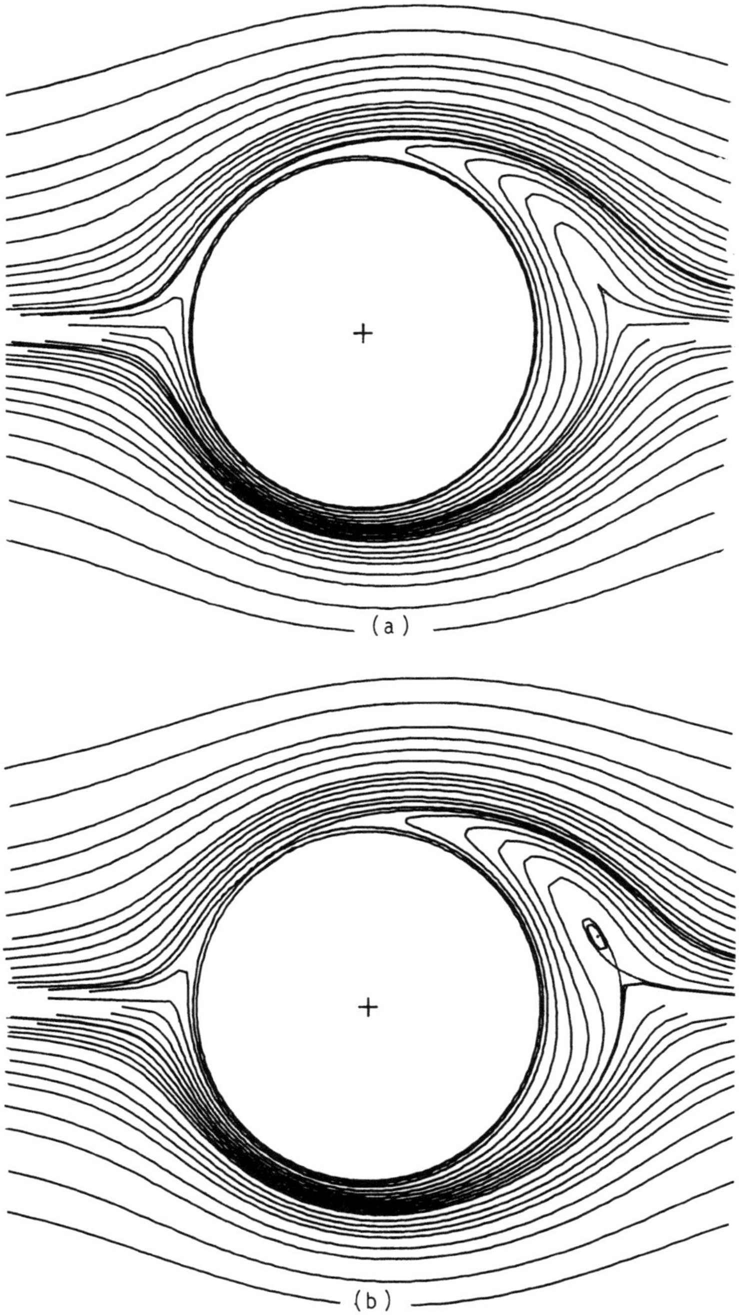

Fig. 3. Vortex formation at Re=100;
(a) initial formation at t=1.4; (b) growth at t=1.5.

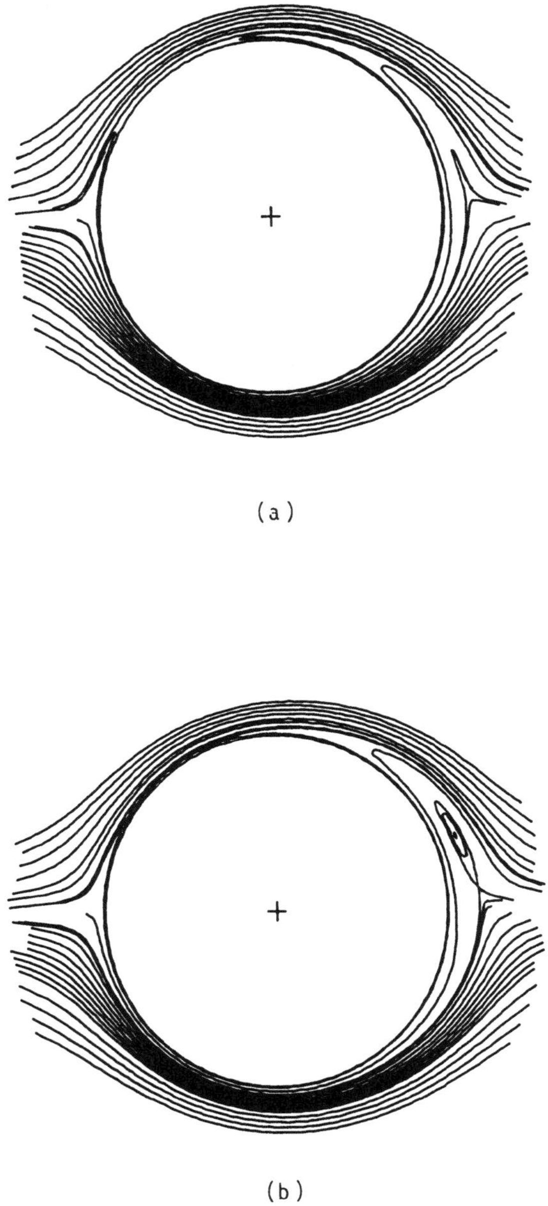

(a)

(b)

Fig. 4 Vortex formation at Re = 500;
(a) initial formation at t=0.87; (b) growth at t=1.0.

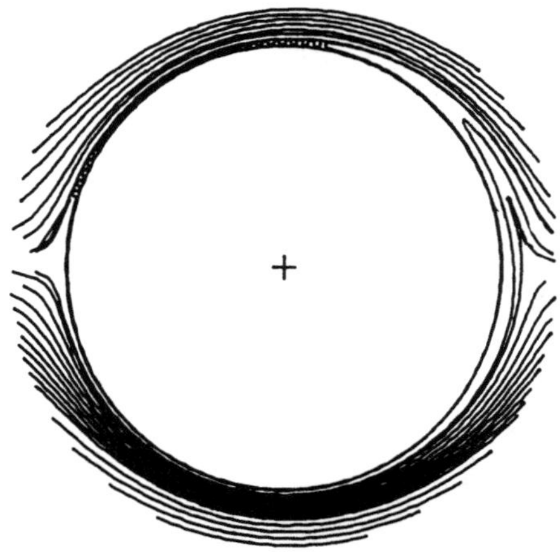

(a)

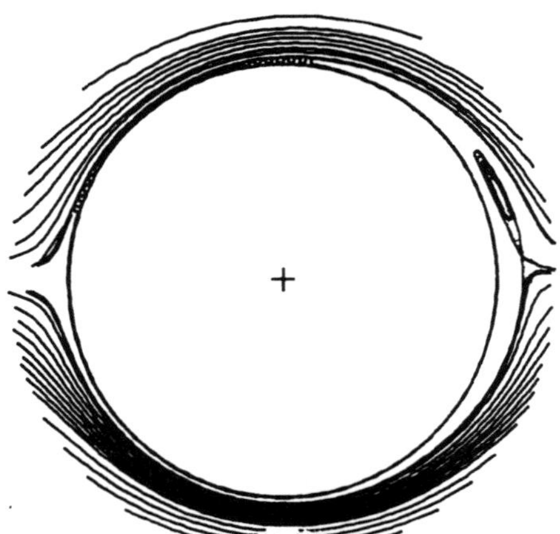

(b)

Fig. 5 Vortex formation at Re = 1000;
(a) initial formation at t=0.79; (b) growth at t=0.9.

formation of the vortex for Re=100, $\alpha=\frac{1}{2}$ showing clearly u=v=0 at this point. It would also be of interest to calculate the corresponding variation along the circular arc $r=r_{cs}$ passing through C=S but we have not yet established the most satisfactory method for this (C=S is not a grid point, in general). There are also many further interesting features to study in the vortex formation; these must await further work.

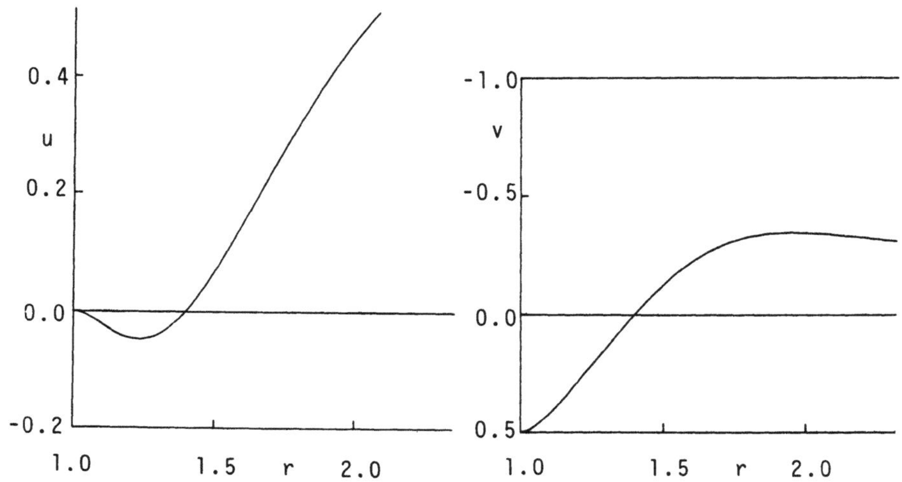

Fig. 6. Radial and transverse velocity components for Re=100.

Vortex Detachment and Merging of Vortices

In [4] an example was given of a phenomenon which was found to occur prior to the detachment of a vortex situated at the rear of the cylinder and its subsequent passage downstream. In that case the illustration was given for the present flow when Re=500, $\alpha=\frac{1}{2}$. It was not possible in [4] to compare the results of calculations with experiment because the experiments were terminated in the case Re=500 before the phenomenon occurs. Yet, actually, the phenomenon was first observed experimentally and reported in [2]. The phenomenon concerned is the transposition between the closure point of the main vortex which is about to detach and that of a secondary counter - rotating vortex which has formed at the surface of the cylinder adjacent to the main vortex. In the case Re=200, $\alpha=\frac{1}{2}$ discussed in [2, p.410 and footnote p. 412] this transposition takes place suddenly at some instant between $\tau=7$ and $\tau=8$ which it is not possible to state precisely in this experimental study.

We have therefore considered this same case in detail and some calculated instantaneous streamlines of the flow are shown in figure 7. Figure 7a shows the flow at $\tau=8.08$. There are two vortices clearly shown, a smaller secondary one with closure point S, through which the streamline is drawn, and a larger one about to detach with closure point S'; here the

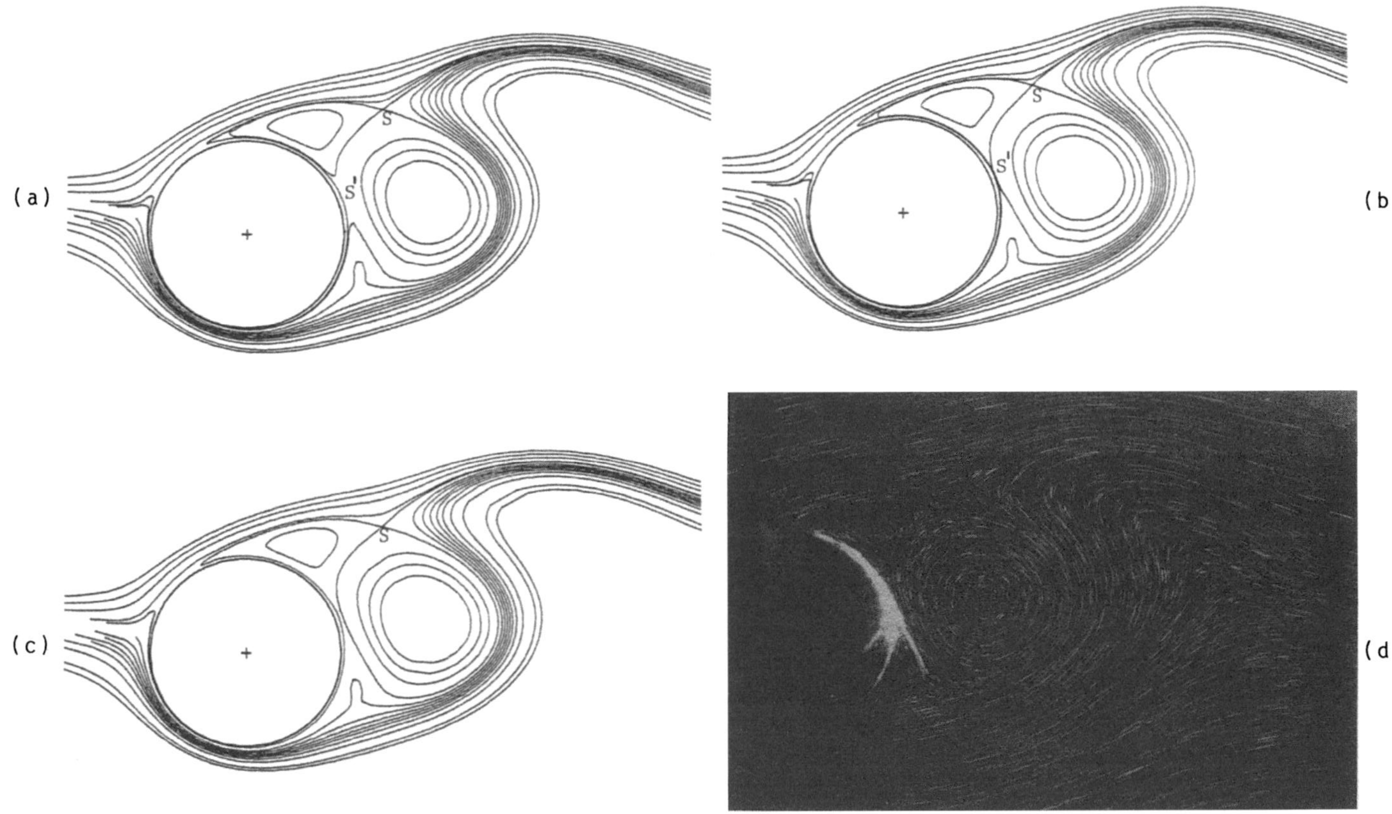

Fig. 7. Transposition of closure points for Re=200, $\alpha=\frac{1}{2}$.
Calculated: (a) τ=8.08; (b) τ=8.10; (c) τ=8.14. Experimental: (d) τ=8.0.

streamline through S' is not drawn, but it encloses the main vortex. In figure 7b, at τ=8.10, we capture the exact instant at which the streamline through S passes also through S' so that each closure point belongs to both vortices. Finally, in figure 7c at τ=8.14 the closure point S belongs to the main vortex and S' (through which the streamline is not drawn) belongs to the secondary vortex. The experimental situation at τ=8 is shown in figure 7d. After the larger vortex has separated from S' (which still continues to exist) it starts to move downstream; by τ=10 (see figure 1(g) of [1] and figure 6(k) of [2]) it has passed well downstream and a new phenomenon, that of coalescence of two secondary co-rotating vortices at the surface of the cylinder, is taking place.

The coalescence phenomenon occurs after the detachment of the main vortex, which thus allows space for it to take place. It is caused by the growth of two secondary vortices which eventually have a common closure point but separate centres until they merge. The process was shown schematically in [4] and had already been shown in [2] experimentally in figures 6(i)-6(n) and in [1] in figures 1(f)-1(i) for the case Re=200, $\alpha=\frac{1}{2}$. The detail of the coalescence process is of some interest because it again embodies a sudden shift of the common closure point of the two vortices as they combine, which moves suddenly to become the closure point of the combined vortex. In figure 8 we have made a detailed study of the process for the case Re=500, $\alpha=\frac{1}{2}$.

Figure 8a indicates the situation at τ=10.25, some time after the time τ=8.06 depicted in figure 3(c) of [4] in which the two secondary vortices are clearly separated by the main vortex which has not yet detached. At τ=10.25 the main vortex has departed downstream and the two secondary vortices now have separate centres but a common closure point; they are rotating in the same direction by virtue of the fluid which passes between them. The situation continues like this for some time as the centres approach each other; figure 8b shows the streamlines at τ=11.25. Soon after, as indicated in figure 8c at τ=11.5, they merge into a single vortex (in effect the upper

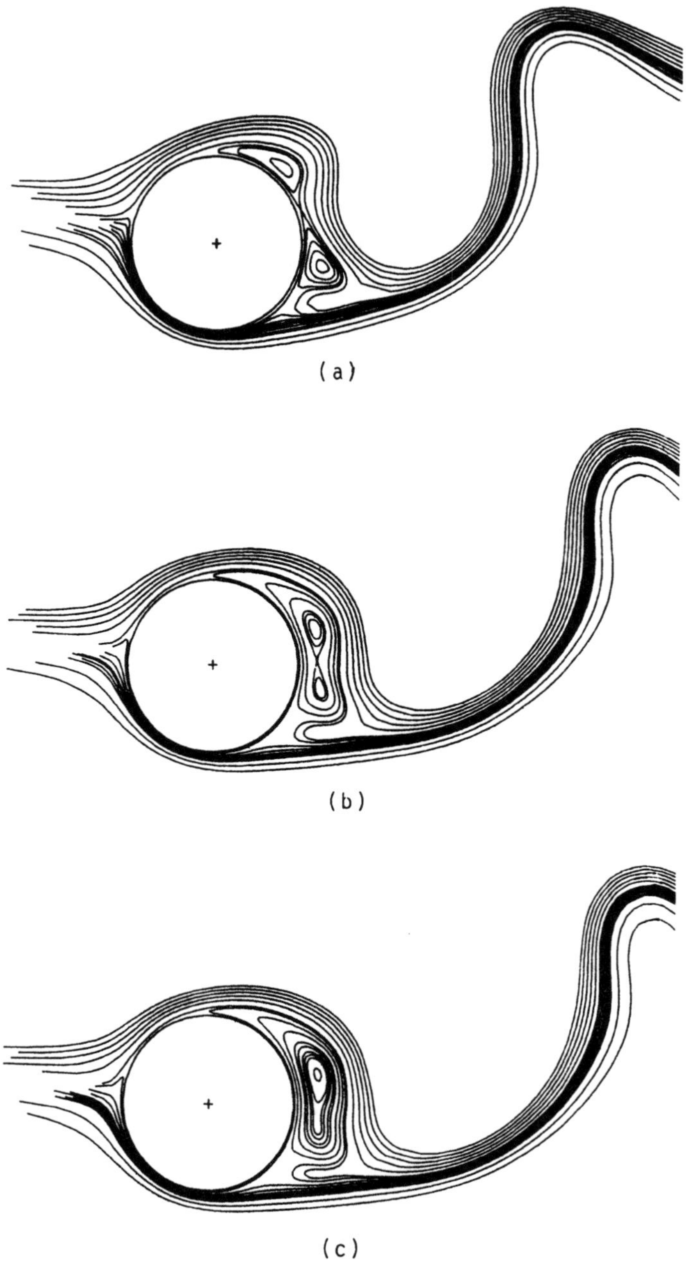

Fig. 8. Vortex coalescence for Re=500, $\alpha=\frac{1}{2}$.
(a) $\tau=10.25$; (b) $\tau=11.25$; (c) $\tau=11.5$.

vortex seems to have swallowed the lower one). The previous common closure point has now been transposed to the region between the cylinder wall and the bulge in the streamline in the lower part of figure 8c. As the single vortex grows it will eventually assume a position somewhat similar to that of the main vortex in figure 7a (or figure 3(a) of [4] in the present case of Re=500). New secondary vortices will appear and the process of transposition of closure points, detachment of the main vortex and combination of the secondary vortices will be repeated as part of a continuing periodic structure. This has not been calculated in the present case but a similar periodic structure of evolution has been calculated in the case of Re=1000.

The provision of experimental diagrams by, and the general collaboration with, Dr. Madeleine Coutanceau and her group at the Laboratoire de Mécanique des Fluides, University of Poitiers, France is gratefully acknowledged. Support for this collaboration by a NATO Scientific Research Grant is also gratefully acknowledged.

References

1. H.M. Badr and S.C.R. Dennis, J. Fluid Mech. 158 (1985) 447-488.

2. M. Coutanceau and C. Ménard, J. Fluid Mech. 158 (1985) 399-446.

3. H.M. Badr, M. Coutanceau, S.C.R. Dennis and C. Ménard, C.R. Acad. Sc. Paris 300, Série II (1985) 529-533.

4. H.M. Badr, M. Coutanceau, S.C.R. Dennis and C. Ménard, C.R. Acad. Sc. Paris 302, Série II (1986) 1127-1130.

5. W.M. Collins and S.C.R. Dennis, Quart. J. Mech. Appl. Math. 26 (1973) 53-78.

6. W.M. Collins and S.C.R. Dennis, J. Fluid Mech. 60 (1973) 105-131.

7. C.A. Koromilas and D.P. Telionis, J. Fluid Mech. 97 (1980) 347-367.

Simulation of Separated Flow Past a Bluff Body Using Navier-Stokes Equations

K.N. Ghia, U. Ghia*, G.A. Osswald and C.A. Liu

Department of Aerospace Engineering and Engineering Mechanics
*Department of Mechanical and Industrial Engineering
University of Cincinnati, Cincinnati, Ohio 45221, U.S.A.

Summary

An analysis is developed for simulating two-dimensional flow past a bluff body, using the incompressible unsteady Navier-Stokes equations in terms of vorticity and stream function. The conservation-law form of the equations is employed, in generalized orthogonal curvilinear coordinates, using the contravariant component of the vorticity. The fully implicit time-marching alternating-direction implicit-block Gaussian elimination (ADI-BGE) method employed is a direct method, with second-order spatial accuracy and, hence, avoids introduction of any artificial viscosity. This unsteady analysis has been carefully applied to simulate flow past a circular cylinder with and without symmetry, requiring the use of either the half or the full cylinder, respectively. For the latter configuration, at early time levels, the evolution of some flow characteristics in the near wake is compared extensively with existing experimental data, in order to partially validate the present analysis. For the half-cylinder configuration, the asymptotic flow structure predicted by the present analysis conforms, to some extent, with the existing numerical and analytical solutions. This analysis has the potential advantage of computing symmetric unstable modes as well as solution bifurcations which may result for flow up to Reynolds number $Re = 10^4$ or even $Re = 10^5$.

Introduction

For flow past airfoils at high angle of attack, the airfoil behaves like a bluff body. The large-scale structure of the separated flow past a bluff body, e.g., a circular cylinder, is not understood well, except at extremely low values of the Reynolds number Re around 120 where Re is based on the cylinder diameter. The understanding of steady separated flow past the model problem of a circular cylinder is crucial, as it may lead to a better understanding of more complex flows such as separated flow past airfoils with or without solution bifurcations associated with lift hysteresis, symmetry breaking, etc. Also, a thorough knowledge of steady separated flow could aid in analyzing unsteady separated flow leading to incipient transition and, eventually, to chaos.

Several asymptotic models have been proposed for high-Re flows past a bluff body. These are based on inviscid-flow considerations and include the Kirchhoff free-streamline model and the Prandtl-Batchelor model. Here, viscosity plays a crucial role in determining which specific inviscid-flow solution represents the correct limit, for large Re, for the flow of

F. T. Smith, S. N. Brown, (Eds.)
Boundary-Layer Separation
IUTAM Symposium London 1986

a real fluid. In 1979, Smith [1] comprehensively studied the model problem of the symmetric flow past a circular cylinder, using asymptotic theories. In this major paper, he proposed an extended Kirchhoff structure with a body-scale flow in the proximity of the cylinder, and a large-scale flow comprising the wake bubble, i.e., the main eddy, and the subsequent wake. The primary eddy was of slender elliptical shape and was characterized by the presence of low velocity and constant pressure. The nondimensional length L and width W of the wake bubble were determined to be of O(Re) and $O(Re^{1/2})$, respectively. Fornberg [2] simulated this flow numerically using the steady Navier-Stokes equations in terms of vorticity ω and stream function ψ. Attention was focussed on arriving at suitable boundary conditions as well as their placement. For Re > 10, the numerical solutions were obtained using the CDC STAR 100 computer. The results agreed well with those of Smith [1] up to Re = 260 but, thereafter, Fornberg's [2] results showed a decrease in growth of the length and increase in the width of the wake bubble with increase in Re, due to convection of vorticity into the interior of the main eddy from the fluid downstream of it. This was in direct contrast to the results of Smith [1] who had shown the length of the bubble to increase with Re and to possess no vorticity in its interior.

In an attempt to resolve the anamoly between the numerical and theoretical results, Fornberg [3] performed an improved numerical simulation in 1985. He provided extremely fine-grid calculations, using a grid of (541, 109) points, for symmetric flow past a cylinder up to Re = 600, using the CDC CYBER 205 computer system. The new results showed that L increases as O(Re) but they continue to conform with the earlier prediction that the width W increases beyond $O(Re^{1/2})$ and becomes nearly linear, i.e., O(Re), around Re = 450. The total wake consisted of a pair of bubbles with low but uniform vorticity within them. At about the same time, Peregrine [4] provided a possible vorticity distribution in the wake bubble with the aid of an exact solution of the Navier-Stokes equations, near the rear stagnation point in the wake. His approach suggests the possibility of W increasing as O(Re) as $Re \to \infty$; this is in conformity with the results of Fornberg [3]. Smith [5] also proposed a new laminar-flow structure for higher Re, based on a massive eddy of size (both L and W) of O(Re). This proposed model supersedes the extended Kirchhoff model discussed in his earlier study. The proposed large-Re structure appears to resolve the questions related to the flow-structure for the primary eddy;

however, the description of the rear-closure of the eddy still appears incomplete. The new proposed model provides a structure for the massive eddy in conformity with the Navier-Stokes results of Fornberg [3].

Recently, K. Ghia, Osswald and U. Ghia [6] have carefully studied self-induced unsteady motion in the near wake of a Joukowski airfoil at angles of attack sufficiently high, so that this airfoil configuration can be looked upon as a bluff body. This led to the initiation of the simulation of flow in the wake of a circular cylinder without symmetry, the results of which will be presented by the authors at a forthcoming meeting. Also, Osswald, K. Ghia and U. Ghia [7] simulated the buffetting-stall phenomenon for the corresponding cambered Joukowski airfoil. In all of these studies, the unsteady Navier-Stokes equations are solved using clustered conformal coordinates where, for the Joukowski airfoil, the use of Joukowski transformation maps the airfoil into a circle. The fully implicit numerical method of the authors experiences no difficulties in their studies for Re = 10^3 and for Re = 10^4 in Ref. [8]. Thus, the present study was motivated, by the need to develop a method which overcomes the numerical difficulties which several investigators have faced (see Fornberg [2]) in accurately capturing the diverse length scales of this apparently simple model flow configuration of symmetric flow past a circular cylinder, and, thereby, contribute towards an accurate description of the large-Re flow structure. Also, if time permitted, it was desired to attempt the solution for Re = 10^3 where there exists a possibility of symmetric unstable modes in this flow.

Analysis and Numerical Methods

The conservation-law form of the unsteady Navier-Stokes equations in terms of vorticity and stream function, in generalized curvilinear coordinates (ξ^1, ξ^2), is used. The far-field boundary conditions of uniform flow are, strictly, valid only at infinity. To circumvent the large values of ψ occurring in the far-field and also facilitate the numerical implementation of the far-field boundary conditions, the stream function ψ is decomposed into two parts such that $\psi = \psi_{in}+\psi_v$, neither of which need to be small in comparison to the other. Here, ψ_{in} is a solution of the stream function equation with zero vorticity and satisfies the free-stream condition at infinity and the zero normal-velocity condition at the surface of the cylinder. This decomposition is used so as to enable the far-field boundary condition to be placed at infinity for the flow calculations.

The Reynolds number used in this study is defined as Re = $U_\infty d/\nu$ where d is the diameter of the cylinder. The characteristic time is defined as $t = t^*/(d/U_\infty)$, with U_∞ being the undisturbed free-stream velocity. The boundary conditions correspond to uniform flow at infinity, together with the no-slip condition along the body surface. For the half-cylinder configurations, symmetry conditions are imposed along the wake centerline aft of the rear stagnation point. The initial conditions for the viscous flow are taken as the corresponding steady-state inviscid solution, and represent an impulsive start from rest.

Clustered conformal transformations are used to obtain appropriate surface-oriented coordinates for a circular cylinder. Figure 1 shows the inviscid flow problem considered first in the physical plane and then transformed to the complex-potential W-plane, where the cylinder is

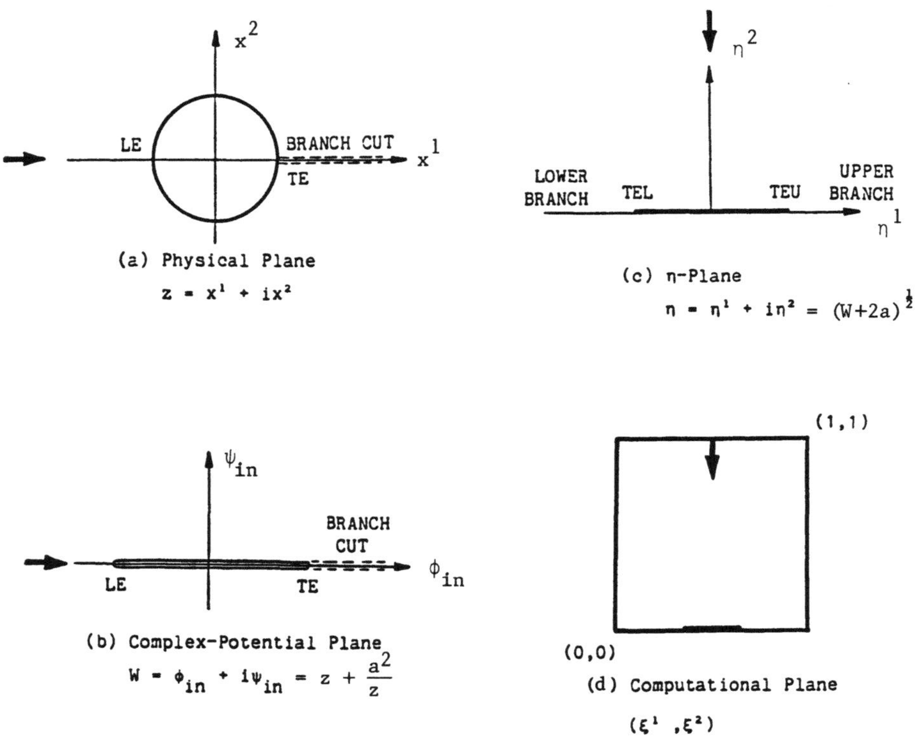

Fig. 1. Flow Past Circular Cylinder in Physical and Transformed Planes.

collapsed to a straight-line segment. The doubly-infinite W-plane could constitute the computational plane, however it has certain disadvantages. The flow domain is infinite in extent and provides an H-grid in the physical plane, which makes the specification of the far-field boundary condition difficult and, also, the lines of ψ_{in} = constant experience large turning near the leading stagnation point. In view of these disadvantages, a power-law conformal transformation is implemented. This has some distinct advantages, namely, it leads to a C-grid in the physical plane, with inherent clustering near the leading stagnation point and also the entire flow domain occupies only the upper-half of this η-plane. This transformation introduces a singularity in the vorticity at the trailing edge and requires special treatment of the wall-vorticity boundary condition in the flow configuration for a circular cylinder without symmetry. Suitable one-dimensional grid clustering transformations are finally applied to obtain bounded computational coordinates (ξ^1,ξ^2), simultaneously attempting to resolve the multiple disparate length scales of the unsteady flow with massive separation, while maintaining the transformation metrics to be smooth and continuous in the entire flow field. Typical clustered conformal grids consisting of (241, 45) and (241, 49) points for a circular cylinder without and with symmetry, respectively, are shown in Fig. 2.

A fully implicit time-marching method is developed such that all spatial derivatives are approximated using central differences and no use is made of any artificial dissipation. The numerical method solves the discretized equations using the alternating-direction implicit-block Gaussian elimination (ADI-BGE) method and has overall $O[\Delta t, (\Delta\xi^1)^2, (\Delta\xi^2)^2]$ accuracy. The boundary conditions on the vorticity are implemented so as to maintain second-order spatial accuracy.

Results and Discussion

Comments on Issues Related to Discretized Problem

The unsteady Navier-Stokes analysis and numerical method developed are applied to symmetric flow past a circular cylinder. Existing computational results [3] for Re up to 600 show that, for Re ≥ 450, the wake bubble size (L,W) grows as [O(Re), O(Re)]. In this range of Re, only the demonstrated symmetric stable mode prevails and the relaxation time for the massive trailing eddy to reach asymptotic state is very long. The present authors are well aware of the fact that, if only the asymptotic state of the flow

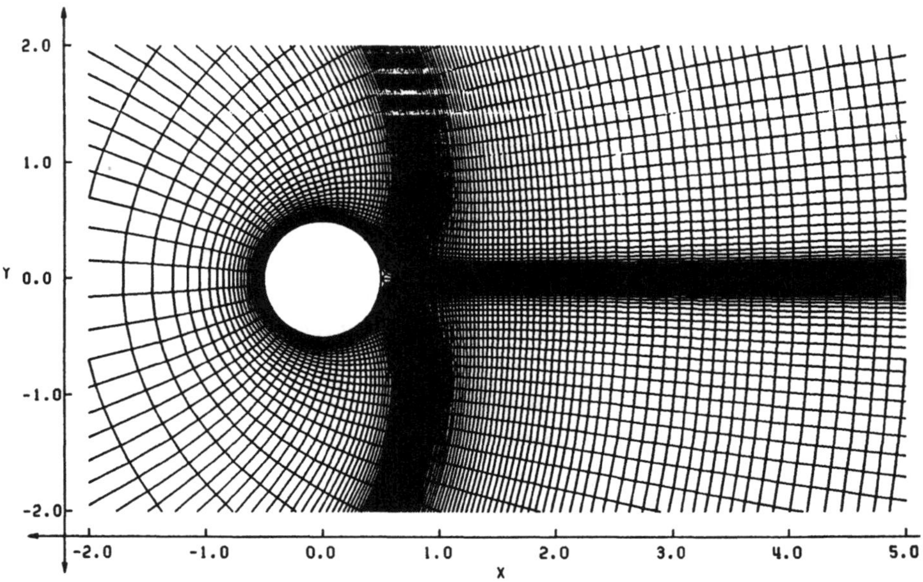

FIG. 2a. TYPICAL GRID WITH (241,45) POINTS, FOR FLOW PAST ENTIRE CYLINDER.

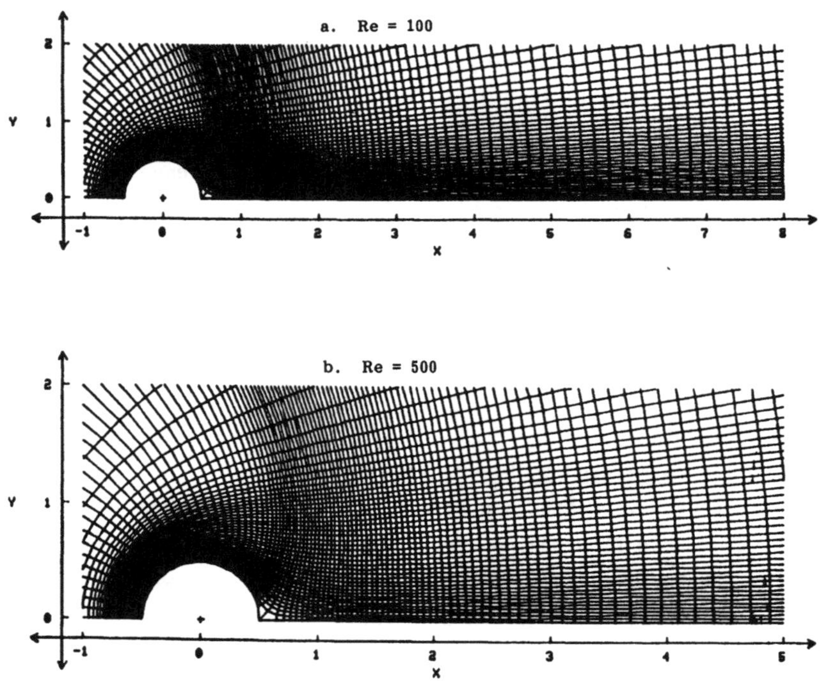

FIG. 2b. TYPICAL GRIDS WITH (241,49) POINTS FOR UPPER-HALF PLANE OF SYMMETRIC FLOW PAST CYLINDER.

is desired, artificial means may be employed for urging the numerical solution to attain the prevailing asymptotic state expeditiously. However, this artificially suppresses the instability modes which directly influence the high-Re viscous solutions. Consequently, only a time-accurate solution procedure is employed in the present study.

The accurate determination of the flow structure in the large primary eddy, including the rear closure, as well as the flow structure in the shear layer separating from the surface of the cylinder, places a tremendous burden on the grid employed. The grid must provide the necessary resolution of the scales of the problem, while keeping the total number of grid points in the domain to a minimum. The choice of a computational grid of (241, 49) points was selected, partly because it was used in Ref. [3] as the coarsest grid, but mainly because it represents nearly the maximum grid that can be accommodated within the storage capacity of the Concurrent 3250 MPS computer system employed. The grid distribution is optimally arrived at to resolve the multiple scales of this problem. Central differences are used throughout the flow field to prevent the computations from becoming contaminated by numerical dissipation. However, for Re = 500, the grid distribution used here is only marginally adequate in the region between the massive eddy and the quiescent zone near the cylinder rear stagnation point; hence, the results are considered only qualitative. The relative computational efficiency of the present algorithm is measured in terms of the CPU time τ required to advance the solution by one time step per spatial grid point. The value of this 'computational effort' index τ is 1.91×10^{-5} seconds for the NASA-Langley Research Center CDC CYBER 205 VPS 32 vector machine. It is difficult to provide the corresponding values of τ for the Concurrent 3250 MPS scalar supermini-computer, but, approximately it would be about 35 to 50 times that for the Langley System. Additional remarks regarding the efficiency of the scheme will be made later in this section.

Circular Cylinder Without Symmetry

In order to validate the present analysis, the flow past a circular cylinder was computed, without assuming symmetry. For this configuration, Bouard and Coutanceau [9] have provided comprehensive experimental results for the flow in the recirculating regions, during the initial phase of the flow development [$t \leq 3.0$] for Re up to 10^4. For Re $\leq$ 550, their results showed the development of two or more eddies which remain stable, symmetrical and attached to the cylinder. The comparative study performed here is also limited to flow in the separation regions for $t \leq 3.5$. Figure 4 presents the characteristics of the flow past a cylinder at early time for both Re = 200 and 550. In Figs. 3(a) and 3(c), the normalized velocity on the axis of the cylinder, aft of the trailing edge, for $0.5 \leq t \leq 3.0$, is compared satisfactorily. It should be noted that the portion of the curves for which the velocity is negative represents a measure of the length of the eddy where the flow is returning from the wake towards the cylinder. Also shown in these figures as an inset are the drag histories for the two cases investigated. The exterior boundaries of the closed wake for $0.5 \leq t \leq 3.0$ for Re = 200 and 550 are compared in Figs. 3(b) and 3(d), respectively, and the agreement is quite good. The instantaneous streamlines and vorticity contours are presented in Fig. 4 for Re = 200. The occurrence of the bulge phenomenon for Re = 200 and $t \geq 1.0$, first reported in Ref. [9], is also computed satisfactorily. It leads to an alteration in the vorticity contours close to the surface of the cylinder in the separated zone. The isolated secondary separation phenomenon reported for Re = 550 was also observed in the present calculation but, due to brevity of this paper, these results are not shown here. The results presented here and compared with the experimental data are intended to establish credence to the present analysis.

Symmetric Cylinder

The original objective was to provide the accurate flow structure for the three symmetric configurations with Re = 100, 500 and 1,000. However, since the present grid of (241, 49) proved to be only marginally adequate for Re = 500, computations were not performed for the configuration with Re = 1,000. Considerable difficulty was experienced from the downtime of the computer system when the actual runs were made and, hence, the results for Re = 500 are still far from the time-asymptotic state. The instantaneous stream-function and vorticity contours for Re = 100 are depicted in Fig. 5 for t = 9, 20, 30 and 50. The flow has reached the asymptotic state considerably prior to t = 50; the computations were continued up to t = 50 in order to establish that the results did not change with additional advancement in time. The vorticity contours show a smooth variation; closer examination of the overall steady state results at t = 50 shows that they compare favorably with those of Fornberg [3]. For Re = 500, the instantaneous stream-function and vorticity contours are

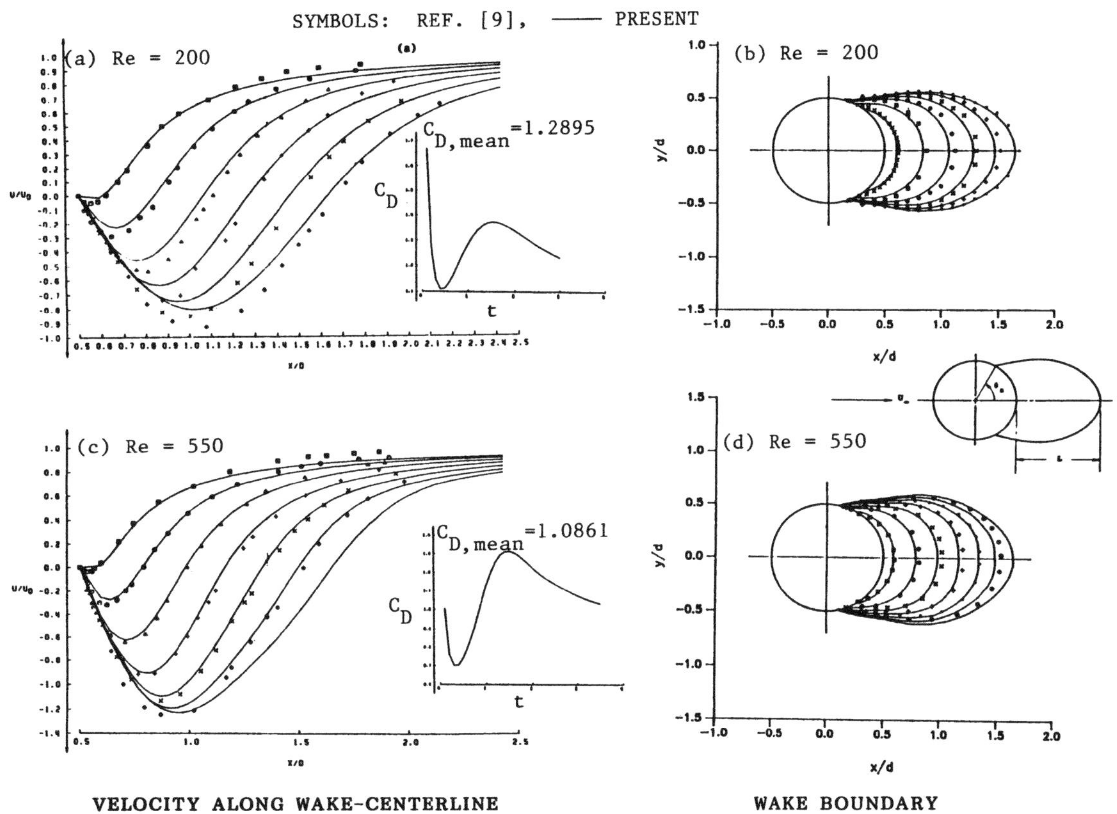

FIG. 3. EARLY-TIME FLOW CHARACTERISTICS FOR CYLINDER
(a),(b): Re = 200; (c),(d): Re = 550.

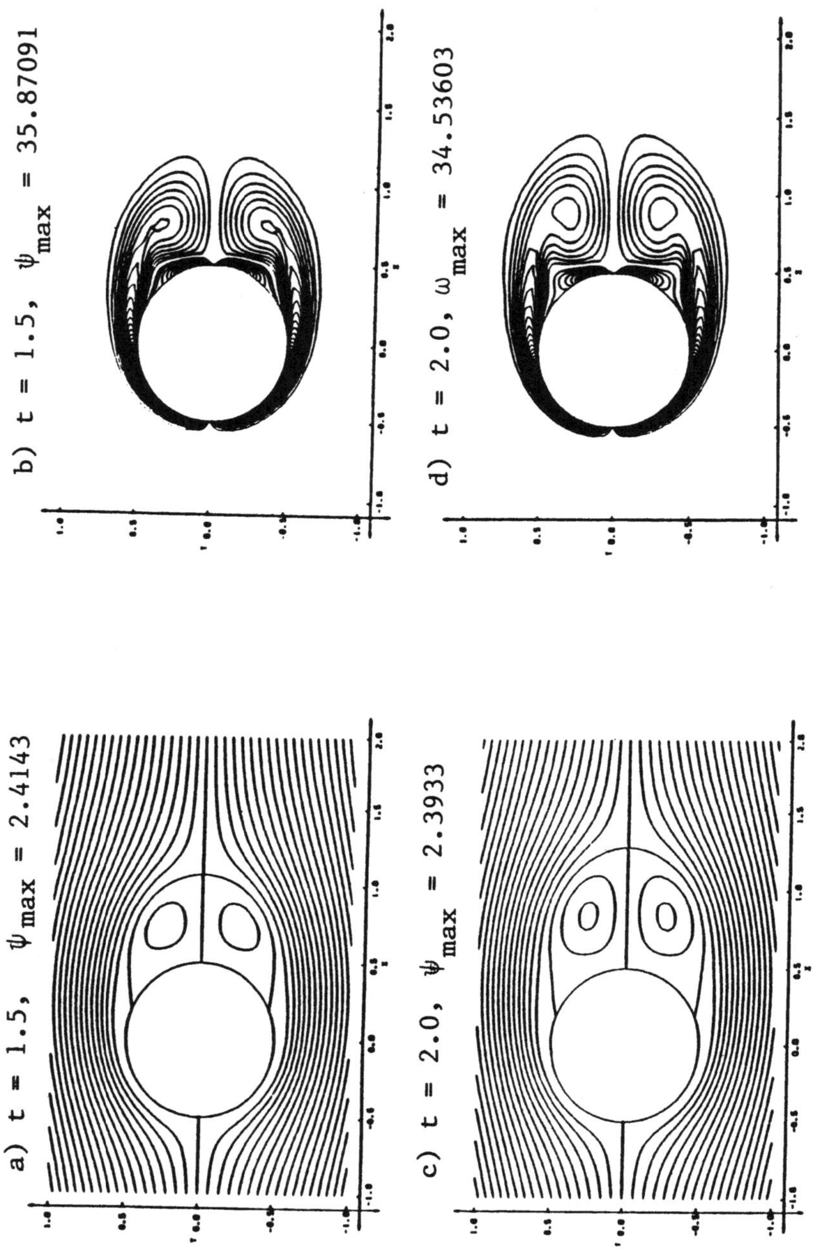
a) t = 1.5, ψmax = 2.4143
b) t = 1.5, ψmax = 35.87091
c) t = 2.0, ψmax = 2.3933
d) t = 2.0, ωmax = 34.53603

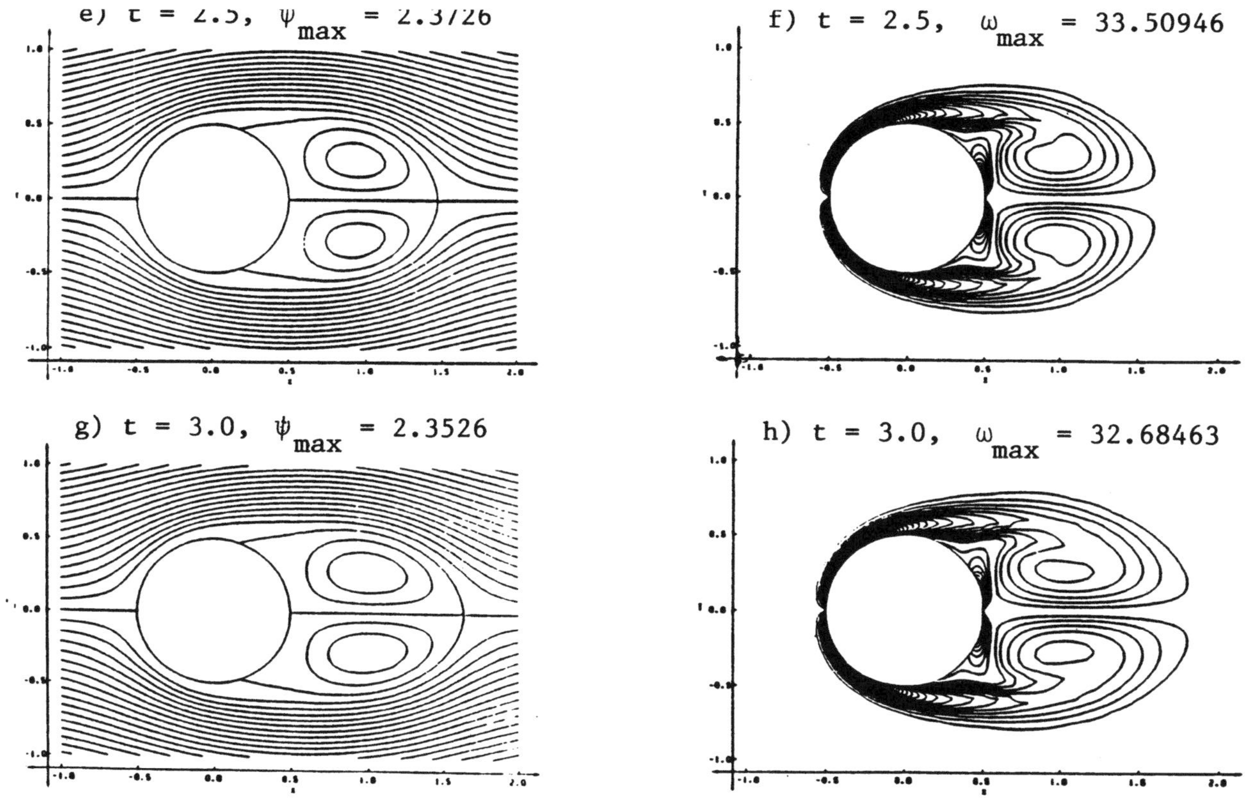

$\Delta\psi = 0.05$

INSTANTANEOUS STREAM-FUNCTION CONTOURS

$\Delta_1 \omega = 1.0$, $\omega_{max1} = 50.0$, $\Delta_2 \omega = 10.0$

VORTICITY CONTOURS

FIG. 4. DEVELOPMENT OF FLOW PAST CYLINDER AT Re = 200; SYMMETRY NOT ASSUMED.

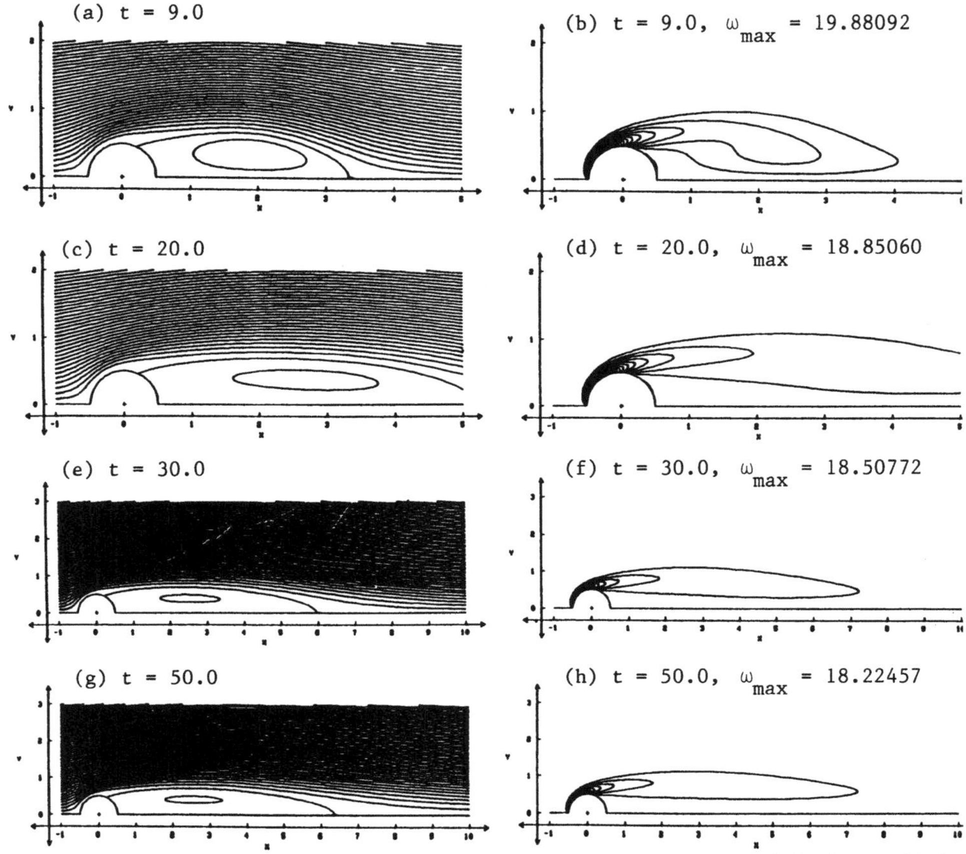

$\Delta\psi = 0.05$ $\Delta_1\omega = 1.0$, $\omega_{max_1} = 50.0$, $\Delta_2\omega = 50.0$

INSTANTANEOUS STREAM-FUNCTION CONTOURS VORTICITY CONTOURS

Fig. 5. Development of Symmetric Flow Past Cylinder at Re = 100.

delineated in Fig. 6 at t = 100, 225, 800 and 1600. The eddy length L, as seen from the stream-function contours, increases from approximately 6 diameters at t=100 to 27 diameters at t=1600; the asymptotic results of Fornberg [3] show this length to be 36 diameters. Further, the region between the cylinder and the main eddy grows from initially about 1 diameter to 5 1/2 diameters by t = 1600; the corresponding asymptotic results of Fornberg [3] show this distance to be approximately 8 diameters. This region can be considered as consisting of two parts: the front part where the separating streamline grows in width as $O(Re^{1/2})$, followed by a 'transition' region in which the growth rate adjusts such that the width

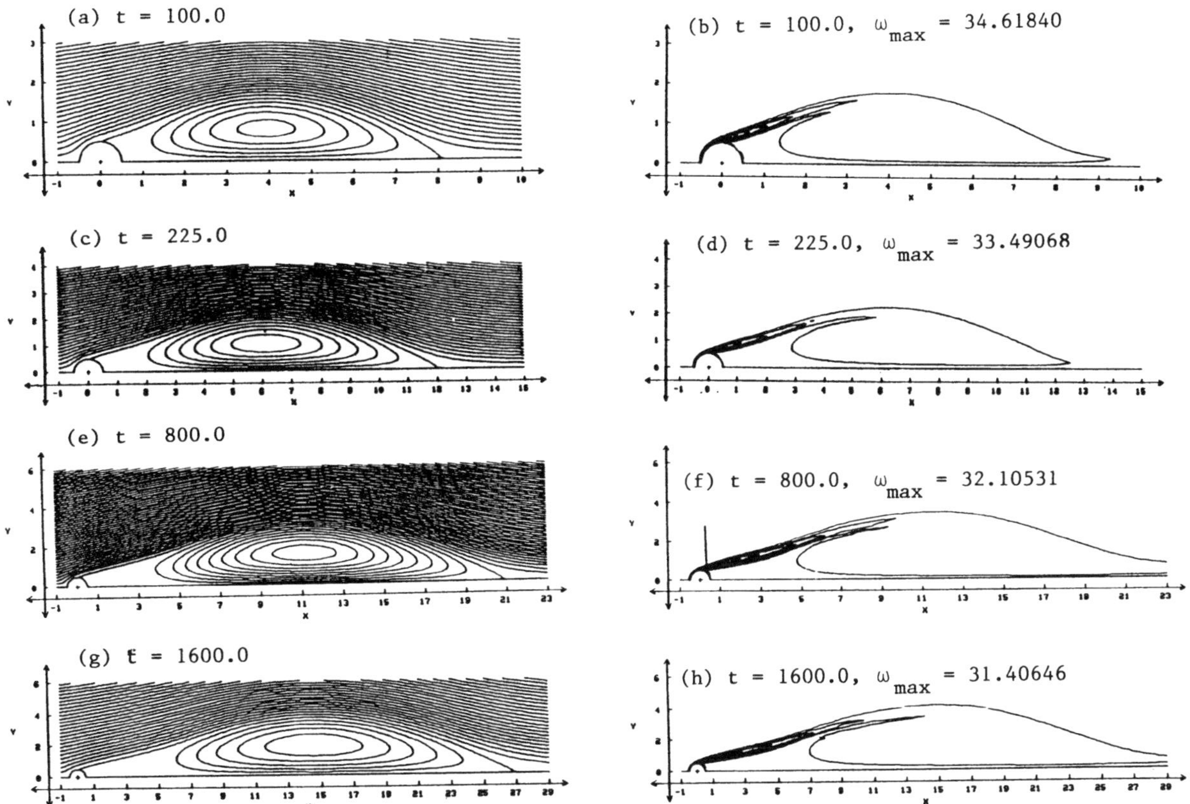

$\Delta\psi = 0.1$

INSTANTANEOUS STREAM-FUNCTION CONTOURS

$\Delta_1\omega=1.0$, $\omega_{max1}=50.0$, $\Delta_2\omega=50.0$

VORTICITY CONTOURS

FIG. 6. DEVELOPMENT OF SYMMETRIC FLOW PAST CYLINDER AT Re = 500.

can grow as O(Re) in the main eddy. Also, in this transition region, the vorticity increases from zero to the level of nearly uniform vorticity present in the main eddy. The kink in the contours of the vorticity is conjectured to be related to this change in the growth rate of the width of the eddy. Figure 6(h) shows a very small wiggle in the inner most contour of $\omega = 0.5$ near a distance of 7 diameters; this wiggle results from inadequate resolution of the normal scales and may have the possibility of growing with time. This suggests that further refinement in the normal grid is essential for obtaining an accurate flow structure computed for the asymptotic state for $Re \geq 500$.

The surface vorticity for Re = 100 and Re = 500 is presented in Fig. 7; the results of Fornberg [3] are also included. For Re = 100, the agreement is good whereas, for Re = 500, the present results do not correspond to the time-asymptotic state, leading to a deviation in the minimum surface vorticity. Fortunately, the minimum surface vorticity is slowly decreasing with time and the deviation may vanish if the present computations are continued for larger time t. Next, the coefficient of surface pressure is plotted in Fig. 8 for these two configurations; also shown here is the inviscid C_p distribution. The associated table shows the coefficient of drag C_D for both Re = 100 and 500 for the present study as well as those obtained by Fornberg [3]. The results from the two sources conform for Re = 100, if the inviscid contribution of 0.009, which is indicative of the accuracy of the numerical integration for the drag determination, is taken into account. For Re = 500, the present results show a smaller value for C_D, with a possibility of further reduction in this value for $t > 225$.

In order to shed light into the large-scale characteristics for flow past a cylinder, the velocity vectors are plotted in Fig. 9 at various streamwise locations for both the configurations investigated, as the differences in the flow structures of the two configurations then become apparent. For Re = 500, the large eddy, with its uniform vorticity, and velocity of O(1), can be seen very clearly. The results presented here do not show O(1) thickness for both the shear layer past the center of the main eddy and the return jet, as proposed by Smith [5]; the present results need to be scrutinized further to provide the accurate flow structure. The Navier-Stokes calculations show that, near Re = 500, the translating cylinder drags with it a massive eddy. At this juncture, a few questions that arise and which need to be clarified are listed. What is the exact

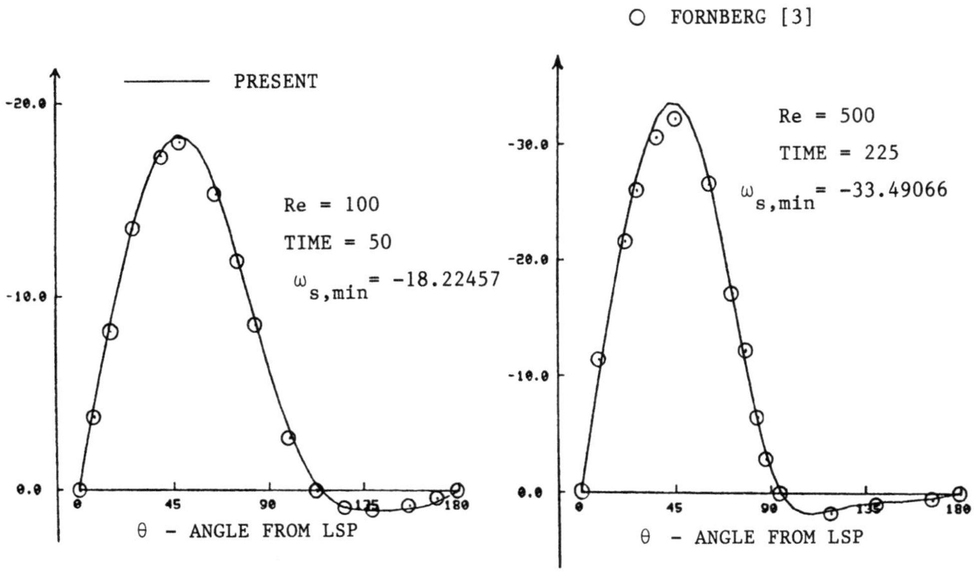

FIG. 7. SURFACE VORTICITY FOR SYMMETRIC FLOW PAST CYLINDER.

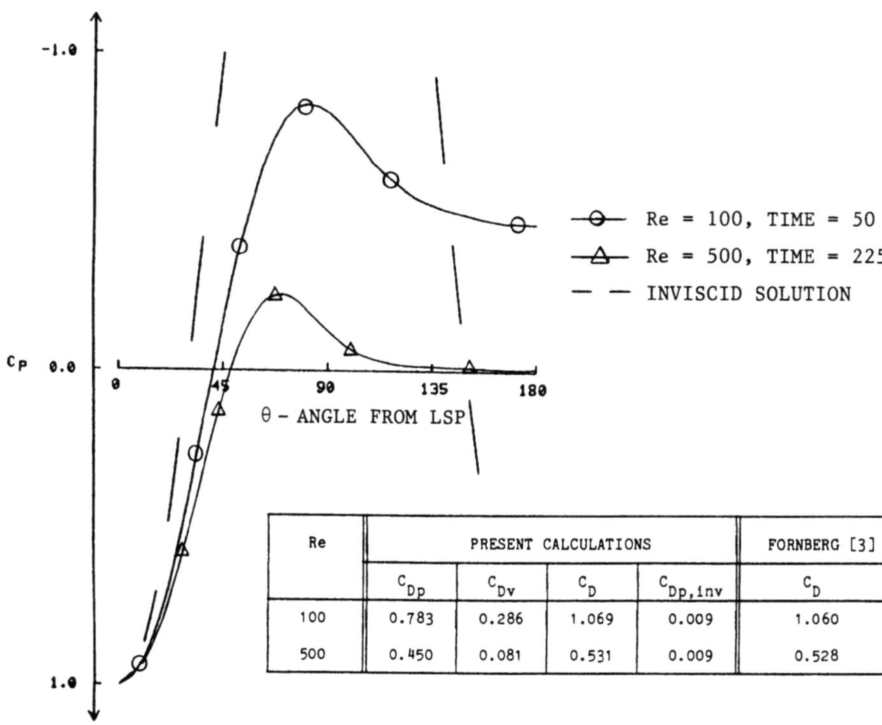

Re	PRESENT CALCULATIONS				FORNBERG [3]
	C_{Dp}	C_{Dv}	C_D	$C_{Dp,inv}$	C_D
100	0.783	0.286	1.069	0.009	1.060
500	0.450	0.081	0.531	0.009	0.528

FIG. 8. COEFFICIENTS OF PRESSURE AND DRAG FOR SYMMETRIC FLOW PAST CYLINDER.

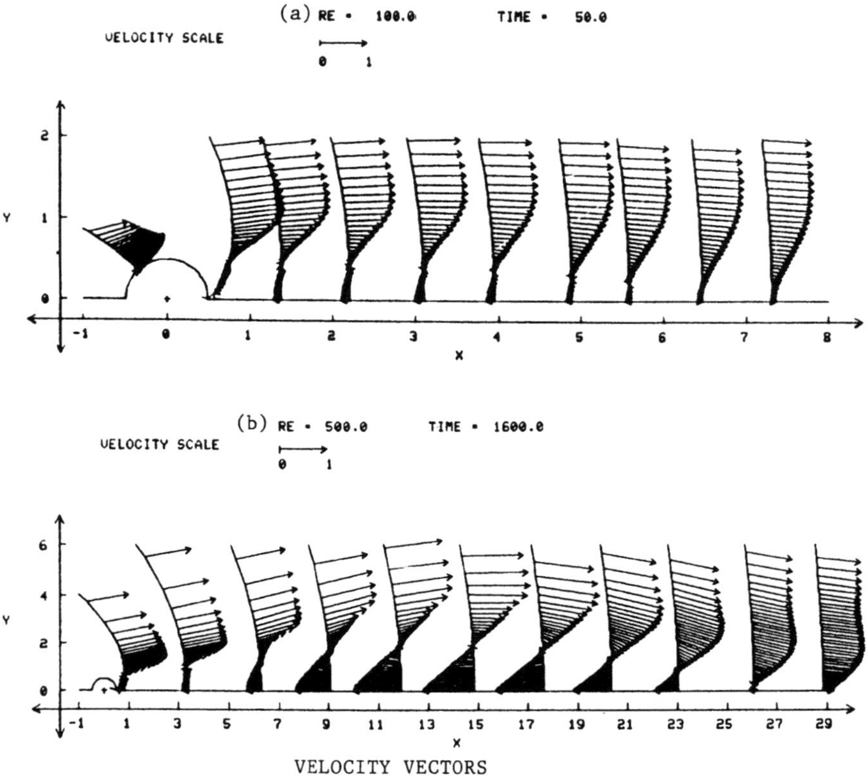

Fig. 9. Large-Scale Characteristics for Flow Past Cylinder.

flow structure near the closure of the eddy at the rear? Will the massive eddy break away from the translating cylinder at higher Re due to instability modes? What will be the large-scale flow structure if the main eddy breaks away and is convected towards infinity? Is it possible to characterize this unsteady evolution as leading to a new asymptotic state? Answers to some of these questions can be addressed by a systematic study, using this unsteady analysis, if proper supercomputer resources are made available. The existing nonlinear stability theory may indicate the existence of only an asymmetric self-sustaining instability mode, while the unstable symmetric mode, being convective in nature, may exist only if externally forced. Hence, the quest for determination of the accurate flow structure for higher Re must continue as many questions can be answered with the help of the unsteady Navier-Stokes analysis.

Acknowledgements

The first two authors greatly appreciate the stimulating discussions they had with Professor H.K. Cheng during their stay at the University of Southern California in Summer 1986. This work was supported, in part, by the AFOSR Grant 85-0231, and, in part, by the Office of Naval Research, Contract N00014-85-K-0049. Part of the computer time used was supported by the NASA Grant NAG-1-465.

References

1. Smith, F.T.: Laminar flow of an incompressible fluid past a bluff body: the separation, reattachment, eddy properties and drag. J. Fluid Mech. 92 (1979) 171-205.

2. Fornberg, B.: A numerical study of viscous flow past a circular cylinder. J. Fluid Mech. 98 (1980) 819-855.

3. Fornberg, B.: Steady viscous flow past a circular cylinder up to Reynolds number 600. J. Computational Physics 61 (1985) 297-320.

4. Peregrine, D.H.: A note on the steady high-Reynolds-number flow about a circular cylinder. J. Fluid Mech. 157 (1985) 493-500.

5. Smith, F.T.: A structure for laminar flow past a bluff body at high Reynolds number. J. Fluid Mech. 155 (1985) 175-191.

6. Ghia, K.N., Osswald, G.A. and Ghia, U.: Simulation of self-induced motion in the near-wake of a Joukowski airfoil. High Reynolds Number Flow Computations, Editor: K. Kuwahara, Springer-Verlag, New York (1986).

7. Osswald, G.A., Ghia, K.N. and Ghia, U.: Simulation of buffetting stall for a cambered Joukowski airfoil using a fully implicit method. Lecture Notes in Physics 264, Editors: F.G. Zhuang and Y.L. Zhu, Springer-Verlag, New York (1986) 516-522.

8. Ghia, K.N., Osswald, G.A. and Ghia, U.: Analysis of two-dimensional incompressible flow past airfoils using unsteady Navier-Stokes equations. Numerical and Physical Aspects of Aerodynamic Flows: Vol. III, Editor: T. Cebeci, Springer-Verlag, New York (1986).

9. Bouard, R. and Coutanceau, M.: The early stage of development of the wake behind an impulsively started cylinder for $40 < Re \; 10^4$. J. Fluid Mech. 101 (1980) 583-607.

Some Features of the Transcritical Boundary Layer Interaction and Separation

V.Ja. NEILAND U.S.S.R. THE CENTRAL INSTITUTE OF AERO-
-AND HYDRODYNAMICS, 140160 ZHUKOWSKY-TOWN,3

1. The asymptotic theory of the inviscid flow boundary layer interaction is an important part of the viscous gas dynamics at high Reynolds numbers. It is based on the fundamental idea of L. Prandtl about a possible division of the total flow into the inviscid flow and thin boundary layer [1]. This idea is due to the attempt to explain rationally the flow separation from a body surface. The idea of Prandtl appeared extremely fruitful not only for the viscous flow dynamics but also for many other problems of the applied mathematics. The initial theory formulation assumes the possibility to get at first the solution for the external inviscid gas flow and then for the boundary layer at a given pressure distribution. Later L. Prandtl [2] showed the possibility to refine the solution allowing for the effect of the boundary layer displacement on the external flow. In the next approximation it is necessary to take into account the impact of the external flow change on the boundary layer flow, etc. Actually, the concept of the weak interaction theory was formulated.

However, the following experimental and theoretical studies indicated that in the most cases taking into account the weak interaction does not allow all the problems to be solved, in particular, the flow at the separation point cannot be described. For example, the study of the boundary layer solutions at a given external pressure distribution results in an "impassable" singular point [3, 4, etc]. The crucial importance of the strong local external flow-boundary layer interaction eliminating the solution singularity and the possibility to continue the solution beyond the separation point was demon-

F. T. Smith, S. N. Brown, (Eds.)
Boundary-Layer Separation
IUTAM Symposium London 1986

strated in [5-7]. Near the laminar separation point in a supersonic gas flow the Navier-Stokes solution is presented asymptotically in a triple-deck form. The use of the triple-deck limiting structure of the solution proved to be very fruitful. For example, in [8] it was used to solve the laminar separation problem for an incompressible fluid, in [9-10] it was extrapolated for unsteady flow studies.

However, the asymptotic approach to the problems of the supersonic viscous flow dynamics allowed us to study a wide range of problems which are not necessarily reduced to the triple-deck scheme and were not described by the classical boundary layer theory: for instance, for flows near the point of the shock incidence on the boundary layer; flows about angular points and steps at the boundary layer bottom. These are problems in which the interaction is strong along the whole body surface which may result in the upstream propagation of disturbances for a body in a hypersonic flow. It is likely that the investigations in these directions for supersonic gas flows are most completely reviewed in [11].

Thus, the range of problems described by searching for the limiting Navier-Stokes solutions at the limit transfer $Re \to \infty$ was considerably expanded. Now it includes a great number of problems important for the practice and theory which are not described by the classical boundary layer theory. This is also of particular importance since the numerical solutions at high Reynolds numbers present some difficulties. Besides, the asymptotic approach explains rationally the physical flow properties and provides convenient approximate similarity laws.

The structure of the supersonic flow-boundary layer interaction regions, the direction of the disturbance propagation and the scale of distances for which the disturbances propagate depend on the Mach number profile in an undisturbed boundary layer. L. Crocco [12] was the first who discovered this phenomenon. He introduced the notion of the subcritical and supercritical boundary layers which, interacting with the shock, are able to transfer upstream the pressure disturbances for distances significant compared with the boundary layer thickness, or which are unable to do it. Crocco noted that for the supercritical boundary layer to acquire this property the Mach number profile in the boundary layer must be changed drastically at very short length comparable with the boundary layer thickness. In terms of [12] it is necessary that the supercritical jump take place. These representations have been justified by general physical arguments and the integral method to describe the interaction.

However, the early studies (before [13] which appeared in 1973) on the asymptotic theory of the separation and interaction considered actually only subcritical conditions. And there arose even the assumption that the supercriticality is not a physical flow property, but results from an inaccurate description of the phenomenon based on the integral boundary layer equations [14]. But in [13] and [15] the asymptotic theory of two- and three-dimensional supercritical flows is developed and a deep analogy is found between the properties of the sub- and supersonic inviscid flows, on the one hand, and the sub- and supercritical boundary layers in a supersonic external flow, on the other hand. The description of the transcritical flows analogous to the transonic condition in the conventional gasdynamics was of great interest.

This paper is devoted to the description of transcritical flows and possible ways of the boundary layer transition from the subcritical condition to the supercritical one and vice versa. To investigate the transcritical interaction consider some typical simplest problems of this type.

2. Consider at first the hypersonic viscous flow about an angular point of the simplest body contour (see fig.1).

Let us discuss the weak interaction on the main part of the body ahead of the region near angular point O. As it is known the parameter distribution profiles in an undisturbed boundary layer are determined by the relations (see [16]):

$$u=u_e f'(\xi,\eta)\ , \quad \bar{u}_0=f'\ , \quad H=(u_e^2/2)g(\xi,\eta)$$

$$\xi=\int_{-l}^{0}\rho_w\mu_w u_e dx \qquad , \ \eta=\left(u_e/\sqrt{2\xi}\right)\int_0^y \rho dy \quad , \qquad (2.1)$$

$$f'''+ff''=0\ , \quad g''+fg'=0\ , \quad f(\xi,0)=f'(\xi,0)=0\ , \ g(\xi,0)=g_w$$
$$f'(\xi,\infty)=1\ .$$

Here the Cartesian coordinates are shown in fig.1, the origin is at the angular point, $x=-l$ corresponds to the boundary layer origin, that is, l is the length of the forward part of a plate, u , H , ρ , μ are the longitudinal velocity component, stagnation enthalpy, density and dynamic viscosity, respectively; subscripts e , w denote the parameter values on the external boundary and near the wall in the boundary layer.

To reduce the computations the assumptions are taken which are not significant in principle: the linear variation of μ with enthalpy and $P_r=1$.

Now, before obtaining the limiting asymptotic solutions of the Navier-Stokes equations by the method of matched asymptotic expansions [17], estimate characteristic scales of quantities in the region of a strong local interaction of the external hypersonic flow with the boundary layer at point O.

The respective estimates for the undisturbed boundary layer ahead of point O are by (2.1) as follows:

$$u\sim u_e\ , \ \rho\sim\rho_0\sim\rho_e/M_e^2 , \ H\sim\frac{u_e^2}{2}\ , \ \delta_0\sim l Re_0^{-1/2}, \ Re_0\sim\rho_0 u_e l/\mu_0 , \ \rho_0 u_e^2\sim p \quad (2.2)$$

where δ_0 is the thickness of the undisturbed boundary layer at $x=l$, subscript e is the parameter value on the outer boundary of the undisturbed boundary layer, subscript 0 are the values of the density and viscosity at the stagnation temperature, respectively.

By the hypersonic small disturbance theory at $M_e \gg 1$, $\theta \ll 1$, $M_e \theta \ll 1$

$$M_e(p - p_e) = \rho_e u_e^2 (d\delta^*/dx + \theta) \tag{2.3}$$

where δ^* is the boundary layer displacement thickness.

If δ^* =0, there is no boundary layer and the body pressure jump occurs in the centred expansion wave. The boundary layer smoothes the discontinuity in the pressure distribution along the body, but the pressure gradients remain high although the pressure disturbance at $M_e\theta \ll 1$ remains low by (2.3):

$$\frac{\Delta p}{p} \sim M_e \theta \quad . \tag{2.4}$$

For the main part of the boundary layer where $u \sim u_e$ we obtain conventional estimates using the equations of momentum and state:

$$\Delta u \sim u_e(\Delta p/p)\ , \quad \Delta\rho \sim \rho_o(\Delta p/p), \quad \Delta\delta/\delta_o \sim \Delta p/p \quad . \tag{2.5}$$

This part of the boundary layer is denoted in fig.1 by subscript 2. However, there is always region 3 where

$$\rho u^2 \sim \Delta p \quad . \tag{2.6}$$

Therefore, if it is proved that the viscous terms in the Navier-Stokes equations are small in this region, the velocity disturbances there occur in the main order [5-7]. In the general case region 3 with low dynamic pressures requires its own asymptotic expansions to be constructed.

Thus, the Navier-Stokes solutions are searched in the following limiting transition:

$$Re_o \to \infty\ , \quad M_e \to \infty\ , \quad \theta \to 0 \text{ at } M_e\theta \to 0\ , \quad q_w \leqslant O(1) \quad . \tag{2.7}$$

The solution for the external region 1 is evidently described by the hypersonic small disturbance theory. In the following the formula is only required for the pressure distribution on the boundary formed by the boundary layer displacement thickness or region 2 (2.3).

It is convenient to introduce Mises variables

$$\rho u\frac{\partial u}{\partial x}+\frac{\partial p}{\partial x}-\rho v\frac{\partial p}{\partial \psi}=\rho u\frac{\partial}{\partial \psi}\left[\rho u\mu\frac{\partial u}{\partial \psi}+\mu\left(\frac{\partial v}{\partial x}-\rho v\frac{\partial v}{\partial \psi}\right)\right]+$$
$$+\left(\frac{\partial}{\partial x}-\rho v\frac{\partial}{\partial \psi}\right)\left[\frac{4}{3}\mu\left(\frac{\partial u}{\partial x}-\rho v\frac{\partial u}{\partial \psi}\right)-\frac{2}{3}\rho u\mu\frac{\partial u}{\partial \psi}\right]; \qquad (2.8)$$

$$\rho u\left(\frac{\partial v}{\partial x}+\frac{\partial p}{\partial \psi}\right)=\rho u\frac{\partial}{\partial \psi}\left[\frac{4}{3}\rho u\mu\frac{\partial v}{\partial \psi}-\frac{2}{3}\mu\left(\frac{\partial u}{\partial x}-\rho v\frac{\partial u}{\partial \psi}\right)\right]+$$
$$+\left(\frac{\partial}{\partial x}-\rho v\frac{\partial}{\partial \psi}\right)\left[\rho u\mu\frac{\partial u}{\partial \psi}+\mu\left(\frac{\partial v}{\partial x}-\rho u\frac{\partial v}{\partial \psi}\right)\right]; \qquad (2.8.1)$$

$$\frac{\partial n}{\partial x}=\frac{v}{u}\ , \qquad \frac{\partial n}{\partial \psi}=\frac{1}{\rho u}\ ; \qquad (2.8.2)$$

$$\rho u\frac{\partial h}{\partial x}-u\frac{\partial p}{\partial x}=\rho u\frac{\partial}{\partial \psi}\left[\rho\mu u\frac{\partial h}{\partial \psi}\right]+ \qquad (2.8.3)$$

$$+\left(\frac{\partial}{\partial x}-\rho v\frac{\partial}{\partial \psi}\right)\left(\mu\frac{\partial h}{\partial x}-\rho v\mu\frac{\partial h}{\partial \psi}\right)+\mu\left[\left(\rho u\frac{\partial u}{\partial \psi}+\frac{\partial v}{\partial x}-\rho v\frac{\partial v}{\partial \psi}\right)^2+\right. \qquad (2.8.4)$$
$$\left.+2\left(\frac{\partial u}{\partial x}-\rho v\frac{\partial u}{\partial \psi}\right)^2+2\left(\rho u\frac{\partial v}{\partial \psi}\right)^2-\frac{2}{3}\left(\frac{\partial u}{\partial x}-\rho v\frac{\partial u}{\partial \psi}+\rho u\frac{\partial v}{\partial \psi}\right)^2\right]$$

in region 2 of the vortex flow with small disturbances including most of the streamtubes of the undisturbed initial boundary layer.

Introduce the following variables for independent variables and functions in region 2:

$$x=l\overline{\Delta x}_2 x_2\ , \qquad \psi=\rho_0 u_e l\overline{\Delta\psi}_2\cdot\psi_2$$

$$p=\rho u_e^2(p_{20}+\overline{\Delta p}_2\, p_{21}+\cdots)\ , \quad \rho=\rho_0(\rho_{20}+\overline{\Delta\rho}_2\cdot\rho_{21}+\cdots)$$

$$u=u_e(u_{20}+\overline{\Delta u}_2\cdot u_{21}+\cdots)\ , \quad v=u_e\overline{\Delta v}_2\cdot(v_{21}+\cdots) \qquad (2.9)$$

$$n=l\mathrm{Re}_0^{-1/2}\left[(n_{20}+\overline{\Delta n}_2\cdot n_{21}+\cdots)+\overline{\Delta n}_3(n_3-n_{30})_\infty\right]$$

where p_{20} , p_{21} , ρ_{20} , ρ_{21} , ... are the nondimensional functions $O(1)$; $\overline{\Delta p}_2$, $\overline{\Delta\rho}_2$, $\overline{\Delta u}_2$, ... are the nondimensional scales which are functions of the limiting transition type (2.7):

$$P_{20}=(\gamma-1)/2\gamma\ , \qquad \rho_0=\rho_e[2/(\gamma-1)M_e^2\,] \tag{2.10}$$
$$\overline{\Delta p}_2=\overline{\Delta u}_2=\overline{\Delta \rho}_2=\overline{\Delta n}_2=M_e\theta\ , \qquad \overline{\Delta\psi}_2=Re_0^{-1/2}\ .$$

$\overline{\Delta x}_2$ is determined for each case but it will be shown that $Re_0^{-1/2}\ll\overline{\Delta x}_2\ll 1$.

After substituting (2.9) into the Navier-Stokes equations (2.8) and performing the limiting transition (2.7) we obtain the following set of the equations:

$$\rho_{20}u_{20}\frac{\partial u_{21}}{\partial x_2}+\frac{\partial p_{21}}{\partial x_2}=0\ , \quad \frac{\partial p_{21}}{\partial \psi_2}=0\ , \quad \frac{\partial}{\partial x}\left(\frac{p_2}{\rho^\gamma}\right)_2=0$$
$$\frac{\partial n_{20}}{\partial \psi_2}=\frac{1}{\rho_{20}u_{20}}\ , \qquad \frac{\partial n_{21}}{\partial \psi_2}=\frac{1}{\rho_{20}u_{20}}\left(-\frac{\rho_{21}}{\rho_{20}}-\frac{u_{21}}{u_{20}}\right)\ . \tag{2.11}$$

Then match the solution in region 2 at $x_2\to-\infty$ with that in the undisturbed boundary layer. This defines the function profiles $\rho_{20}(\psi_2)$, $u_{20}(\psi_2)$ and the value $p_{21}(x_{21}\to-\infty)\to 0$. The set (2.11) is easily integrated and allows us to define the contribution of region 2 to the variable part of the displacement thickness δ^* in the considered flow region at g_w =0:

$$n_{21}(x_2)=\frac{1}{\gamma}\,\frac{p_{21}(x_2)}{p_{20}}\int_0^{n_{20e}}\left(\frac{1}{M_0^2}-1\right)dn_{20}\ . \tag{2.12}$$

It is obvious, that n_{21} is the linear function of the pressure disturbance p_{21} , and the nature of the variation is defined by the Mach number profile M_0 in the undisturbed boundary layer upstream of the initial part of the interaction region. It is worthy to note, that M_0 for the boundary layer on a flat plate at the temperature factor g_w =0 is such that the integral in (2.12) is negative. It means that region 2 behaves like a supersonic streamline, that is, as the pressure decreases ($p_{21}<$ 0) its thickness rises ($n_{21}>$ 0).

3. Now consider the flow in region 3 where (2.6) is met and the disturbances may be nonlinear.

Introduce the following scales of arguments and functions:

$$x = l\overline{\Delta x_3}\,x_3\ , \qquad \psi = \rho_o u_e l Re_o^{-1/2}\overline{\Delta\psi_3}\cdot\psi_3$$

$$p = \rho_o u_e^2 (p_{30} + \overline{\Delta p_3}\cdot p_{31} + \cdots)\,, \qquad \rho = \rho_o\overline{\Delta\rho_3}\cdot(\rho_3 + \cdots) \tag{3.1}$$

$$u = u_e\,\overline{\Delta u_3}(u_3 + \cdots)\ , \qquad n = l Re_o^{-1/2}\,\overline{\Delta n_3}(n_3 + \cdots)$$

$$v = u_e\,\overline{\Delta v}\,(v_3 + \cdots)\,, \qquad \mu = \mu_o\overline{\Delta\mu_3}(\mu_3 + \cdots)$$

$$h = (u_e^2/2)\ \overline{\Delta h_3}\ (h_3 + \cdots)$$

where

$$\overline{\Delta p_3} = \overline{\Delta p_2} = \overline{\Delta\rho_3}\cdot\overline{\Delta u_3}^2\,, \quad p_{30} = p_{20}\,, \quad p_{31} = p_{21}\,, \overline{\Delta\psi_3} = \overline{\Delta\rho_3}\,\overline{\Delta u_3}\,\overline{\Delta n_3}\ .$$

Substituting (3.1) into (2.8) and carrying out the limiting transition (2.7) gives us the following set of equations:

$$\rho_3 u_3 \frac{\partial u_3}{\partial x_3} = -\frac{\partial p_3}{\partial x_3} + K\cdot\rho_3 u_3 \frac{\partial}{\partial\psi_3}\left(\rho_3 u_3 \mu_3 \frac{\partial u_3}{\partial\psi_3}\right)$$

$$\frac{\partial n_3}{\partial\psi_3} = \frac{1}{\rho_3 u_3}\ , \quad \rho_3 h_3 = \frac{2\gamma}{\gamma - 1}p_{30}\,, \quad K = \frac{\overline{\Delta\mu_3}\cdot\overline{\Delta x_3}}{\overline{\Delta n_3}^2\cdot\overline{\Delta\rho_3}\cdot\overline{\Delta u_3}} \tag{3.2}$$

$$\frac{\partial h_3}{\partial x_3} = K\frac{\partial}{\partial\psi_3}\left(\rho_3 u_3 \mu_3 \frac{\partial h_3}{\partial\psi_3}\right)\ , \qquad \frac{\partial p_{31}}{\partial\psi_3} = 0\ .$$

The relative dimensionless scales of variables Δx , Δp , $\Delta\rho$, ... may take on different values depending on the relation of small quantities g_w , $M_e\theta$. When the limiting transition (2.7) including possibility $g_w \to 0$ is performed, the equations of different types are obtained from the set (3.2).

4. Estimate the orders of magnitudes of the functions according to the conventional procedure and classify the flow conditions on its basis.

For these purpose it is convenient to return (only in this section of the paper) to the dimensional variables, denoting by subcripts 0, 1, 2, 3 their characteristic values in the undisturbed boundary layer and in regions 1, 2, 3 of the disturbed flow.

Using (2.1) we obtain the following estimates for the behavior of the nondimensional velocity $u = u/u_e$ and enthalpy $\bar{u}=u/u_e$ $g=2h/u_e^2$ at the bottom of the undisturbed boundary layer near the wall with the temperature defined by the value $g_w \lesssim 1$:

$$g \sim \sqrt{\bar{y} + g_w^2} \quad , \quad \bar{u} \sim \sqrt{\bar{y}+g_w^2} - g_w \ . \tag{4.1}$$

The factors O(1) in (4.1) are omitted as being of no importance for estimating the order of magnitude of $\bar{y} = y/\delta_0$ functions.

Consider at first the flow in which region 3 characterized by the thickness $\bar{y} \sim \bar{\delta}\,(\bar{\delta}_3 \equiv \delta_3/\delta_0)$ is almost isothermal, that is

$$1 \gtrsim g_w^2 \gg \bar{\delta}_3 \ . \tag{4.2}$$

Then, according to (4.1), the estimates are obtained:

$$g_3 \sim g_w \, , \quad \rho_3 \sim \rho_w \sim \rho_0/g_w \, , \quad \mu_3 \sim \mu_0 g_w \, , \quad \bar{u}_3 \sim \bar{\delta}_3/g_w \tag{4.3}$$

where ρ_0 and μ_0 are the density and viscosity at the stagnation temperature. The last estimate (4.3) is easily obtained from (4.1) when (4.2) is accounted for.

By the definition in region 3

$$\Delta p_3 \sim \rho_3 u_3^2 \sim \rho_0 u_e^2 \bar{\delta}_3^2 / g_w^3 \ . \tag{4.4}$$

Then the thickness of this layer is of the order of magnitude:

$$\bar{\delta}_3 \sim g_w^{3/2} \left(\frac{\Delta p}{p}\right)^{1/2} . \tag{4.5}$$

If the coefficient in the dimensionless scales in the right part of the momentum equation of the set (3.2) $K \to 0$ or remains O(1) when the limiting transition occurs (2.7), the velocity changes in the main order. Hence,

$$\Delta\delta_3 \sim \delta_3 \sim \delta_0 \left(\frac{\Delta p}{p}\right)^{1/2} g_w^{3/2} \quad . \tag{4.6}$$

If the limiting condition occurs in such a way that $K \to \infty$, the disturbances in region 3 are $\Delta u_3 \ll u_3$, that is the flow is described by the linearized equations with viscous terms. In the following the linear conditions are not considered, being of less interest.

Thus, consider for the present the flow with nonlinear disturbances in region 3. Various flows are possible depending on the relation of g_w and $\Delta p/p$ values.

If g_w is not too small or in general $g_w = O(1)$ the relation is met:

$$\bar{\delta}_3 >> \Delta\bar{\delta}_2 \sim (\Delta p/p) \tag{4.7}$$

that is, the main part of the derivative of the boundary layer displacement thickness δ^* (2.3) is due to the flow in region 3 which is subsonic. Then the flow is inevitably subcritical as in the classical triple-deck scheme.

The sufficient subcriticality condition follows immediately from the comparison of (4.5), (4.6),(4.7) and (2.5).

$$\frac{\Delta p}{p} << g_w^3 \quad . \tag{4.8}$$

If it is assumed that the layer 2 is supercritical, that is, the integral in (2.12) (call it L) converges (due to the limiting transition $g_w \to 0$) and is negative (as it occurs in almost all practically important cases), the flow as a whole may become supercritical at

$$|\Delta\delta_2| >> \Delta\delta_3 \sim \delta_3 \quad . \tag{4.9}$$

So, at $L < 0$ the flow is supercritical if

$$\frac{\Delta p}{p} >> g_w^3 \quad . \tag{4.10}$$

Thus, the transcritical flows with sub- or supercritical boundary layer in different regions occur if

$$\frac{\Delta p}{p} \sim g_w^3 \quad , \qquad \Delta\delta_2 \sim \delta_3 \ . \qquad (4.11)$$

The radical differences of sub- and supercritical flow properties are similar to those of sub- and supersonic flows and are illustrated below when solving the specific problems.

Considering obviously subcritical conditions ($\Delta\delta_3 \gg \Delta\delta_2$) and possible transcritical flows ($\Delta\delta_3 \sim \delta_3 \sim \Delta\delta_2$, $L < 0$) and using the Ackeret formula (2.3) to obtain estimates yields the scale Δx of the region where the disturbances propagate upstream:

$$\frac{\Delta x}{\ell} \sim \left(M_e \frac{\delta_0}{\ell}\right) g_w^{3/2} \left(\frac{\Delta p}{p}\right)^{-1/2} \ . \qquad (4.12)$$

Comparison of (4.11) and (4.12) shows that for all sub- and transcritical interactions $\Delta x \gg \delta_0$. This means that in the main approximation there are no transverse pressure differences in regions 2 and 3 except for singular points of the solution which are considered below.

Now compare the characteristic values of the viscous and inertial terms relation in the momentum equation. For the flow in region 3 to remain nonlinear it is necessary and sufficient to meet the condition:

$$\rho_3 u_3^2 \delta_3^2 \,/\, \Delta x \,\mu_3 u_3 \gtrsim O(1) \ . \qquad (4.13)$$

Using the above estimates from (4.13) we obtain

$$\left(\frac{\Delta p}{p}\right)^2 \gtrsim O\left(M_e Re_0^{-1/2}\right) \ . \qquad (4.14)$$

Thus, the flow in layer 3 is described by the complete Prandtl equations (that is, it is "viscous and nonlinear") when the pressure disturbances are by the order of magnitude equal to

$$\frac{\Delta p}{p} \sim M_e^{1/2} Re_0^{-1/4} \ . \qquad (4.15)$$

This complies with the extrapolation of the results [5-7] for hypersonic flows (see [18]).

The pressure disturbances being lower by the order of magnitude, the disturbances in region 3 are linear. If

$$1 \gg \frac{\Delta p}{p} \gg M_e Re_o^{-1/2} \qquad (4.16)$$

the flow in this region should be nonlinear and inviscid in the main approximation. (To meet the nonslip conditions a more thin viscous sublayer has to be introduced).

In what follows more emphasis is placed on the transcritical conditions with a nonlinear inviscid flow in region 3 for which (4.11) and (4.16) are satisfied. In fig.2 these conditions are represented by the line AB in the plane $O(\frac{\Delta p}{p})$, $O(g_w)$. The subcritical flows with nonlinear disturbances in region 3 fall within the region ABC. Point C corresponds to the conventional triple-deck flow model [5-7], and the disturbances to the right of AB and above AC (that is, with inviscid nonlinear layer 3) are discussed in [19, 20].

In all sub- and transcritical nonlinear problems which are within the triangle ABC (fig.2) the flow in region 3 remains isothermal in the main approximation, that is $g_3 \sim g_w$.

Therefore, it may be expected that to the left of AB the supercritical interactions with the same properties (4.3) (4.4) may be found (at $L < 0$).

For the supercritical interactions to appear, the main part of the δ^* variation must be due to the supersonic streamline in region 2. For this purpose the condition (4.10) has to be met. It follows from the Ackeret formula and the estimate $\Delta\bar{\delta}_2$ in (4.7) the estimate

$$\Delta x \sim M_e \delta_o \quad . \qquad (4.17)$$

Hence, if g_w decreases at a given pressure disturbance the scale Δx in ABC (fig.2) decreases to AB by (4.12), but intersecting this line of transcriticality it stops varying by the order of magnitude (4.17) since layer 3 no longer affects it in the main approximation.

Using (4.13) let us determine how the critical pressure difference varies by the order of magnitude to the left of AB:

$$\frac{\Delta p}{p} \sim (M_e Re_o^{-1/2})^{2/3} g_w^{-1} \quad . \qquad (4.18)$$

It is easy to see that point A (fig.2) where the transcriticality condition is met (4.11) the critical pressure difference value satisfies (4.15).

When moving in the plot of fig.2 in accordance with (4.18) in the direction of g_w decreasing the thickness of the nonlinear viscous disturbed layer in region 3 continues to drop, and the pressure disturbance increases till the condition (4.2) (by which the estimates in layer 3 have been obtained so far) is violated. Using (4.5) it may be shown that (4.2) is violated as the temperature factor values are:

$$g_w \sim \frac{\Delta p}{p} << \left(\frac{\Delta p}{p}\right)^{1/3} \quad . \qquad (4.19)$$

In fig.3 the transition to the left across the line BD on which (4.19) is met corresponds to the transition to the nonisothermal region 3. By (4.1) at lower g_w the parameters in region 3 with the nonlinear disturbances are as follow:

$$g_3 \sim \bar{\delta}_3^{1/2}, \quad \bar{u}_3 \sim \bar{\delta}_3^{1/2}, \quad \bar{\mu}_3 \sim \bar{\delta}^{1/2}, \quad \bar{\rho}_3 \sim \bar{\delta}_3^{-1/2} . (4.20)$$

Comparison of the orders of magnitude of the main viscous and inertial terms (as above) gives the order of magnitude for the critical pressure difference which may induce the separation in compressed flows:

$$\frac{\Delta p}{p} \sim (M_e Re_o^{-1/2})^{1/3} \quad . \qquad (4.21)$$

The line ED in fig.2 corresponds to (4.21). So, fig.2 gives a full picture of the possible flows with a strong local interaction up to the condition that results in a separation. Most of the available investigations are devoted to subcritical flows of the ABC and, in particular, at the point C.There are a few number of papers discussing the supercritical conditions [13, 15]. Therefore, the present paper investigates specific features of the transcritical flow similar, in a way, to the transonic flow in the conventional gasdynamics.

5. Now consider in detail the properties of the transcritical flows corresponding to the line AB in fig.2. According to the results of section 4 flow 3 should be described by the inviscid boundary layer equations, and

$$\Delta\delta_2 \sim \delta_3 . \quad (5.1)$$

The simplest way of defining the variable part of the boundary layer displacement thickness $\Delta\delta^*$ is to use the unified formula for regions 2 and 3 resulting from the mass conservation equation

$$\Delta\delta^* = \int_0^{\delta_0} \frac{\rho_0}{\rho} \frac{u_0}{u} dn_0 \quad (5.2)$$

where subscript 0 denotes the parameter values upstream of the interaction zone. As the flows in both regions are locally inviscid the expression

$$\Delta\delta^* = \int_0^{\delta_0} \left[\frac{(p_0/p)^{1/\gamma}}{\left\{1 + \frac{2}{(\gamma-1)M_0^2}\left[1-\left(\frac{p}{p_0}\right)^{\frac{\gamma-1}{\gamma}}\right]\right\}^{1/2}} - 1 \right] dy_0 \quad (5.3)$$

may be obtained using the condition of isentropic flow and the Bernoulli equation. Thus, the profile of an initial undisturbed boundary layer is introduced as $M_0(y_0)$ relationship where y_0 is the coordinate of a respective streamline in an undisturbed boundary layer.

Using (2.1) it is possible to obtain:

$$M_0 = [2/(\gamma-1)]^{1/2}\, \bar{u}_0 / g^{1/2} . \quad (5.4)$$

At the considered condition $\Delta p/p = O(g_w^3)$ the velocity and stagnation enthalphy profiles $\bar{u}_0(\bar{y}_0)$ and $g(\bar{y}_0)$ near a body can be represented as follows:

$$\begin{aligned} \bar{u}_0 &= (\tau_w u_e / 2q_w)\left[\sqrt{g_w^2 + (4q_w/\mu_0 u_e^2)\, y_0} - g_w\right] + \cdots \\ g_w &= \sqrt{g_w^2 + (4g_w/\mu_0 u_e^2)\, y_0} + \cdots \end{aligned} \quad (5.5)$$

where τ_w , g_w are the shear stress and heat transfer rate in an undisturbed boundary layer ahead of the local interaction region.

Since $\frac{p}{p_0} = 1 + \frac{\Delta p}{p_0}$ where $\frac{\Delta p}{p_0} \ll 1$, the second term under the radical in the denominator of (5.3) may be expanded in $\Delta p/p_0 << 1$ everywhere except for the zone where M_0 is also low. The region $M_0 = O(1)$ is region 2, and the nonlinear zone 3 corresponds to low y_0 for which the second term is no longer small.

In sublayer 3 we have:

$$\bar{u}_0 \approx (A\bar{B}/2)(\bar{y}_0/g_w) + \cdots, \quad g \approx g_w + \cdots, \qquad A = (\tau_w u_e/2g_w), \quad B = (4g_w/\mu_0 u_e^2) . \tag{5.6}$$

Now introduce the intermediate integration limit:

$$0 < \delta_1(\varepsilon) << \delta_0 . \tag{5.7}$$

The integral in the limits from 0 to δ_1 is taken as

$$\Delta\delta_3^* = \left(\frac{p_0}{p}\right)^{1/\gamma}\left[\sqrt{\delta_1^2 + \frac{\gamma-1}{\gamma}\frac{4g_w^3}{A^2B^2}\varepsilon} - \sqrt{\frac{\gamma-1}{\gamma}\frac{4g_w^3}{A^2B^2}\varepsilon}\right], \quad \varepsilon = -\frac{\Delta p}{p} . \tag{5.8}$$

In the "linear" region 2, that is, for $\delta_1(\varepsilon) \leq y_0 \leq \delta_0$ we obtain:

$$\Delta\delta_2^* = \left[\left(\frac{p_0}{p}\right)^{1/\gamma} - 1\right](\delta_0 - \delta_1) - \left(\frac{p_0}{p}\right)^{1/\gamma}\left(\frac{\gamma-1}{2\gamma}\right)\varepsilon \int_{\delta_1(\varepsilon)}^{\delta_0} \frac{g\,dy_0}{\bar{u}_0^2} . \tag{5.9}$$

It should be remembered that at $\varepsilon \to 0$, $\delta_1(\varepsilon) \to 0$, and the integral (5.9) diverges. Therefore, "matching" of the results has to be performed carefully accounting for the following relationships:

$$\varepsilon \sim g_w^3 , \quad \varepsilon << \delta_1(\varepsilon) << 1 \tag{5.10}$$

and at $\varepsilon \to 0$ the expression is obtained:

$$\Delta\delta^* = \lim_{\varepsilon \to 0}(\Delta\delta_2^* + \Delta\delta_3^*) = \varepsilon\left[\frac{1}{\gamma}\delta_0 - \left(\frac{\gamma-1}{\gamma}\right)\delta_0 \int_0^1 \left(\frac{g}{\bar{u}_0^2} - \frac{4g_w^3}{A^2\bar{B}^2\bar{y}_0^2}\right)d\bar{y}_0\right] - \sqrt{\frac{\gamma-1}{\gamma}}\,\frac{2\delta_0}{A\bar{B}}(\varepsilon g_w^3)^{1/2}, \quad y_0 = \delta_0\bar{y}_0 , \quad \bar{B} = \delta_0 B = O(1) . \tag{5.11}$$

The integrand in (5.11) is such that the integral at $y_0 \to 0$ converges since $\bar{u}_0'(0) = \bar{u}_0''(0) = 0$ due to (2.1). To understand this result it should be first of all noted that by (5.6) A is nondimensional and B has a dimension inverse of the length.

Secondly, it follows from (5.11) that $\Delta\delta^*/\delta_0 \sim \varepsilon \sim g_w^3$. The second term appears when integrating along a nonlinear always subsonic sublayer. Hence, for the expanding flows, that is for $\varepsilon > 0$, the second term is always negative. The first term sign depends on the form of the Mach number profile in the main part of the boundary layer, but its supersonic part dominates almost in all cases, that is, the first term in square brackets at $g_w \ll 1$ is positive. It should be noted that for the inviscid flow in region 3 the case $\varepsilon < 0$ is of no physical sense due to the Bernoulli equation.

Let us use the Ackeret formula (2.3) and (5.11) to obtain the equation describing the pressure distribution:

$$\frac{dP}{d\tau} = \frac{P^{1/2}(P^{1/2} + \tilde{\theta})}{P_*^{1/2} - P^{1/2}} \tag{5.12}$$

where the designations are introduced:

$$\tilde{\theta} = \gamma M_e \theta / g_w^3 \ , \quad \varepsilon = g_w^3 P \ , \quad \tau = x/(\gamma M_e \delta_0 T), \quad P_*^{1/2} = R/2T$$

$$T = (1/\gamma)\left[1 - \frac{\gamma-1}{2}\int_0^1 \left(\frac{g}{\bar{u}_0^2} - \frac{4 g_w^3}{A^2 \bar{B}^2 \bar{y}_0^2}\right) d\bar{y}_0\right] \ , \quad R = \sqrt{\frac{\gamma-1}{\gamma}}\,\frac{2}{A\bar{B}} \ . \tag{5.13}$$

For a sufficiently full profile of the nondisturbed boundary layer when the transition to the supercritical flow is possible, $T > 0$.

6. Analysing (5.12) a family of integral curves can be plotted for surfaces $\tilde{\theta}(\tau)$ at various critical pressure differences. It will be recalled that the transcritical flows are considered which are locally inviscid in region 3 and correspond to the curve AB in fig.2. Therefore, only $P > 0$ are of sense. The relation of the variable part of the displacement thickness $\Delta\delta^*$ (5.11) in variables (5.13) takes on the form:

$$\Delta\delta^* = \delta_0 g_w^3 (TP - RP^{1/2}), \quad T > 0 \ , \quad R > 0 \ . \tag{6.1}$$

It is easy to see that at very low expansions sublayer 3 dominates and $\Delta\delta^*$ decreases as the expansion is raised. This is subsonic or subcritical behaviour. At $P > P_*$, $\Delta\delta^*$ starts to increase as in onedimensional supersonic flow. The critical point similar to sonic point M=1 corresponds to:

$$\left(\frac{d\delta^*}{dP}\right) = 0 \quad , \quad P = P_* = (R/2T)^2 . \qquad (6.2)$$

The explicit dependence of P_* on g_w is cancelled by the transformation (5.13), while the dependence on the shape of the undisturbed boundary layer profile is obvious from (5.11) and (5.13).

To analyse qualitatively the possible solutions of the problem we present in fig.3 using (5.12) a scheme of integral curves for the areas of bodies $\tilde{\theta}$ =const at various relationships of $\tilde{\theta}$ and P_* .

If the expanding flow with an angular point, that is, $\tilde{\theta}(\zeta<0)=0$, $\tilde{\theta}(\zeta>0)=\tilde{\theta}_w<0$, is considered the solution is easily obtained in analytical form.

However, even before the interaction the essential flow properties are easily available from figs.3a, 3b, 3c. Indeed the pressure $\zeta>0$ behind the angular point P has to tend to $-\tilde{\theta}$. Two cases are possible: the first $P_*>-\tilde{\theta}$ is a turn by a subcritical angle. As fig.3b shows, at $\zeta>0$ there are no integral curves coming to the line $P=-\tilde{\theta}$. Thus, the whole turn to $P=-\tilde{\theta}$ is realized up to the angular point $\zeta=0$ in the flow region $\zeta<0$ where $\tilde{\theta}=0$ along one of the integral curves (fig.3a).

The second case of the supercritical turn is $P_*<-\tilde{\theta}$. Fig.3c shows that the integral curves issuing from the critical point come to the line $P=-\tilde{\theta}$ at $\zeta\to+\infty$. Since at $\zeta<0$ the flow is described by the integral curves (fig.3a), a flow with the "sonic" pont corresponding to $P(0)=P_*$ (that is, this point is angular) is only possible.

Thus, in the flow with a large turn angle above the critical value $|\tilde{\theta}|>P_*$, the whole subsonic part is realized ahead of the angular point and the whole supersonic part behind it. In particular, for the supercritical initial state when $R \ll T$, essentially the whole turn occurs downstream of the angular point.

The solution of some problems requires discontinuous solutions of the supercritical jump type proposed by L. Crocco [12].

It should be noted that these solutions are possible within the asymptotic theory. Remember, that for super- and transcritical flows we have $\Delta x \sim M_e \delta_0$. So, the variations on lengths shorter by the scale are discontinuities on basic integral curves. A detailed study of the flow on $\Delta x \sim \delta_0$ lengths with rapidly varying pressures is, evidently, a special problem that is beyond the scope of the paper. It may be only noted that on $\Delta x \sim \delta_0$ there should be no variations in the main term:

$$[\Delta \delta^*] = 0 \quad . \tag{6.3}$$

The solution for such regions can be written but for the purposes of the paper the flow structure is of no importance, just as the Becker solution for structure of short wave is not essential either for the conventional gas dynamics. Using (6.1) we write the Nugoniot condition analog following from (6.3):

$$P_2^{1/2} = 2P_*^{1/2} - P_1^{1/2} \tag{6.4}$$

where subscripts 1, 2 denote the function values upstream of a discontinuity and downstream of it respectively, when (6.3) is met. Fig.4 shows the relationship $\Delta\delta^*(P)$. A transfer through critical point O of the continuous expanding flow is possible from point A to B. Point A may be reached from B through a supercritical jump defined by (6.4).

Using continuous integral curves (fig.3) and the condition in the discontinuity (6.4) (fig.4) some fairly complex flows may be constructed with characteristic transcritical properties similar, to some degree, to those of transonic inviscid flows.

Consider the flow around a swept step (fig.5). Let initially be $|\tilde{\theta}| < P_*$. The flow area $\zeta > \zeta_B$ corresponds to fig.3a. It is seen that there are no integral curves where $P \to 0$. This requires $P(\zeta_B) = 0$. Near the line AB the flow is described by integral curves (fig.3b). All integral curves for $P < -\tilde{\theta}$ are defined to additive constant shift along axis ζ . As $P(\zeta_B)=0$ any one of these curves may be taken, and having passed along it the distance $\zeta_B-\zeta_A$ corresponding in similarity variables to AB length (fig.5), we obtain $P(\zeta_A)<P_*$. It remains to describe the disturbances decay on $\zeta<\zeta_A$ by integral curves (fig.3a). Thus, the solution is completely constructed and does not require discontinuities.

It is a more complicated case when dealing with $|\tilde{\theta}|>P_*$. As in the first case, $P(\zeta_B)=0$. The solution of AB is described by fig.3c. Denote by $\Delta\zeta_*$ the distance between points $P=P_*$ and $P=0$. Let initially the length AB in similarity variables be $\Delta\zeta_{AB}\leq\Delta\zeta_*$. In this case, by fig.3c, find $P(\zeta_A)$ laying off the length $\Delta\zeta_{AB}$ along an integral curve from point P =0. This value is, evidently, $P<P_*$. Then, using fig.3a, construct a part of the solution for $\zeta<\zeta_A$. Thus, for a small length the solution can be continuous and subcritical even at a supercritical angle $|\tilde{\theta}|>P_*$.

However, there is no such a possibility at $\Delta\zeta_{AB}>\Delta\zeta_*$. In this case the flow in which $P(\zeta_A)=P_*$ should be considered when the supercritical properties arise even on the first portion. Having passed the whole integral curve in fig.3a we move from point A upwards along the integral curve in fig.3c. To reach point B where P =0 a jump (6.4) is required. Since in this case we have $\Delta\zeta>\Delta\zeta_*$ a jump position can be chosen that allows us to reach such a point on the subcritical integral curve from which the pressure at point B vanishes throughout the remaining distance up to B. It is clear that this is always possible since when moving along the integral curve $P_*<P<-\tilde{\theta}$ (fig.3c) any distance can be passed. The curve reaches the line $P=-\tilde{\theta}$ only asymptotically at $\zeta\to\infty$. Thus, we have obtained an interesting result. If the turn angle is supercritical $-\tilde{\theta}>P_*$, the flow becomes supercritical only for a sufficient length of AB and it should be decelerated across the jump (6.4). The length of AB being in similarity variables lower than a critical value $\Delta\zeta_{AB}<\Delta\zeta_*$, the flow is smooth and subcritical throughout.

As the last example let us discuss the flow around a "profile" presented in fig.6. In this case we have $P(\tau_D)=0$ again. At $-\tilde{\theta} < P_*$ the construction of the solution is of no difficulties. Let us measure the pressure in fig.3b from point $P=0$ on the integral curve at a distance $\Delta\tau=\Delta\tau_{CD}$. Thus $P(\tau_C)$ is obtained. Then, in fig.3a we obtain $P(\tau_B)$ for a known $\Delta\tau_{BC}$. Using fig.3d at a known τ it is possible to find $P(\tau_A)$. Then by fig.3a we obtain the solution for the length $\tau < \tau_A$.

Now let be $-\tilde{\theta} > P_*$. As in the above case we have $P(\tau_D)=0$. The length of CD is represented by curves in fig.3c. Two cases are possible. $\Delta\tau_{CD} < \Delta\tau_*$ is more simple. We find the value $P(\tau_C) < P_*$ in fig.3c. The length of BC corresponds to fig.3a. When $\Delta\tau_{BC}$ is known we find $P(\tau_B) < P_*$. By fig.3c we define $P(\tau_A)$, and the pressure distribution upstream of the point A is found using fig.3a. Thus, if the lengths of the swept parts AB and CD in similarity variables remain shorter than a critical value $\Delta\tau_*$ the flow remains subcritical even at $-\tilde{\theta} > P_*$.

The flow in fig.6 is more complex if $-\tilde{\theta} > P_*$ and $\Delta\tau_{CD} > \Delta\tau_*$. $P(\tau_{CD})=0$ on CD. According to 3c there should be a jump (6.4) due to $\Delta\tau_{CD} > \Delta\tau_*$. Let us denote the coordinate of the jump (6.4) as τ_1 and the upstream pressure as P_1; the downstream pressure P_2 is defined from the relation (6.4) as $P_2(P_1)$. Fig.3c shows that there is a relationship $P_2(\tau_D-\tau_1)$ and $P_1(\tau_1-\tau_C)$ as well, since at point C we have necessarily $P=P_*$. Hence, to define three unknowns P_1, P_2, τ_1 we have three relations which are always solved at $\Delta\tau_{CD} > \Delta\tau_*$. The flow for BC at a known $\Delta\tau_{BC}$ may be calculated using integral curves in fig.3a, and then $P(B) < P_*$ is obtained. Now find $P(\tau_A)$ from fig.3d at a known length $\Delta\tau_{AB}$, and the disturbance decay upstream of A from fig.3a.

The flow is fully described, and it is evident again that at turn angles $-\tilde{\theta} > P_*$ the supercritical flow occurs only at sufficient lengths of acceleration zones and terminates decelerated by jumps of (6.4) type. The flow structure inside the jump region (6.4) at the length $\Delta x \sim \delta_0$ will be further studied in detail. It is of particular interest since a thin viscous sublayer formed at the bottom of a local inviscid flow in region c may separate locally in the jump area $\Delta x \sim \delta_0$ (6.4) which is also possible at points B (fig.5) and D (fig.6).

References

1. Prandtl L. Über Flüssigkeitsbewegung bei sehr kleiner Reibung. d.III. Intern.Math.Kongr., Heidelberg, 1904.
2. Prandtl L. Mekhanika vyaskikh zhidkostei/Red. Dürend V.F. Aerodinamika t.III, razdel Zh., Oborongiz, 1939, p.47-237.
3. Landau L.D., Lifshitz E.M. Mekhanika sploshnykh sred. Gostekhizdat, Moskva, 1944.
4. Goldstein S. Quart.J.Mech.Appl.Math., 1948, N 1, p.43-69.
5. Neiland V.Ja. Doklad na III Fcesoyuznom sjezde po teoreticheskoi i prikladnoi mekhanike. 25.01 + 1.02.1968. Sb. annotatsii dikladov sjezda, "Nauka", Moskva, 1968.
6. Neiland V.Ja. Izv. AN SSSR, MZhG, 1969, N 4.
7. Stewartson K., Williams P. Proc. Roy. Soc. A, 1969, 312.
8. Sychev V.V. Izv. AN SSSR, MZhG, 1972, N 3.
9. Ryzhov A.S. DAN SSSR, t.234, N 4, 1977.
10. Schneider W. J.Fluid Mech., 1974, vol.63, p.463.
11. Neiland V.Ja. Asymtoticheskaya teoriya otryva i vzaimodeistviya pogranichnogo sloya so sverkhzvukovym potokom gaza. Uspekhi mekhaniki, t.4, vyp.2, 1981, p.3-62.
12. Crocco L. Proc. Conf. on High-Speed Aeron. Brooklin,1955.
13. Neiland V.Ja. Izv. AN SSSR, MZhG, 1973, N 6.
14. Brown S., Stewartson K. Ann. Rev. of Fluid Mech., 1961, N 1.
15. Neiland V.Ja. Uchyonye zapiski TsAGI, 1974, t.V, N 2, p.70-79.
16. Heiss U.D., Probstein P.F. Teoriya giperzvukovykh techenii. M., Izd. inostr. lit. 1962.
17. Van Dyke M. Metody vozmushchenii v mekhanike zhidkosti. M., "Mir", 1967.

18. Neiland V.Ja. Izv. AN SSSR, MZhG, 1970, N 4.

19. Neiland V.Ja. Izv. AN SSSR, MZhG, 1967, N 4.

20. Olsson G.R., Messiter A.F. AIAA J., 1969, vol.7, N 7.

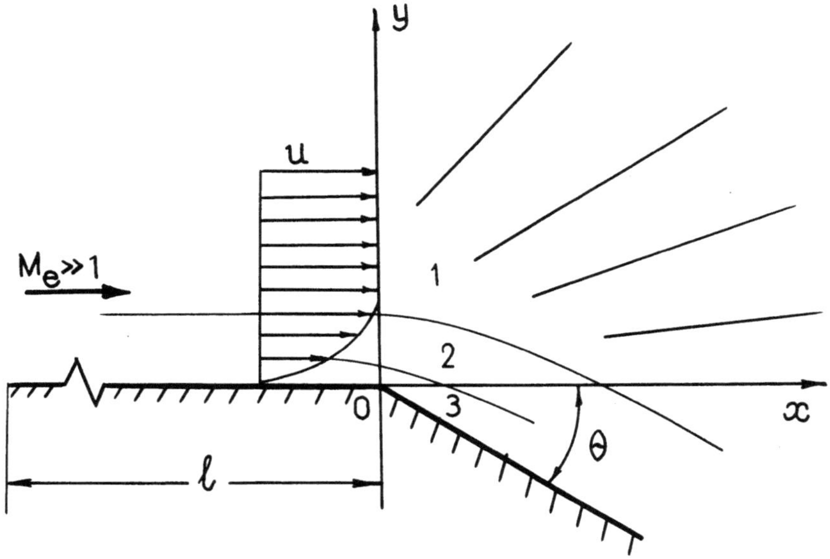
y
u
$M_e \gg 1$
1
2
3
0
x
Θ
ℓ

Fig. 1

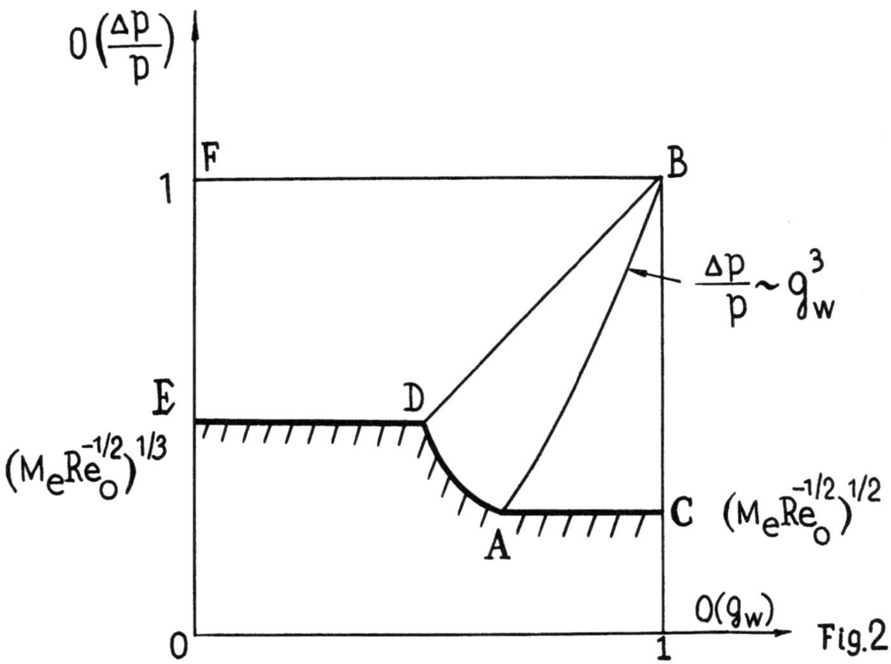
$O\left(\frac{\Delta p}{p}\right)$
F
B
1
$\frac{\Delta p}{p} \sim g_w^3$
E
D
$(M_e Re_o^{-1/2})^{1/3}$
C
$(M_e Re_o^{-1/2})^{1/2}$
A
$O(g_w)$
0
1

Fig.2

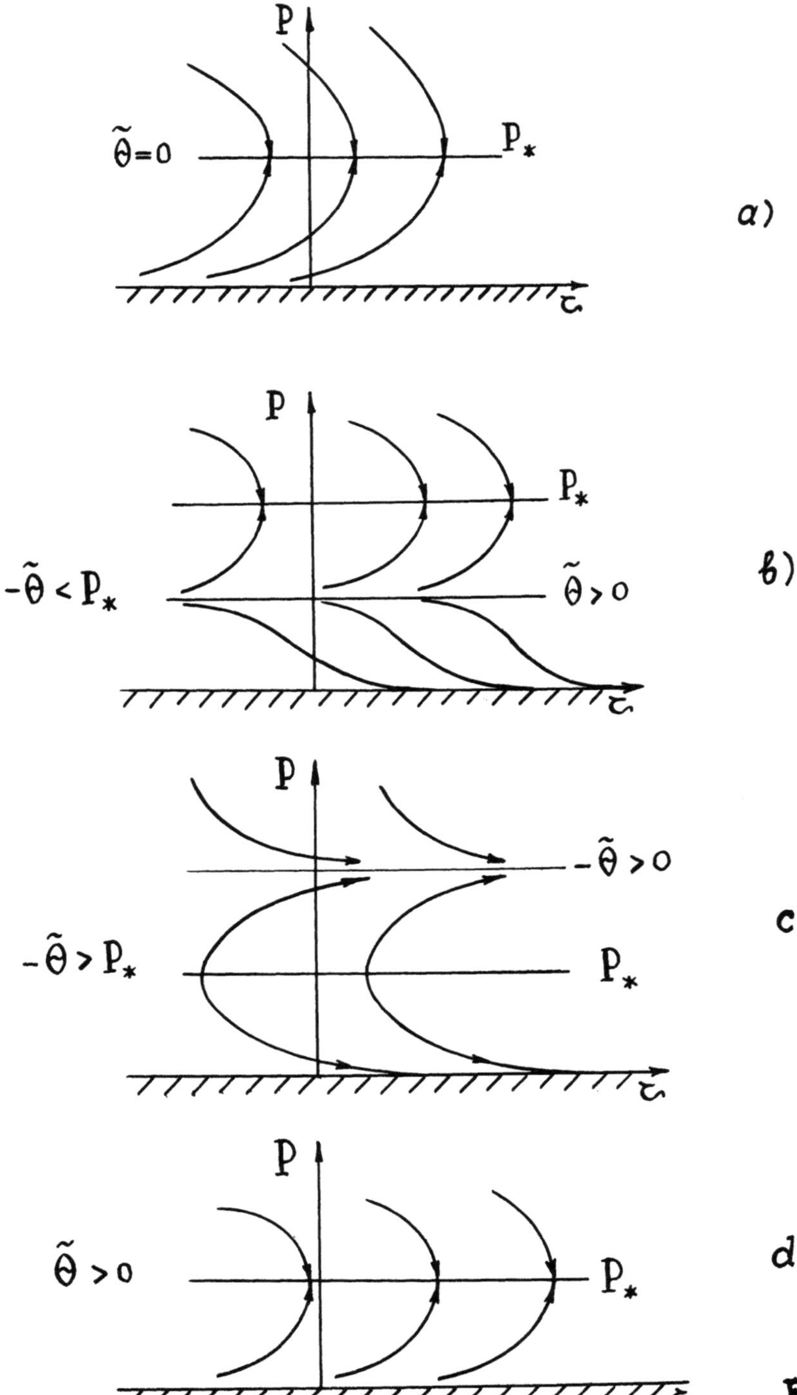
P
$\tilde{\Theta}=0$
P_*
a)
P
P_*
$-\tilde{\Theta} < P_*$
$\tilde{\Theta} > 0$
b)
P
$-\tilde{\Theta} > 0$
$-\tilde{\Theta} > P_*$
P_*
c)
P
$\tilde{\Theta} > 0$
P_*
d)
τ

Fig. 3

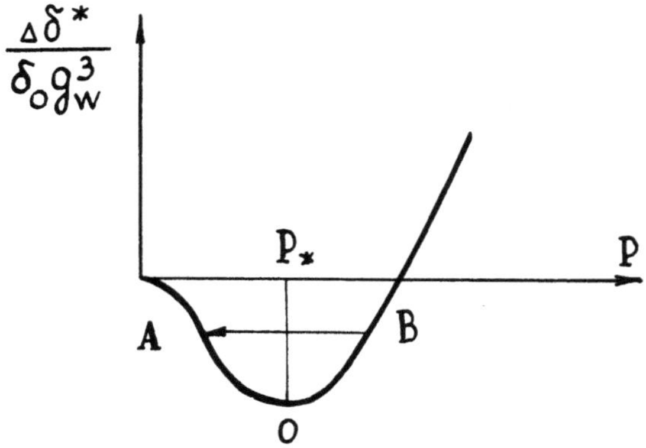

Fig. 4

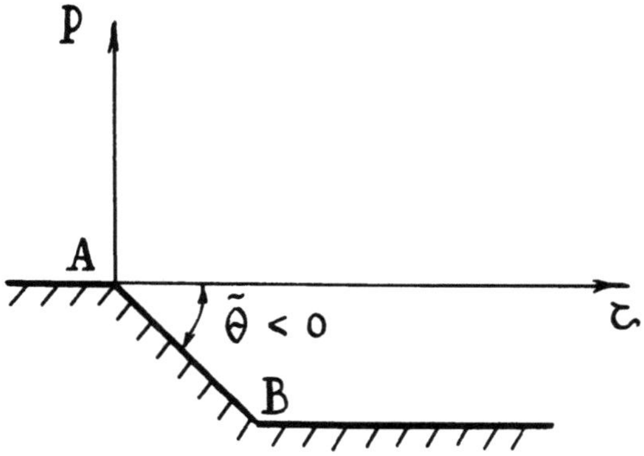

Fig. 5

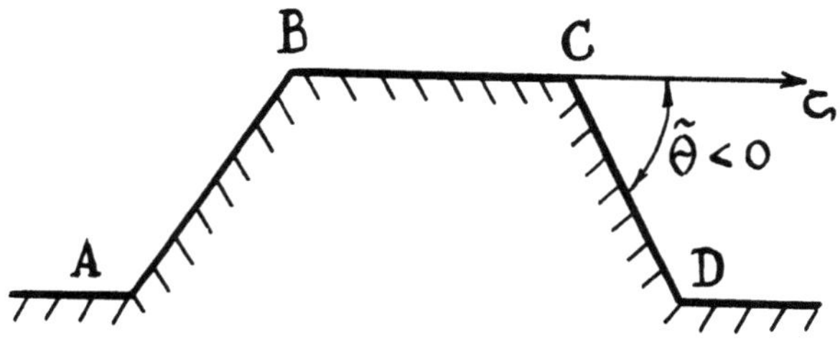

Fig. 6

Session 5

Unsteady Triple-deck Flows Leading to Instabilities

P.W. DUCK

Department of Mathematics,
University of Manchester,
Manchester, England.

Summary

A number of two-dimensional supersonic and incompressible non-linear unsteady triple-deck flows are investigated pertaining to Blasius type flows over small unsteady wall distortions. The method of solution is that of the pseudo-spectral method of Burggraf and Duck [16], in which the solution is (partly) determined in Fourier space.

In a number of the computations it is found that an apparent breakdown of the governing equations occurs, which seems to be related to a large wavenumber (inviscid) instability (and also to a possible catastrophic boundary layer separation). In the incompressible case, Tollmien-Schlichting waves are clearly seen to be present, and our results suggest that non-linearity may well enhance their amplitude growth.

We also present a set of (linearised) results for incompressible flow over an unsteady three-dimensional wall distortion.

1. Introduction

It is now known (Smith [1]) that the (subsonic) triple-deck of Stewartson [2] and Messiter [3] can provide a rational description of the lower branch of the neutral stability curve of the Blasius (and similar) boundary layers (including non-parallel effects). Indeed, many of the elements of triple-deck theory are to be found in the classical hydrodynamic stability work of Lin [4].

Further evidence of this link is seen in the work of a number of Russian authors (see for example Bogdanova and Ryzhov [5], Terent'ev [6]). Very recent results suggest that triple-deck theory can go even further in describing not just the formation, but also the evolution of Tollmien-Schlichting waves, and provide the basis for a rational description of the early stages of some transition processes. This has been illustrated by Smith and Burggraf [7] and Smith [8], where high frequency asymptotic analysis was exploited to enable analytic progress to be made within the framework of the triple-deck. Incidentally the corresponding supersonic triple-deck does not admit

F. T. Smith, S. N. Brown, (Eds.)
Boundary-Layer Separation
IUTAM Symposium London 1986

Tollmien-Schlichting instabilities (at least when subject to small disturbances). Duck [9] and Smith [8] have integrated the full unsteady triple-deck equations, and this paper is primarily concerned with this aspect.

In addition to Tollmien-Schlichting instabilities, other forms of instability have been observed within the triple-deck formulation. In [9] it was found that for sufficiently large boundary layer disturbances, the governing equations appear to fail; a spectral analysis indicates that this is accompanied by a rapid growth in the solution at large wavenumbers. The work of Smith and Bodonyi [10] and of Tutty and Cowley [11] (who also carried out a study of a form of the full boundary layer equations), indicates that this feature is as a result of an inviscid, large wavenumber, Rayleigh type instability, linked to inflexional type profiles. At the same time there exists a possible link between this form of instability and a catastrophic unsteady separation. Further examples of this type of phenomenon have been presented by Duck [12] (for internal pulsatile flows) and Duck [13] (for external pulsatile flows).

This paper is concerned with, in particular, the solution of the non-linear unsteady triple-deck equations, pertaining to viscous flows over small disturbances of the wall boundary (lying within the scale of the lower deck). We assume throughout this paper that unsteadiness is introduced into the problem by means of a normal motion of the wall distortion, and as such may be considered to be a crude model of the famous vibrating ribbon experiments of Schubauer and Skramstad [14].

2. Two-dimensional formulation and solution technique

We suppose that we have a laminar flow over a flat plate on which is situated a small distortion, at a distance L from the leading edge of the plate. The flow far from the plate has kinematic viscosity ν, density ρ, and velocity components $u/U_\infty = 1$, $v/U_\infty = 0$ in the xL and yL directions respectively, where U_∞ is constant; the pressure is written as $\rho U^2_\infty p$. We are now able to define a Reynolds number Re, and a related dimensionless parameter ϵ by

$$Re = \epsilon^{-8} = U_\infty L/\nu \quad . \tag{1}$$

Throughout we assume $Re \gg 1$ (and hence $\epsilon \ll 1$). If the characteristic timescale of the unsteadiness of the problem is Ω^{-1}, then a further dimensionless parameter may be defined namely

$$\beta_0 = (U_\infty/\Omega L)^{1/2}. \tag{2}$$

We require that β_0 corresponds to the so called triple-deck frequency, i.e. the acceleration term appears explicitly in the lower deck equations, and this then requires that

$$\beta_0 = \epsilon\beta \quad , \quad \beta = O(1) \tag{3}$$

(other orders of β_0 have been considered in some detail by Duck [15]).

As noted in the previous section, we choose the wall distortion to lie entirely within the "lower deck" of the triple-deck, and so the scale of the distortion is $O(\epsilon^3 L)$ in the streamwise direction by $O(\epsilon^5 L)$ in the transverse direction.

The resulting problem in the lower deck is as follows (having carried out the appropriate scalings, and invoked the Prandtl transposition theorem, details of which may be found in [9] and [15])

$$\frac{1}{\beta^2} U_t + UU_X + VU_Y = U_{YY} - P_X, \tag{4a}$$

$$U_X + V_Y = 0 \ , \tag{4b}$$

$$U = V = 0 \quad \text{on} \quad Y = 0, \tag{4c}$$

$$U \longrightarrow Y + hF\,(X,t) + A(X,t) \quad \text{as} \quad Y \rightarrow \infty \ , \tag{4d}$$

$$P = \mathcal{L}\{A\} \ , \tag{4e}$$

$$\text{and} \qquad U \longrightarrow Y \quad \text{as} \quad X \longrightarrow -\infty. \tag{4f}$$

Here $X = (x-1)/\epsilon^3$, $Y = y/\epsilon^5 - hF$, $U = u/\epsilon$, $V = v/\epsilon^3 - UhF_X - hF_t/\beta^2$, $P = p/\epsilon^2$. The function $F(X,t)$ in (4d) denotes the wall perturbation shape (generally $F(X,t) = O(1)$), whilst h is the scale of the distortion (which again is generally taken to be $O(1)$. Equation (4e) provides a link between the pressure $P(X,t)$ and the displacement function $A(X,t)$ where $\mathcal{L}\{A\}$ denotes some (linear) operator acting on A; the precise details of $\mathcal{L}$ are generally determined by considering the "upper deck" equation. Particular examples are

$$P = -A_X \quad \text{for supersonic flows,} \tag{5a}$$

$$P = \frac{1}{\pi} \int_{-\infty}^{\infty} \frac{A_\xi}{X-\xi}\, d\xi \quad \text{for incompressible flows.} \tag{5b}$$

We now turn to the general solution procedure to be followed, which is based on that of Burggraf and Duck [16] and involves computing the flow (partly) in Fourier space. Briefly (a fuller account is given in Burggraf and Duck [16] and Duck [9]) the technique is as follows. Equation (4a) is differentiated with respect to Y to eliminate the pressure term, and the solution variables are split into two components, namely that corresponding to the uniform shear (U=Y) plus a perturbation component (denoted by a tilde). We also transform from Y to η, where $Y = f(\eta)$. This is for the purposes of numerical expedience, and for this investigation we took

$$f(\eta) = \frac{\eta}{1-\eta} , \tag{6}$$

where $0 \leqslant Y < \infty$ as $0 \leqslant \eta < 1$; note that a uniform distribution of points in η space gives a non-uniform distribution of points in Y space, with the generally advantageous property that points are concentrated close to $Y = 0$. In addition the infinite (Y) space is mapped on to a bounded (η) space.

The following system then results:

$$\frac{1}{\beta^2} \tilde{\tau}_t + f(\eta)\, \tilde{\tau}_X - \frac{1}{(f'(\eta))^2} \tilde{\tau}_{\eta\eta} + \frac{f''(\eta)}{(f'(\eta))^3} \tilde{\tau}_\eta = - [\tilde{U}\, \tilde{\tau}_X + \frac{\tilde{V}}{f'(\eta)} \tilde{\tau}_\eta]$$

$$= \tilde{R} \tag{7}$$

where $\tilde{\tau} = \tilde{U}_Y = \frac{\tilde{U}_\eta}{f'(\eta)}$ is the (perturbation) shear.

The next step is to take Fourier Transforms in X, and this yields

$$\frac{1}{\beta^2} \tilde{\tau}^*_t + ik\, f(\eta)\, \tilde{\tau}^* - \frac{1}{[f'(\eta)]^2} \tilde{\tau}^*{}_{\eta\eta} + \frac{f''(\eta)}{[f'(\eta)]^3} \tilde{\tau}^*_\eta = \tilde{R}^*, \tag{8}$$

where an asterisk denotes a transformed variable, e.g.

$$\tilde{\tau}^*(k,\ \eta,\ t) = \int_{-\infty}^{\infty} \tilde{\tau}\ (X,\eta,t)\ e^{-ikX}\, dX\ . \tag{9}$$

Two conditions must be applied to the system (8), namely that

$$\tilde{\tau}^* \longrightarrow 0 \quad \text{as} \quad \eta \longrightarrow 1, \tag{10}$$

together with

$$\left.\frac{\tilde{\tau}^*{}_\eta}{f'(\eta)}\right|_{\eta\to 0} = ik\, \Lambda \left\{\left[\int_0^\eta f'(\eta)\, \tau^*\, d\eta\right]_{\eta\to 1} - hF^*\right\}, \tag{11}$$

where the transform of the pressure-displacement relation (4e) is written as

$$P^* = \Lambda\ (k)\ A^*. \tag{12}$$

Equation (11) arises from a combination of equations (4d), (12), and (4a) evaluated at $Y = 0$, and serves as a solubility condition.

The system is now in a form suitable for numerical treatment. Equation (8) (and its associated conditions) will be considered at a finite number of k points ($-k_{min} \leqslant k \leqslant 0$ in steps of Δk), noting that we are able to use the following property since $\tau(X,\eta,t)$ must be a real function:

$$\tilde{\tau}^*\ (k,\eta,t) = \text{complex conjugate}\ \left\{\tilde{\tau}^*\ (-k,\eta,t)\right\}\ . \tag{13}$$

Crank-Nicholson differencing is applied in the timewise direction (stepsize Δt), whilst standard central differencing is applied on η derivatives ($0 \leqslant \eta \leqslant \eta_\infty < 1$ in steps of $\Delta\eta$, where $1 - \eta_\infty << 1$). Quadratures (including those to evaluate $\tilde{U}$ and $\tilde{V}$) are evaluated using the trapezoidal rule.

The quantity $\tilde{R}^*$ in (8) is not evaluated directly (by means of a convolution) but instead the following procedure is followed. The four components $\tilde{U}$, $\tilde{\tau}_X$, $\tilde{V}$ and $\tilde{\tau}_\eta$ are evaluated from their analogous Fourier transformed quantities, by means of the Fast Fourier Transform (F.F.T.) of Cooley and Tukey [17]; the non-linear quantity $\tilde{R}$ in (7) is then evaluated in physical space, and then $\tilde{R}^*$ is obtained, again using the F.F.T. This particular, somewhat roundabout, approach can be shown to be considerably faster than the more direct convolution approach.

For the first iteration at each time step, $\tilde{R}^*$ was guessed at all points. At each k,t station we consider the discretized form of (8) at $\Delta\eta \leqslant \eta \leqslant \eta_\infty - \Delta\eta$ in steps of $\Delta\eta$ and there then results (retaining the $\tilde{R}^*$ on the "right-hand-side" of the system) a linear, almost tridiagonal set of algebraic equations (the discretized form of (11) results in a full bottom row), which is amenable to standard Gaussian-elimination procedures. The solution is obtained at all k, η points, and the $\tilde{R}^*$ is re-evaluated and the procedure repeated. The procedure is continued until the maximum change in any of the calculated $|P^*\ (k,t)|$ is less than (typically) 10^{-6}, at which point the computation is advanced in time by an amount Δt.

Certain so called linearised solutions are also calculated. These correspond to diminishingly small heights of wall distortion, i.e.

$h \longrightarrow 0$, in which limit we may properly neglect the non-linear terms in (7), i.e. $\tilde{R}$. Consequently we may set $\tilde{R}^* \equiv 0$ for all k, η, t in (8), and the solution may be advanced forward in time without the need for any form of iteration.

This so called "pseudo-spectral" scheme has a number of distinct advantages over more conventional, purely finite-difference methods. In particular, reversed flows need no special treatment; diverse pressure-displacement relationships may all be handled in a similar fashion; and as noted above, the linearised computations require no iteration.

For the ensuing results, we took the following numerical parameters: $\Delta t = .005$, $\Delta\eta = .0396$, $\eta_\infty = .95$, $k_{min} = 14.055$, $\Delta k = .11$. Rarely were more than four iterations required at each timestep. Results are presented in the following section.

3. Two-dimensional results

A number of results of this general class have appeared, already, in the literature [9], where the emphasis was on wall distributions with maintained oscillations. In this section, however, we pay particular attention to wall perturbations of the form

$$F(X,t) = \begin{cases} 0 & t < 0, \\ \frac{1}{1+X^2}\tanh^2 t & t > 0, \end{cases} \tag{14}$$

giving

$$F^*(k,t) = \begin{cases} 0 & t < 0, \\ \pi e^{-|k|}\tanh^2 t & t > 0. \end{cases} \tag{15}$$

All the results presented here are for $\beta = 1$ (other frequency parameters were considered in [9]).

The first set of results (shown in figure 1) relate to the supersonic (non-linear) case with $h = -5$; This corresponds to a cavity type of wall distortion. The (perturbation) wall shear $\tilde{\tau}_w$ is seen to grow in magnitude (notice that at the two later times shown, reversed flow is present, where $\tilde{\tau}_w < -1$), and seems to be approaching the corresponding steady state results (see [16]).

The second set of results are presented in figure 2, and correspond to $h = -10$ (other details remain unchanged from the first example), a rather more severe distortion. The wall shear is seen to grow in magnitude but after $t \approx 1.3$, just downstream (relative to the freestream flow) of $X = 0$, this growth increases rapidly. In fact after $t \approx 1.5$ this particular computation became unreliable, due to the slow decay of the (spectral) solution variables as $|k| \longrightarrow \infty$, suggesting the likely occurence of some form of singularity. This, of course, is confirmed in the physical solution. This type of phenomenon has been observed previously (see Section 1), and seems to be related to a large wavenumber, inviscid (Rayleigh type) instability. At the same time it seems that this breakdown of solution may be related to perhaps a catastrophic boundary layer separation, involving a burst of vorticity into the main body of the fluid. Indeed, finite time singularities have been observed in quite a number of unsteady boundary layer calculations, in a range of diverse situations previously (see van Dommelen and Shen [18] for an example in a "non-interacting" boundary layer computation).

We next consider a number of incompressible flow examples. The first of these is the linearised example (corresponding to $h \longrightarrow 0$), and figure 3 shows the (normalised) perturbation wall shear behaviour at two times. These distributions clearly show the development of Tollmien-Schlichting waves downstream, with increasing amplitude; although $\beta = 1$ is below the critical value for sustained wall oscillations (for which $\beta_c = 0.660$ - see [9]), the startup process is responsible for triggering all time modes. Close to the peak of the distortion ($X = 0$), and also upstream, by $t = 3.5$ we see that a steady state is practically achieved whilst further downstream the Tollmien-Schlichting waves are being propagated downstream as a wave packet, although as time progresses the steady state regime moves progressively further downstream.

The next example we consider here is the non-linear incompressible computation for $h = -5$. The important feature to be observed in these results is the effect of non-linearity on the solution, in particular on the Tollmien-Schlichting waves. Perturbation wall shear distortions at these times are shown in figures 4 and 5. If we follow the wave "trough" centered around $X \approx 6.5$ at $t = 2$, we see that its amplitude is enhanced by non-linearity (compare figure 4 with figure 3 at $t = 2.5$). Figure 5 shows the perturbation wall shear distribution at $t = 2.75$. However, soon after this time, the computation became unreliable due to a similar lack of convergence of the spectral solution as $|k| \to \infty$, as that observed in the $h = -10$ supersonic example described perviously. Figure 6 shows the distribution of Real $(\tilde{\tau}_w^*)$ (which is typical of all spectral variables) with k at $t = 2.75$, and indicates the slow decay of the spectral solution as $|k|$ increases. The large amplitude oscillations at smaller values of $|k|$ correspond to the primary Tollmien-Schlichting instability. Taken as a whole, these results suggest that in this example the non-linearity is causing the Tollmien-Schlichting waves themselves to be subject to a secondary inviscid type of instability.

The final example we present in this section is with $h = -10$ (other details remain unchanged from the previous example); perturbation wall shear distributions are shown in figure 7. Again a rapid growth is observed in $|\tilde{\tau}_w|$, this time close to the bottom of the hump, after $t \doteq 1.0$, and again the computation ultimately became unreliable due to the poor convergence of the solution as $|k| \to \infty$. However the situation is somewhat different fromthat of the previous example $(h = -5)$. In this more severe case the rapid growth in $|\tilde{\tau}_w|$ does not involve the Tollmien-Schlichting waves directly which do not appear to have been properly established by $t \doteq 1$.

4. Three-dimensional formulation and solution technique

We now consider the interaction between a two-dimensional oncoming boundary layer and a three-dimensional wall distortion. Scales are unchanged for Section 2 although we must introduce a new cross-flow scale $Z = z/\epsilon^3$, where zL is the dimensional cross-flow coordinate and $w\, U_\infty$ is the dimensional cross flow component of velocity. The shape function F is now a function of X, Z and t.

Having followed the same procedures as in Section 2, we obtain the following problem (with $\tilde{W} = w/\epsilon$)

$$\frac{1}{\beta^2}\,\tilde{\tau}_t + Y\,\tilde{\tau}_X - \tilde{W}_Z - \tilde{\tau}_{YY}$$

$$= -(\tilde{U}\,\tilde{\tau}_X + \tilde{V}\,\tilde{\tau}_Y + \tilde{W}\,\tilde{\tau}_Z - \tilde{\tau}\tilde{W}_Z + \tilde{W}_Y\tilde{U}_Z) = \tilde{R}_1, \tag{16}$$

together with

$$\frac{1}{\beta^2}\,\tilde{W}_t + Y\,\tilde{W}_X - \tilde{W}_{YY} + P_Z = -\,(\tilde{U}\,\tilde{W}_X + \tilde{V}\,\tilde{W}_Y + \tilde{W}\,\tilde{W}_Z) = \tilde{R}_2 \tag{17}$$

and $\tilde{U}_X + \tilde{V}_Y + \tilde{W}_Z = 0.$ (18)

The boundary conditions are

$$\tilde{W} \longrightarrow 0, \ \tilde{U} \longrightarrow A(X, Z, t) + hF(X, Z, t) \ \text{as} \ Y \longrightarrow \infty,$$

and $\tilde{U} = \tilde{V} = \tilde{W} = 0$ on $Y = 0$. (19)

We restrict attention to just incompressible flows, and so the pressure - displacement relationship is

$$P(X, Z, t) = -\frac{1}{2\pi} \int_{-\infty}^{\infty} \int_{-\infty}^{\infty} \frac{A_{\xi\xi} \, d\xi \, d\zeta}{[(X-\xi)^2 + (Z-\zeta)^2]^{\frac{1}{2}}} . \qquad (20)$$

A full numerical study of the steady problem (equivalent to $\beta \longrightarrow \infty$) has been undertaken by Duck and Burggraf [19] and by Smith [20], whilst the limit of $\beta \longrightarrow 0$ has been considered by Duck [21]. Our technique now follows that of [19]; we take Fourier Transforms in the X and Z directions, where (for example)

$$\tilde{\tau}^{**} (k, \ell, Y, t)) = \int_{-\infty}^{\infty} \int_{-\infty}^{\infty} \tilde{\tau} (X, Z, Y, t) \, e^{-ikX-i\ell Z} dXdZ \quad . \qquad (21)$$

We now make the following change of variables, from $\tilde{\tau}^{**}$ to σ^{**}, where

$$\sigma^{**} = \tilde{\tau}^{**} + \frac{\ell}{k} \tilde{W}_Y^{**} \qquad (22)$$

(this was suggested in [19]); we also use the transformation (6) and then (16)-(17) become

$$\frac{1}{\beta^2} \sigma_t^{**} + ikf(\eta) \sigma^{**} - \frac{1}{[f'(\eta)]^2} \sigma^{**}_{\eta\eta}$$

$$+ \frac{f''(\eta)}{[f'(\eta)]^3} \sigma_\eta^{**} = \tilde{R}_1^{**} + \frac{\ell}{kf'(\eta)} \tilde{R}^{**}_{2\eta} , \qquad (23)$$

$$\frac{1}{\beta^2} \tilde{W}_t^{**} + ik f(\eta) \tilde{W}^{**} - \frac{1}{[f'(\eta)]^2} \tilde{W}_{\eta\eta}^{**}$$

$$+ \frac{f''(\eta)}{[f'(\eta)]^3} \tilde{W}_\eta^{*} + i\ell P^{**} = \tilde{R}_2^{**}. \qquad (24)$$

The conditions to be applied are that

$W^{**} = 0$ on $\eta = 0$ together with W^{**}, $\sigma^{**} \longrightarrow 0$ as $\eta \longrightarrow 1$, and (25)

$$\left.\frac{\sigma_\eta^{**}}{f'(\eta)}\right|_{\eta=0} = ik\,(k^2 + \ell^2)^{\frac{1}{2}} \left\{\left[\int_0^\eta f'(\eta)\,\sigma^{**}\,d\eta\right]_{\eta\to 1} - h\,F^{**}\right\} . \qquad (26)$$

Note that the problem for σ^{**} is very similar to that for $\tilde{\tau}^*$ in Section 2; the motivation in using (22) is to effectively enable the determination of σ^{**} independently of $\tilde{W}^{**}$ (although terms involving $\tilde{W}^{**}$ are of course included in $\tilde{R}_1^{**}$ and $\tilde{R}_{2\eta}^{**}$). Consequently the procedure used in Section 2 to determine $\tilde{\tau}^*$ may be used to determine σ^{**}. Here the solution is determined at a finite number of k <u>and</u> ℓ points (if the distortion is symmetrical with respect to $Z = 0$, then we need to consider just half the range of ℓ, and we take $-\ell_{min} < \ell < 0$ in steps of $\Delta\ell$). P^{**} is simply obtained by means of

$$P^{**} = \frac{-\,ik}{(k^2+\ell^2)} \left.\frac{\sigma_\eta^{**}}{f'(\eta)}\right|_{\eta=0} , \qquad (27)$$

and then $\tilde{W}^{**}$ may be obtained from (24). Linearised $(h \longrightarrow 0)$ solutions require no iteration and are obtained by setting

$$\tilde{R}_1^{**} \equiv \tilde{R}_2^{**} \equiv 0 . \qquad (28)$$

The three-dimensional results to be presented here are of this linearised form.

5. Three-dimensional results

Here we present a set of linearised three-dimensional results, using the method described in the previous section. The particular example we take is

$$F(X,\ Z,\ t) = \begin{cases} 0 & t < 0, \\ \dfrac{\tanh^2 t}{(1+X^2)(1+Z^2)} & t > 0. \end{cases} \qquad (29)$$

We took the following numerical parameters: $\Delta t = 0.01$, $\Delta\eta = 0.095$, $\eta_\infty = 0.95$, $k_{min} = 7.554$, $\Delta k = 0.118$, $\ell_{min} = 5.034$, $\Delta\ell = 0.0787$.

Figure 8 shows the perturbation wall shear along $Z = 0$ (normalised with respect to h) at two times. These distributions show clearly the development of the Tollmien-Schlichting wave packet observed in our two-dimensional results. Indeed, these distributions may be compared directly with the two-dimensional results shown in figure 3. There is quite a striking similarity between the two sets of results, although the two-dimensional Tollmien-Schlichting waves do generally seem to be of somewhat larger amplitude. Figure 9 shows the corresponding distributions of the normalised cross-flow component of wall shear (at $Z = -1.2468$). All these distributions indicate that just as in the two-dimensional case, a steady state regime has practically been achieved upstream and close to the distortion, whilst an increasing amplitude wave packet is being propagated downstream.

Figures 10-12 show normalised perturbation wall shear contours at times 2.5, 3.5, 5 respectively. These figures confirm the above observations, whilst at the same time it seems that as $|Z| \longrightarrow \infty$, the waves, whilst reducing is magnitude, become quasi-two dimensional in form. Notice, in particular the patterns of the Tollmien-Schlichting waves as they migrate downstream.

6. Conclusions

This work emphasises the importance of the triple-deck model in boundary layer stability and separation, and indicates that both these areas are strongly inter-related. Much work still remains to be done: the apparent singularity observed in a number of the non-linear results certainly requires definite identification and classification. Indeed, there is also some evidence (not all of which is presented here) that for the class of incompressible flows studied here (i.e. those involving an impulsively started wall motion) that non-linearity will always (ultimately) cause a breakdown in the solution however small $|h|$, due to the enhanced growth of the Tollmien-Schlichting waves, the position of breakdown moving downstream (and occuring at later times) as h reduces in magnitude. Another point worth noting is that in a recent study Duck [13] considered an example of an unsteady oncoming flow. (Blasius together with a Stokes flow) encountering a steady distortion, and found that some linearised ($h \longrightarrow 0$) solutions developed an apparent singularity for large enough unsteady component. Additionally, much remains to be done in the study of three-dimensional disturbances. Work is underway to investigate all these (and other) areas.

References

1. Smith, F.T. Proc. Roy. Soc. Lond. A368 (1979) 573, see also A371 (1980), 439.
2. Stewartson, K. Mathematika 16 (1969) 106.
3. Messiter, A.F. SIAM J. Appl. Math. 18 (1970), 241.
4. Lin, C.C. Theory of Hydrodynamic Stability, (1955) Cambridge University Press.
5. Bogdanova, E.V. and Ryzhov, O.S. Q.J. Mech. Appl. Math. 36 (1983), 271.
6. Terent'ev E.D. IUTAM symposium on Laminar-Turbulent Transition 1984, published by Springer Verlag, Berlin (1985).
7. Smith, F.T. and Burggraf, O.R. Proc. Roy. Soc. Lond. A. 399, (1985), 25.
8. Smith F.T. United Technologies Research Center Report 85-36 (1985).
9. Duck, P.W. J. Fluid Mech. 160 (1985), 465.
10. Smith F.T. and Bodonyi, R.J. Aeronautical J. 36 (1985), 205.
11. Tutty, O.R. and Cowley S.J. J. Fluid Mech (1986) to appear.
12. Duck, P.W. Q. Jl. Mech appl. maths 38 (1985), 621.
13. Duck, P.W. Submitted to J. Fluid Mech. (1986).
14. Schubauer, G.B. and Skramstrad, H.K. NACA Rep. 909 (1948).
15. Duck, P.W. Mathematika 25 (1978) 24.
16. Burggraf, O.R. and Duck, P.W. In Proc. Symp. Phys. Num. Aspects in Aerodyn. Flows. (1981) Springer.
17. Cooley, J.W. and Tukey, J.W. Maths. Comp. 19 (1965) 297.
18. van Dommelen, L.L. and Shen, S.F. J. Comput. Phys. 38 (1980), 325.
19. Duck, P.W. and Burggraf, O.R. J. Fluid. Mech 162 (1986), 1.
20. Smith, F.T. United Technologies Research Center Report 83-46 (1983)
21. Duck P.W. ZAMP 32 (1980) 24.

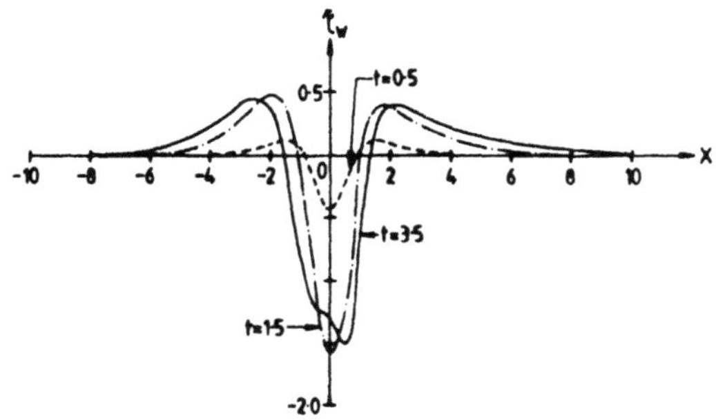

Fig. 1. Surface stress for 2D supersonic case h = -5

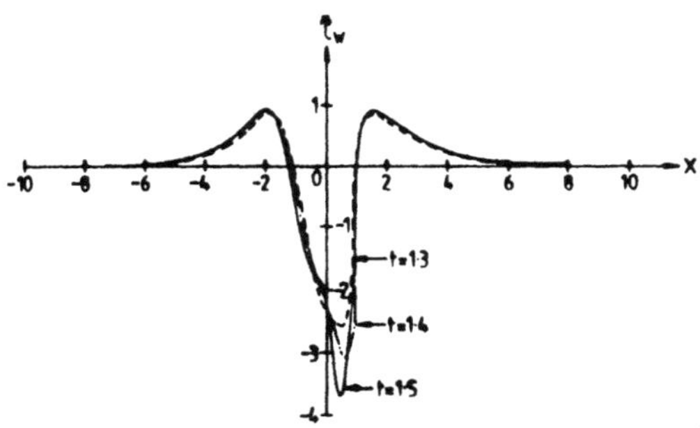

Fig. 2. Surface stress for 2D supersonic case h = -10

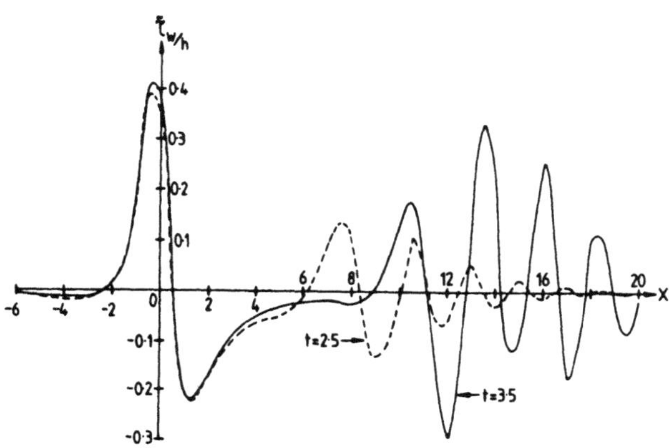

Fig. 3. Normalised surface stress for 2D incompressible case

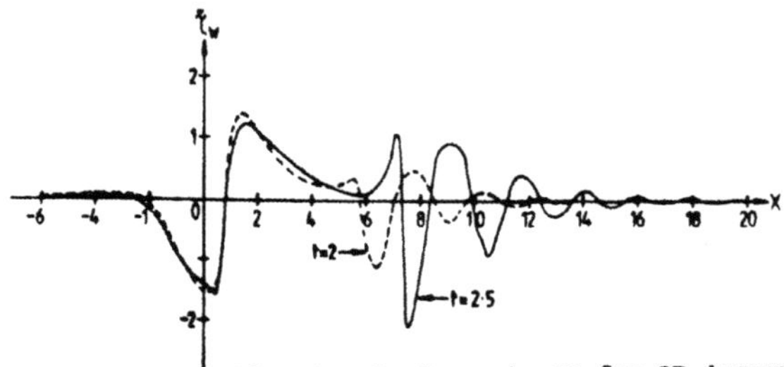

Fig. 4. Surface stress for 2D incompressible case h = -5

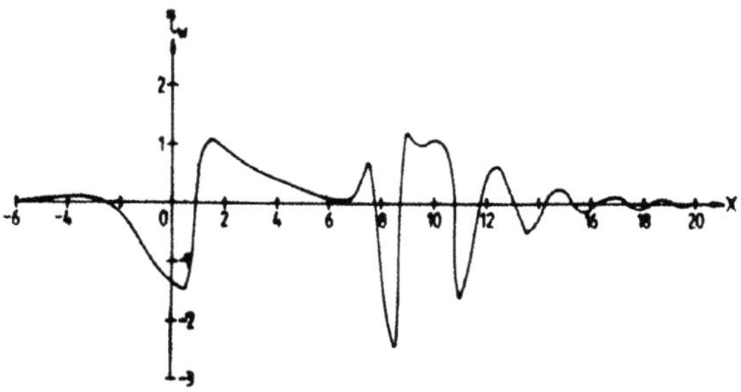

Fig. 5. Surface stress for 2D incompressible case h = -5, t = 2.75

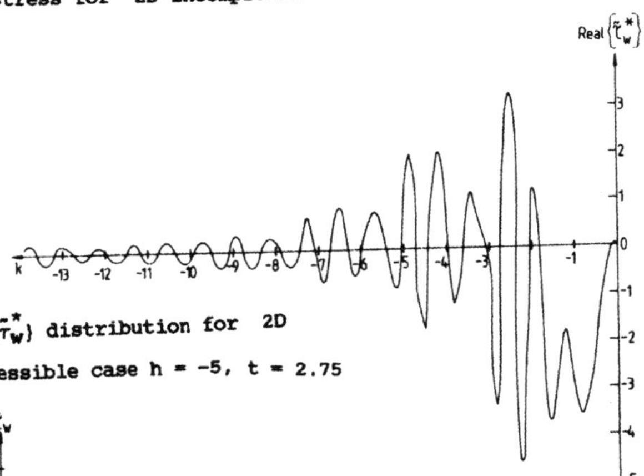

Fig. 6. Real $\{\tilde{\tau}_w^*\}$ distribution for 2D incompressible case h = -5, t = 2.75

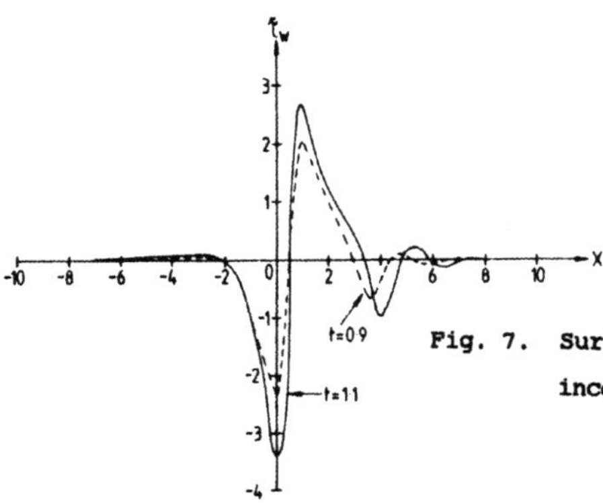

Fig. 7. Surface stress for 2D incompressible case h = -10

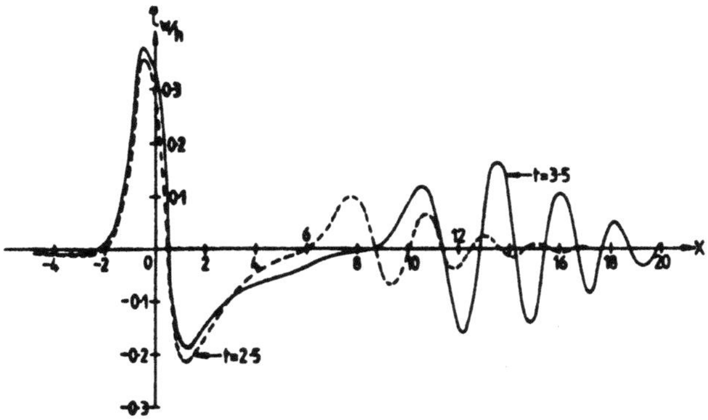

Fig. 8. Normalised surface stress for 3D case Z = 0

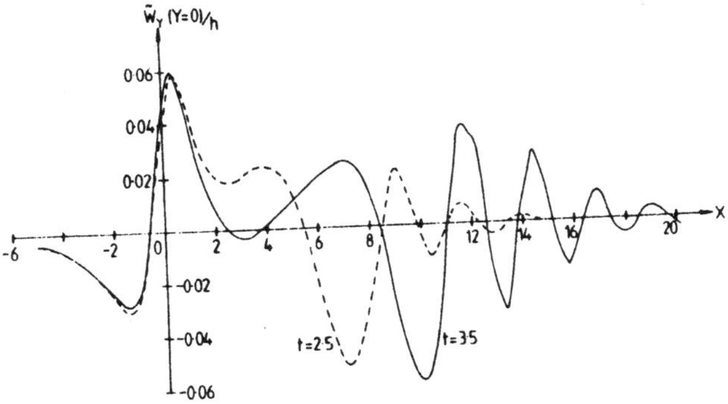

Fig. 9. Normalised cross-flow surface stress for 3D case Z = -1.2468

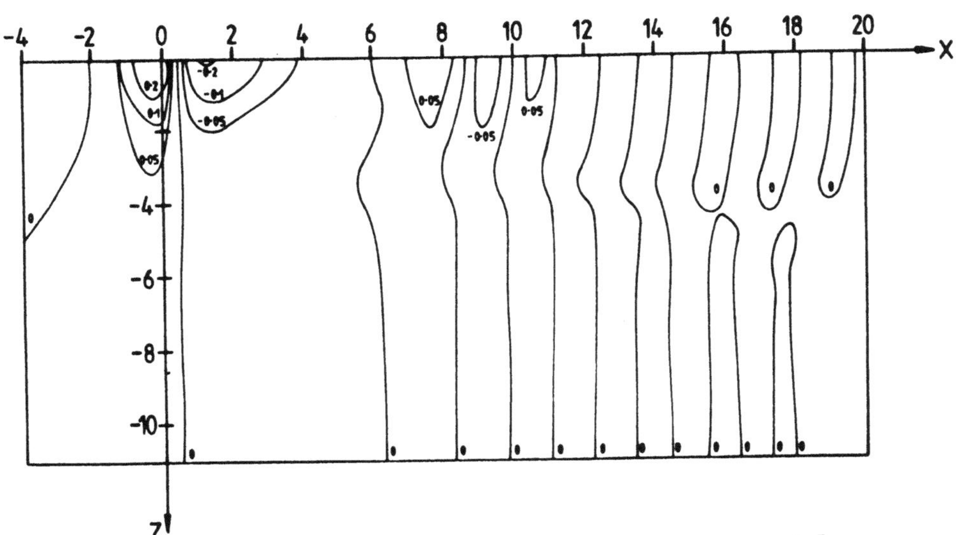

Fig. 10. Normalised surface stress contours for 3D case t = 2.5

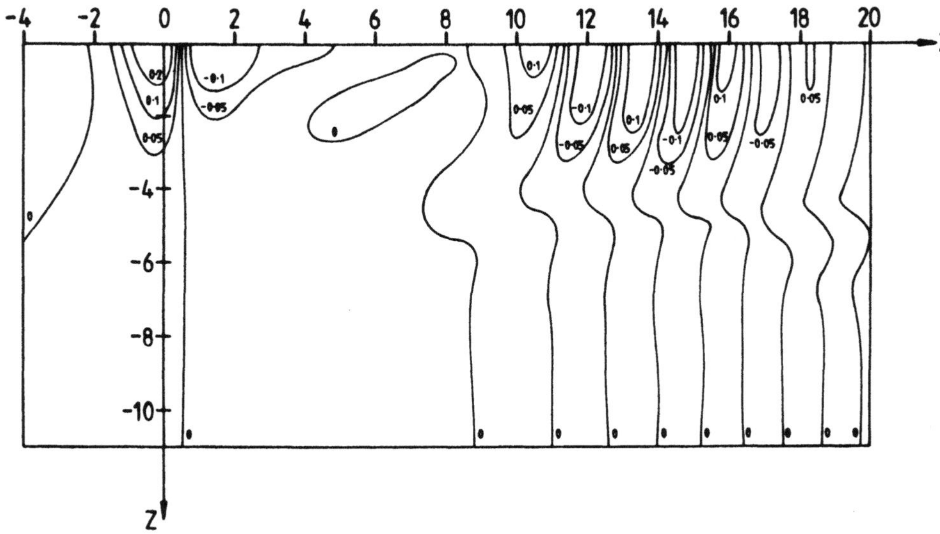

Fig. 11. Normalised surface stress contours for 3D case t = 3.5

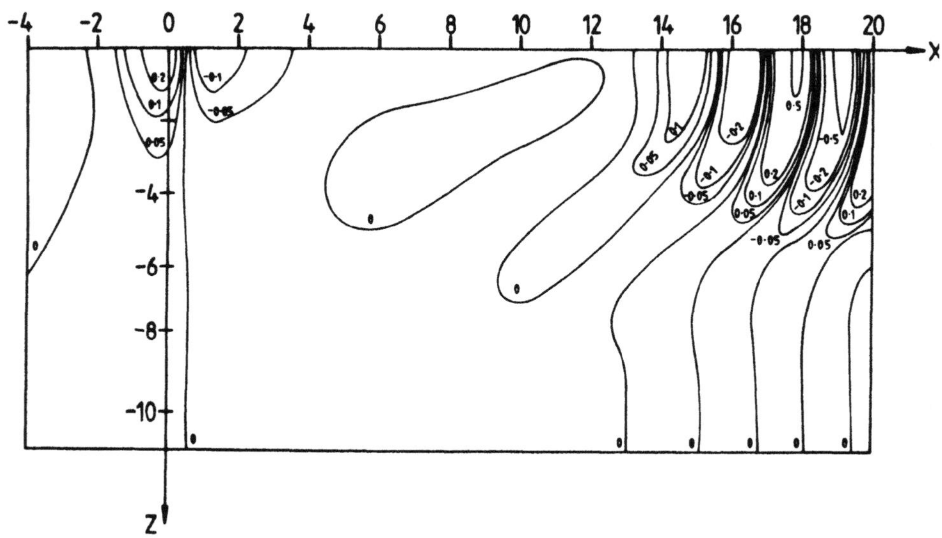

Fig. 12. Normalised surface stress contours for 3D case t = 5.0

Separation and Reattachment near the Leading Edge of a Thin Wing

TUNCER CEBECI, KALLE KAUPS and A. A. KHATTAB

Aerodynamics Research and Technology
Douglas Aircraft Company, Long Beach, California

Abstract

An interactive boundary-layer procedure based on a quasi-three-dimensional approximation is used to calculate separation and reattachment near the leading-edge of a thin wing. Results for a given sweep angle show that, as in two-dimensional flows, reverse flow solutions exist only for a limited range of angles of attack above the critical angle at which the non-interactive boundary layer separates. The solutions for the upper branch behave in the same manner as those predicted by the triple-deck theory for marginal separation in two-dimensional flows. The existence of solutions for the lower branch remains to be investigated.

Introduction

Leading-edge separation bubbles, which are common on thin and sharp-nosed airfoils and wings, play an important role in aerodynamic flows because of their association with the phenomenon of stall. The laminar boundary layer first experiences a favorable pressure gradient which causes the flow to accelerate from the stagnation point and is then subjected to an adverse pressure gradient. On thin airfoils or wings, the adverse pressure gradient is strong enough to cause the flow to separate even at small angles of attack with subsequent transition to turbulence. Reattachment occurs and a turbulent boundary layer extends to the trailing edge. With increasing incidence angle, the separated region moves toward the leading edge and eventually the bubble of the separated flow bursts. Reattachment does not occur and stall sets in.

As an initial step towards the development of a general method for calculating the properties of separation bubbles on airfoils and subsequently on wings, a simple leading-edge separation problem was studied by Cebeci, Stewartson and Williams [1]. A thin ellipse with its axis located at the center of the coordinate system $(\bar{y},\bar{z})$

F. T. Smith, S. N. Brown, (Eds.)
Boundary-Layer Separation
IUTAM Symposium London 1986

$$\frac{\bar{y}^2}{b^2} + \frac{\bar{z}^2}{c^2} = 1 \tag{1}$$

at an angle of attack α and in a uniform stream of speed u_∞ was considered. Attention was directed to its nose region and the ellipse was approximated locally by a "nose-fitting" parabola and an expression was derived for the external velocity distribution $\bar{u}_e$,

$$\bar{u}_e = \frac{u_e}{u_\infty(1 + t)} = \frac{\xi + \xi_o}{\sqrt{1 + \xi^2}} \tag{2}$$

Here t denotes the thickness ratio c/b, ξ_o corresponds to a reduced angle of attack, α/t, and the parameter ξ is the reduced polar coordinate ϕ/t and is related to the $\bar{y}$- and $\bar{z}$-coordinates of the ellipse by $\bar{y} = 1/2\ bt\xi$, $\bar{z} = bt^2\xi/2$, and to the surface distance s by

$$s = bt^2 \int_0^\xi (1 + \xi^2)^{1/2} d\xi \tag{3}$$

The investigation of Cebeci et al. [1] made use of the external velocity distribution given by Eq. (2) and showed that the laminar boundary layer near the leading edge is well behaved and unseparated if $\xi_o < \xi_s = 1.16$, although there was significant adverse pressure gradient. At higher values of ξ_o, however, separation occurred with an associated singularity and required the use of an interactive theory to link the viscous and inviscid flows. With this theory, solutions were obtained for separation bubbles but reattachment occurred in a very limited range of the reduced angle of attack. For $\xi_o > 1.218$, calculations broke down shortly after the flow reversal in the boundary layer and the subsequent studies of Stewartson, Smith and Kaups [2] seemed to imply that a dramatic switch to another separated form of motion can occur. Their results suggest that separated and reattached-flow solutions are not always unique and for certain circumstances a small change in the calculation may cause the reattachment location to move in space or in time by a disproportionately large amount. These conjectures may also have a physical counterpart in that separation bubbles can be unstable with a tendency to form an open region of separation.

In Fig. 1, we show calculated distributions of the reduced skin friction f_w'' at different values of ξ_o for $t = 0.1$ and $R_L(\equiv 2u_\infty b/\nu) = 2 \times 10^6$ and in Fig. 2 we demonstrate the failure of the iteration sequence at a slightly

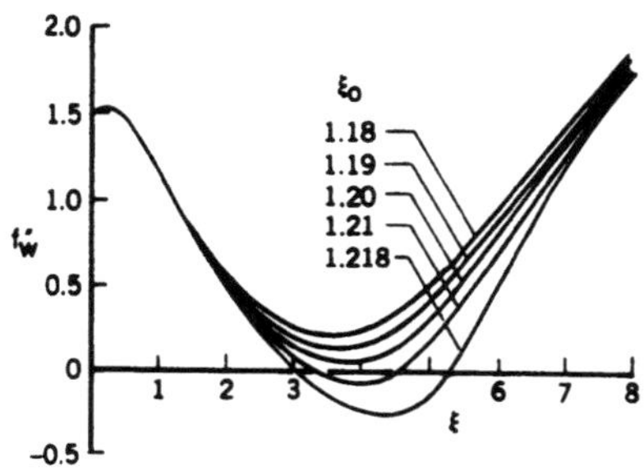

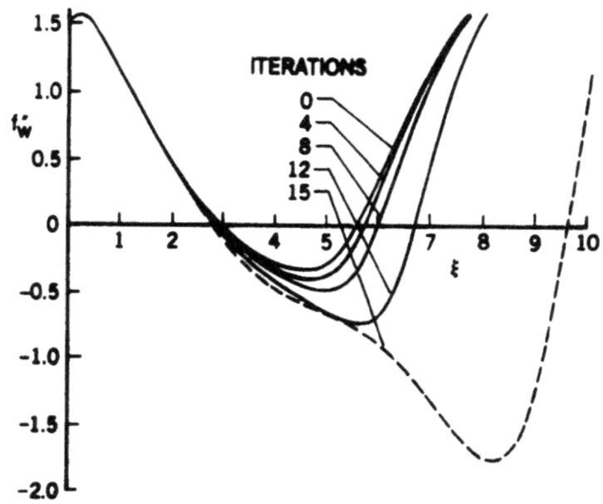

Figure 1. Variation of laminar wall shear parameter f''_w with ξ for various reduced angles of attack, ξ_o.

Figure 2. Variation of wall shear parameter f''_w with several iterations for $\xi_o = 1.220$.

larger value of $\xi_o = 1.218$. It is interesting to note that these observations provide support to the notion of a critical value of ξ_o beyond which attached flow is impossible, as discussed by Stewartson [3]; it should be remembered, however, that transition occurs in flows which have separated and this leads to reattachment, as was discussed in [4,5].

In the present paper we extend the work of Cebeci et al. [1] to the leading-edge separation problem on a wing and, as before, we consider a model problem corresponding to a triaxial ellipse given by

$$\frac{\bar{x}^2}{a^2} + \frac{\bar{y}^2}{b^2} + \frac{\bar{z}^2}{c^2} = 1 \tag{4}$$

Here $\bar{x}, \bar{y}, \bar{z}$ denote the Cartesian coordinates and a,b,c represent the lengths of the principal semi-axes. With the proper choice of the axis ratios a:b:c and the onset flow directions, this model problem provides the extension of the work of Cebeci et al. [1] to three-dimensional separation bubbles on the leading-edge of a swept wing.

In general, the behavior of separation bubbles along the leading edge of a wing is similar to that in two-dimensional flows, at least locally, since the surface pressure gradient acts more or less in the direction normal to the longest axis of the ellipsoid. Qualitative confirmation for this assertion can be found from flow visualization tests on wind-tunnel models. However, data on detailed quantitative measurements is scarce and the most complete and readily available study is by Horton [6]. From a physical point of view, a three-dimensional separation bubble differs from the two-dimensional one by the spanwise inflow or outflow. A theoretical investigation of the influence of inflow or outflow on the marginal separation along the lines of the study by Stewartson, Smith and Kaups [2] was given by Brown [7] who analyzed the separation and reattachment in the plane of symmetry of a prolate spheroid. The principal results that came from her study are that separation and reattachment are related to two parameters, one of which represents the inflow or outflow, and the other the pressure gradient or angle of attack, as in the two-dimensional case. The triaxial ellipsoid is a good model to study the leading-edge separation and reattachment on a thin wing if we take the longest dimension of the ellipsoid $\bar{x}$ in the swept spanwise direction and the $\bar{y}$-axis in the chordwise direction normal to the sweep line (see Fig. 3). The maximum thickness of the sections normal to the sweep line, as before, is given by $t = c/b$ and the wing planform area is πab. The aspect ratio of the wing, with λ denoting the sweep angle, is given by $(4/\pi)(a/b)\cos^2\lambda$.

As in the case of a thin ellipse at incidence α, we can derive expressions for the external velocity components for the nose region of a triaxial ellipse and apply the "nose-fitting" parabolas normal to the $\bar{x}$-axis, as discussed in [8]. With the assumption that $t \ll 1$, this procedure leads to the following expressions for the velocity components:

$$\frac{u_e}{u_\infty} = \frac{H[BH - A\ell(\bar{x}/a)](\xi + \xi_o)(1 + \xi^2)^{1/2}}{H^2(1 + \xi^2) + (\bar{x}/a)^2\ell^2} \tag{5a}$$

$$\frac{w_e}{u_\infty} = \frac{[AH(1 + \xi^2) + \bar{x}/a\ \ell(B - C\xi)][H^2 + (\bar{x}/a)^2\ell^2]^{1/2}}{H^2(1 + \xi^2) + (\bar{x}/a)^2\ell^2} \tag{5b}$$

Here $H = [1 - (\bar{x}/a)^2]^{1/2}$, ℓ denotes the principal axis ratio, b/a, the parameters A, B and C denote the maximum velocities over the ellipsoid induced by the freestream velocity u_∞ in the $\bar{x}$, $\bar{y}$ and $\bar{z}$ directions, respectively, and are given by

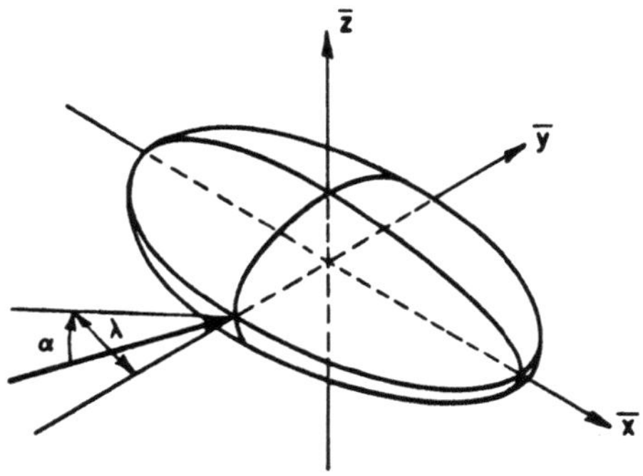

Figure 3. Orientation of the ellipsoid.

$$A = u_\infty V_1 \cos\alpha \sin\lambda \tag{6a}$$

$$B = u_\infty V_2 \cos\alpha \cos\lambda \tag{6b}$$

$$C = u_\infty V_3 \sin\alpha \tag{6c}$$

where $u_\infty \cos\alpha \sin\lambda$, $u_\infty \cos\alpha \cos\lambda$, and $u_\infty \sin\alpha$ are the components of the onset flow, $V_{\bar{x}_\infty}$, $V_{\bar{y}_\infty}$ and $V_{\bar{z}_\infty}$ in the $\bar{x}$, $\bar{y}$, and $\bar{z}$ directions, respectively as shown in Fig. 4. The coefficients V_1, V_2, V_3 are dimensionless maximum velocities induced by unit onset velocities in the $\bar{x}$, $\bar{y}$ and $\bar{z}$ directions, respectively, and are functions of the axis ratios a:b:c. As before ξ_o corresponds to a reduced angle of attack and is given by

$$\xi_o = - \frac{C[H^2 + (\bar{x}/a)^2 \ell^2]}{H[A\ell(\bar{x}/a) - BH]} \tag{7}$$

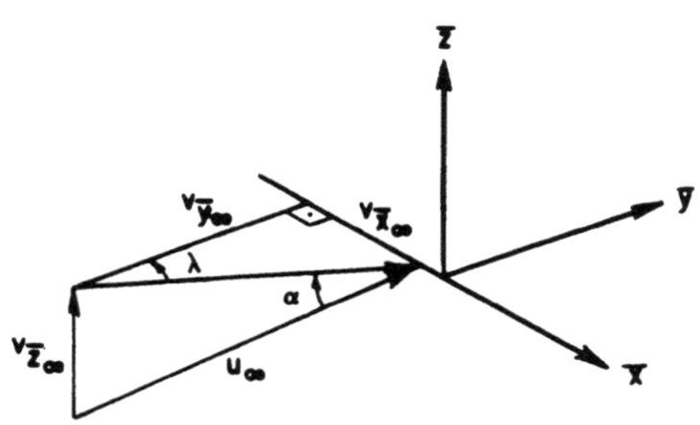

Figure 4. Resolution of the freestream velocity u_∞ along the axis of the ellipsoid.

We note that for an ellipse defined by Eq. (4), the external velocity u_e given by Eq. (5a) is identical to that of Eq. (2) since by letting $a \to \infty$, and keeping b fixed, $\ell \to 0$, $V_2 = 1 + t$. For small angles of attack $\cos\alpha = 1$, $B = u_\infty \cos\lambda$, and B can be thought of as an equivalent freestream velocity.

Equations and Boundary Conditions

The boundary-layer equations and their boundary conditions for a three-dimensional incompressible laminar flow in a curvilinear nonorthogonal coordinate system are given in [9] and can be written as

$$\frac{\partial}{\partial x}(uh_2 \sin\theta) + \frac{\partial}{\partial z}(wh_1 \sin\theta) + \frac{\partial}{\partial y}(vh_1h_2 \sin\theta) = 0 \tag{8}$$

$$\frac{u}{h_1}\frac{\partial u}{\partial x} + \frac{w}{h_2}\frac{\partial u}{\partial z} + v\frac{\partial u}{\partial y} - K_1u^2 \cot\theta + K_2w^2 \csc\theta + K_{12}uw = -\frac{\csc^2\theta}{h_1\rho}\frac{\partial p}{\partial x} + \frac{\cot\theta \csc\theta}{h_2\rho}\frac{\partial p}{\partial z} + \nu\frac{\partial^2 u}{\partial y^2} \tag{9}$$

$$\frac{u}{h_1}\frac{\partial w}{\partial x} + \frac{w}{h_2}\frac{\partial w}{\partial z} + v\frac{\partial w}{\partial y} + K_1u^2 \csc\theta - K_2w^2 \cot\theta + K_{21}uw = \frac{\csc\theta \cot\theta}{h_1\rho}\frac{\partial p}{\partial x} - \frac{\csc^2\theta}{h_2\rho}\frac{\partial p}{\partial z} + \nu\frac{\partial^2 w}{\partial y^2} \tag{10}$$

$$y = 0,\quad u = v = w = 0; \qquad y = \delta:\quad u = u_e(x,z),\quad w = w_e(x,z) \tag{11}$$

Here h_1 and h_2 denote metric coefficients, θ represents the angle between the coordinates x and z, and the parameters K_1 and K_2 are the geodesic curvatures of the curves z = const. and x = const., respectively.

For a flow with small spanwise pressure gradient, the above equations can be simplified by neglecting flow variations with respect to z. The resulting equations from this approximation, which we shall refer to as "quasi-three-dimensional flow equations" can be viewed as providing a better approximation to Eqs. (8) to (10) than those based on the strip theory or infinite swept-wing approximations.

The quasi-three-dimensional flow equations can be written as

$$\frac{\partial}{\partial x}(uh_2 \sin\theta) + \frac{\partial}{\partial y}(vh_1h_2 \sin\theta) = 0 \tag{12}$$

$$\frac{u}{h_1}\frac{\partial u}{\partial x} + v\frac{\partial u}{\partial y} - K_1u^2\cot\theta + K_2w^2\csc\theta + K_{12}uw$$

$$= -\frac{\csc^2\theta}{h_1\rho}\frac{\partial p}{\partial x} + \nu\frac{\partial^2 u}{\partial y^2} \tag{13}$$

$$\frac{u}{h_1}\frac{\partial w}{\partial x} + v\frac{\partial w}{\partial y} + K_1u^2\csc\theta - K_2w^2\cot\theta + K_{21}uw$$

$$= \frac{\csc\theta\cot\theta}{h_1\rho}\frac{\partial p}{\partial x} + \nu\frac{\partial^2 w}{\partial y^2} \tag{14}$$

The first-order boundary-layer theory implies that the metric coefficients h_1 and h_2 and the coordinate angle θ are functions of the surface coordinates x and z only, and can be obtained from the relations between $\bar{x},\bar{y},\bar{z}$ and x,y,z as discussed by Cebeci et al. [8]. The determination of the coordinate system amounts to the choice of the (x,z) net on the body surface. It is defined here by the families of surfaces, $\bar{x}/a$ = const. intersecting the ellipse and we designate

$$z = \frac{\bar{x}}{a} \tag{15a}$$

For the x-coordinate we choose the polar angle ϕ related to coordinates of the ellipse by

$$\bar{y} = -bH\cos\phi, \qquad \bar{z} = cH\sin\phi \tag{15b}$$

where

$$H = \sqrt{1 - z^2} \tag{16}$$

We now restrict our attention to the leading edge where ϕ is small and, as discussed in [8], write the metric coefficients and the coordinate angle as

$$h_1 = bHt(1 + x^2)^{1/2} \tag{17a}$$

$$h_2 = \frac{b(H^2 + \ell^2z^2)^{1/2}}{\ell H} \tag{17b}$$

$$\cos\theta = \frac{\ell xz}{(H^2 + \ell^2z^2)^{1/2}(1 + x^2)^{1/2}} \tag{17c}$$

Here x is equal to ϕ/t and can be viewed as equivalent to ξ in two-dimensional flows.

The geometric parameters K_1, K_2, K_{12}, K_{21} defined by

$$K_1 = \frac{1}{h_1 h_2 \sin\theta}\left[\frac{\partial}{\partial x}(h_2 \cos\theta) - \frac{\partial h_1}{\partial z}\right]$$

$$K_2 = \frac{1}{h_1 h_2 \sin\theta}\left[\frac{\partial}{\partial z}(h_1 \cos\theta) - \frac{\partial h_2}{\partial x}\right]$$

$$K_{12} = \frac{1}{h_1 h_2 \sin^2\theta}\left[(1 + \cos^2\theta)\frac{\partial h_1}{\partial z} - 2\cos\theta\frac{\partial h_2}{\partial x}\right]$$

$$K_{21} = \frac{1}{h_1 h_2 \sin^2\theta}\left[(1 + \cos^2\theta)\frac{\partial h_2}{\partial x} - 2\cos\theta\frac{\partial h_1}{\partial z}\right]$$

can be calculated from the expressions given by Eq. (17).

Solution Procedure

As in our previous studies, we use transformed variables to express the equations of the previous section in a form more suitable for solution. For this purpose, with $ds = h_1 dx$, we define

$$x = x, \qquad d\eta = \sqrt{\frac{u_o}{\nu s}}\,dy \tag{18}$$

and use the definition of stream function $\psi(x,y)$

$$uh_2 \sin\theta = \frac{\partial\psi}{\partial y}, \qquad vh_1 h_2 \sin\theta = -\frac{\partial\psi}{\partial x} \tag{19}$$

so that Eqs. (13) and (14) can be written as

$$f''' + m_1 f f'' + m_3[(f')^2 - e^2] + m_5(f'g' - \bar{w}_e e) + m_4[(g')^2 - (\bar{w}_e)^2]$$

$$= m_9\left(f'\frac{\partial f'}{\partial x} - f''\frac{\partial f}{\partial x} - e\frac{\partial f}{\partial x}\right) \tag{20}$$

$$g''' + m_1 f g'' + m_6[(g')^2 - \bar{w}_e^2] + m_7[(f')^2 - e^2] + m_8(f'g' - \bar{w}_e e)$$

$$= m_9\left(f'\frac{\partial g'}{\partial x} - e\frac{\partial \bar{w}_e}{\partial x} - g''\frac{\partial f}{\partial x}\right) \tag{21}$$

where, with $f' = u/u_o$, $g' = w/u_o$, $w_e = \bar{w}_e/u_o$, $e = u_e/u_o$,

$$m_1 = \frac{\sqrt{s}}{h_1 h_2 \sin\theta} \frac{\partial}{\partial x}(\sqrt{s}\, h_2 \sin\theta), \quad m_3 = K_1 s \cot\theta$$

$$m_4 = -K_2 s \csc\theta, \quad m_5 = -K_{12} s, \quad m_6 = s K_2 \cot\theta \tag{22}$$

$$m_7 = -m_3, \quad m_8 = -K_{21} s, \quad m_9 = s/h_1$$

The boundary conditions follow from Eq. (11) and can be written as

$$\eta = 0, \quad f = f' = 0, \quad g = g' = 0 \tag{23a}$$

$$\eta = \eta_e, \quad f' = e(x), \quad g' = \bar{w}_e(x) \tag{23b}$$

In the so-called standard method $e(x)$ and $w_e(x)$, which represent the external velocity components, are known and as a result the solution of Eqs. (20), (21) can be obtained readily subject to the boundary conditions given by Eqs. (23). In an interactive boundary layer a relationship between the external velocity components and boundary-layer parameters is needed and is provided by an interaction formula as described below.

An Interaction Formula for Quasi-Three-Dimensional Flows

In the current procedures used for two-dimensional flows an interaction formula is obtained through a link between the external velocity and displacement thickness. Two procedures are used for this purpose. In the first, the solutions of the boundary-layer equations are obtained for a prescribed external velocity in order to obtain a displacement-thickness distribution $\delta^{*o}(x)$ [10,11]. If this initial calculation encounters separation, $\delta^{*o}(x)$ is extrapolated to the trailing edge and one complete cycle of the viscous and inviscid calculation is performed. In general, this leads to two external velocity distributions, $u_{ev}(x)$ derived from the boundary-layer solution and $u_{ei}(x)$ derived from the updated approximation to the inviscid velocity past the body with the added displacement thickness. A relaxation formula in the form

$$\delta^{*\nu+1}(x) = \delta^{*\nu}(x)\left[1 + \omega\left(\frac{u_{ev}(x)}{u_{ei}(x)} - 1\right)\right],$$
$$\nu = 0, 1, 2, \ldots \tag{24}$$

where ω denotes a relaxation parameter, is then introduced to define an updated displacement thickness distribution and to obtain new solutions of the boundary-layer equations so that the interactive procedure between inviscid and viscous flow solutions can be carried out until convergence.

The second approach, which is recommended on the grounds of generality and physical basis, treats the external velocity $u_e(x)$ and the displacement thickness $\delta^*(x)$ as unknown quantities [12,13]. The boundary-layer equations are solved simultaneously in an inverse mode and with successive sweeps over the body surface. For each sweep, the external boundary condition is written as the sum of the inviscid velocity $u_e^o(x)$ over the body and a perturbation velocity $\delta u_e(x)$ which is computed from the Hilbert integral, that is,

$$y = \delta, \quad u_e(x) = u_e^o(x) + \delta u_e(x) \tag{25}$$

with $\delta u_e(x)$ computed from the Hilbert integral given by

$$\delta u_e(x) = \frac{1}{\pi} \int_{s_a}^{s_b} \frac{d}{d\sigma} (u_e \delta^*) \frac{d\sigma}{s - \sigma} \tag{26}$$

and the interaction confined between s_a and s_b.

In the present study on three-dimensional flows, we use the second approach and compute the external velocities $u_e(x)$ and $w_e(x)$ as part of the solution. Since the irrotationality condition for an orthogonal system, for example,

$$\frac{\partial}{\partial x} [h_2(w_e^o + \delta w_e)] = \frac{\partial}{\partial z} [h_1(u_e^o + \delta u_e)] \tag{27}$$

provides a relationship between the two velocity components u_e and w_e, the choice of computing the perturbations due to viscous effects is not arbitrary. By replacing the exact solution by a quasi-three-dimensional approach, we have implied that the perturbation quantities δu_e and δw_e are functions of x alone. Thus, from the irrotationality condition

$$\frac{\partial(\delta w_e)}{\partial x} = 0 \tag{28}$$

which implies that

$$\bar{w}_e = \bar{w}_e^o + \text{const.} \tag{29}$$

We have taken the constant to be zero in our calculations. In order to take into account the inevitable variation of w_e, the quasi-three-dimensional calculations will be linked by updating the potential-flow solution after each set of converged inverse boundary-layer calculations.

This provides a new set of u_e and w_e which are functions of x and z, and in this context the quasi-three-dimensional solutions can be viewed as providing the necessary blowing velocity for the global iteration.

In terms of the transformed variables defined by Eqs. (18) and (19), Eqs. (25) and (26) can be written as

$$\eta = \eta_e, \qquad e(x) = \bar{u}_e^o(x) + \frac{1}{\pi} \int_{s_a}^{s_b} \frac{d\Delta}{d\sigma} \frac{d\sigma}{s - \sigma} \tag{30}$$

where

$$\Delta = \frac{s}{\sqrt{R}} (\eta_e e - f_e), \qquad R = \frac{u_o s}{\nu} \tag{31}$$

Following Cebeci et al. [13], we write Eq. (30) as

$$e(x) = \bar{u}_e^o(x) + \sum_{j=1}^{n} c_{ij} \Delta_j \tag{32}$$

Here c_{ij} denotes the interaction-coefficient matrix which is obtained from a discrete approximation to the Hilbert integral in Eq. (30). In this form, Eq. (32) provides an outer boundary condition for the viscous-flow calculation which represents the viscous/inviscid interaction.

Numerical Method

We use the Mechul function formulation of [14] to solve the system given by Eqs. (20), (21), (23) and (30). In essence the solution procedure is similar to that described in [5] for two-dimensional flows. We again use Keller's Box scheme [15], which in this case requires the system to be expressed in terms of seven first-order equations. This is accomplished by introducing the following new variables

$$f' = u \tag{33a}$$

$$u' = v \tag{33b}$$

$$g' = w \tag{33c}$$

$$w' = t \tag{33d}$$

Since e is an unknown in an inverse formulation and is independent of η, we write

$$e' = 0 \tag{33e}$$

In terms of the definitions given by Eqs. (33a-d), Eqs. (20), (21) and their boundary conditions given by Eqs. (23) and (30) become

$$v' + m_1 fv + m_3(u^2 - e^2) + m_5(uw - \bar{w}_e e) + m_4(w^2 - \bar{w}_e^2)$$

$$= m_9(u \frac{\partial u}{\partial x} - v \frac{\partial f}{\partial x} - e \frac{\partial e}{\partial x}) \tag{33f}$$

$$t' + m_1 ft + m_6(w^2 - \bar{w}_e^2) + m_7(u^2 - e^2) + m_8(uw - \bar{w}_e e)$$

$$= m_9(u \frac{\partial w}{\partial x} - e \frac{\partial \bar{w}_e}{\partial x} - t \frac{\partial f}{\partial x}) \tag{33g}$$

$$\eta = 0, \quad f = u = g = w = 0 \tag{33h}$$

$$\eta = \eta_e, \quad u = e(x), \quad w = \bar{w}_e(x) = \bar{w}_e^o(x) \tag{33i}$$

$$e - \tilde{c}_{ii} (\eta_e e - f_e) = g_i \tag{33j}$$

where

$$\tilde{c}_{ii} = c_{ii} \frac{s}{\sqrt{R}}, \qquad g_i = \bar{u}_e^o + \sum_{j \neq i} c_{ij} \Delta_j$$

The solution of the above system is obtained with Keller's box method which is an efficient, second-order finite-difference method extensively used by Cebeci and his associates for a wide range of flows, as discussed in [9]. The general features of the inverse procedure are also described in [9]. As in previous studies the FLARE approximation [16] is employed in which the convective terms $e(\partial e/\partial x)$ and $u(\partial w/\partial x)$ are set equal to zero in the recirculating region and no attempt was made to improve the accuracy of the solutions resulting from this approximation.

Results and Discussion

Calculations were performed for three sweep angles $\lambda = 10°$, $20°$ and $30°$, at a fixed Reynolds number $R_L = 2 \times 10^6$ and with a range of angles of attack including flows with separation and those for which the three-dimensional laminar separation bubble becomes unstable, as previously encountered in two-dimensional flows. Calculations were first made for $\lambda = 20°$ with reduced angles of attack ξ_o ranging from 1.305 to 1.319

and the results are shown in Fig. 5. As can be seen from Fig. 5a, the streamwise wall shear parameter f''_w first becomes negative for $\xi_o = 1.312$ and the calculations proceed smoothly until the reduced angle of attack becomes 1.323 when they change substantially with the number of sweeps as shown in Fig. 5b. Up to this value of ξ_o, the region of separated flow is small and does not change much with each sweep although the rate of convergence of the solutions at a given ξ-location is rather slow.

The distributions of spanwise wall shear parameter g''_w are shown in Fig. 5c and, although they exhibit a sign change, they are well behaved in the range of angles of attack of Fig. 5a. Since the equations are coupled, the problems encountered for $\xi_o = 1.323$, and shown in Fig. 5b for f''_w, are also present in g''_w.

The dimensionless displacement thickness $\tilde{\delta^*}(\equiv (u_e/u_\infty)(\delta^*/b)\sqrt{R_L/2}$ variations of Fig. 5d exhibit the features of flows with separation; away from separation, the displacement thickness is insensitive to reduced angles of attack but in its vicinity shows a rapid variation.

For two-dimensional flows, the critical reduced angle of attack $(\xi_o)_{cr}$ is 1.22 [1], a value which is smaller than 1.312 in three-dimensional flows with $\lambda = 20°$ (Fig. 5). To explore the effect the sweep angle λ has on the critical value of ξ_o, additional calculations were performed for $\lambda = 10°$ and 30°. The results in Figs. 5, 6 and 7 confirm this dependence and show that, as expected, the critical reduced angle increases with increasing sweep angle. For example, for $\lambda = 10°$ (see Fig. 6), $(\xi_o)_{cr}$ is between 1.22 and 1.23, a value which is slightly larger than its corresponding value for two-dimensional flows. With increasing sweep angle, $\lambda = 30°$ (see Fig. 7), $(\xi_o)_{cr}$ exceeds that of $\lambda = 20°$ and becomes 1.481. A comparison between the results in Fig. 1 and Figs. 5, 6 and 7 shows that the recirculation region in three-dimensional flows is even somewhat smaller than in two-dimensional flows. This is partly due to different Reynolds numbers used in the calculations and partly due to the different external velocity distributions.

The studies in [1,2] showed that two-dimensional laminar flows can have very small (marginal) separation bubbles and that the interactive boundary-layer approach yields results which are in agreement with those of the triple-deck theory, although the latter method also predicts multiple solu-

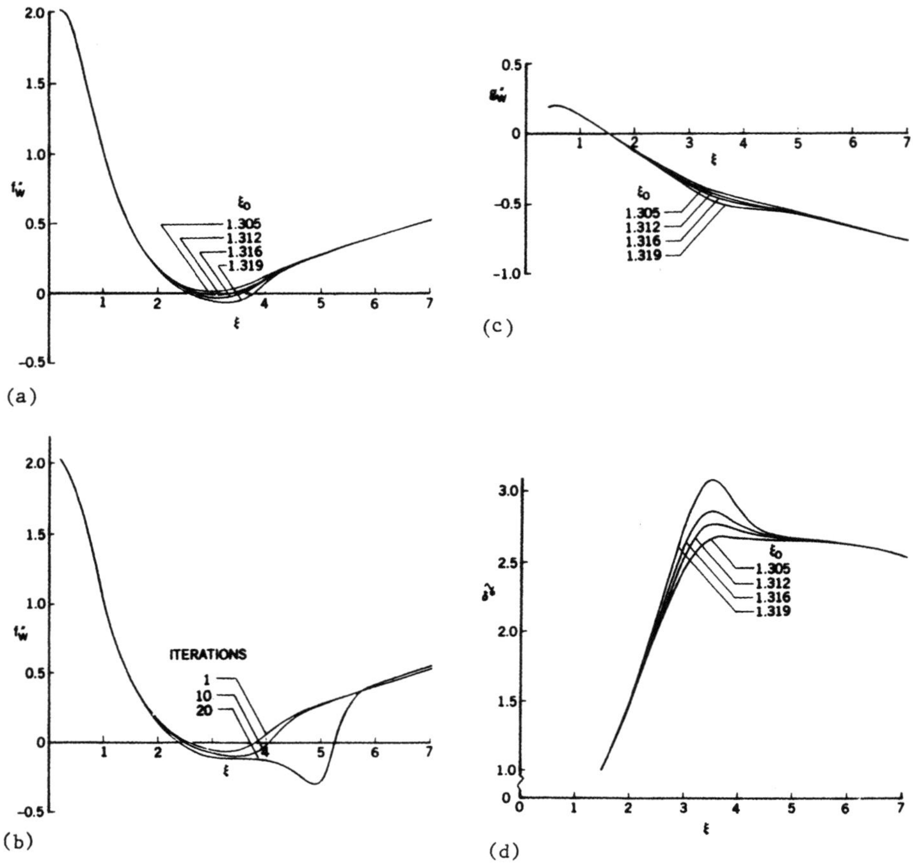

Figure 5. Results for $\lambda = 20°$ for various reduced angles of attack, ξ_o. Variation with ξ of: (a) streamwise wall shear parameter f_w'', (b) streamwise wall shear parameter f_w'' for several iterations for $\xi_o = 1.323$, (c) spanwise wall shear parameter g_w'', (d) dis-displacement thickness $\tilde{\delta}^*$.

tions [2,7]. Figure 8 compares the present results for three sweep angles with the multiple solutions of [2] for two-dimensional flows. In comparing the results in this figure, we have used the results from Fig. 1 of the paper by Stewartson et al. [2], which in its original form presents the variation of A(0) as a function of Γ. Here A(0) represents the wall shear parameter at the location where the noninteractive boundary layer separates (X = 0) and Γ corresponds to an angle of attack. Since A(0) is proportional to the wall shear parameter f_w'' and since Γ is proportional to the angle of attack expressed as $\alpha - \alpha_c$, the results in Fig. 8 are

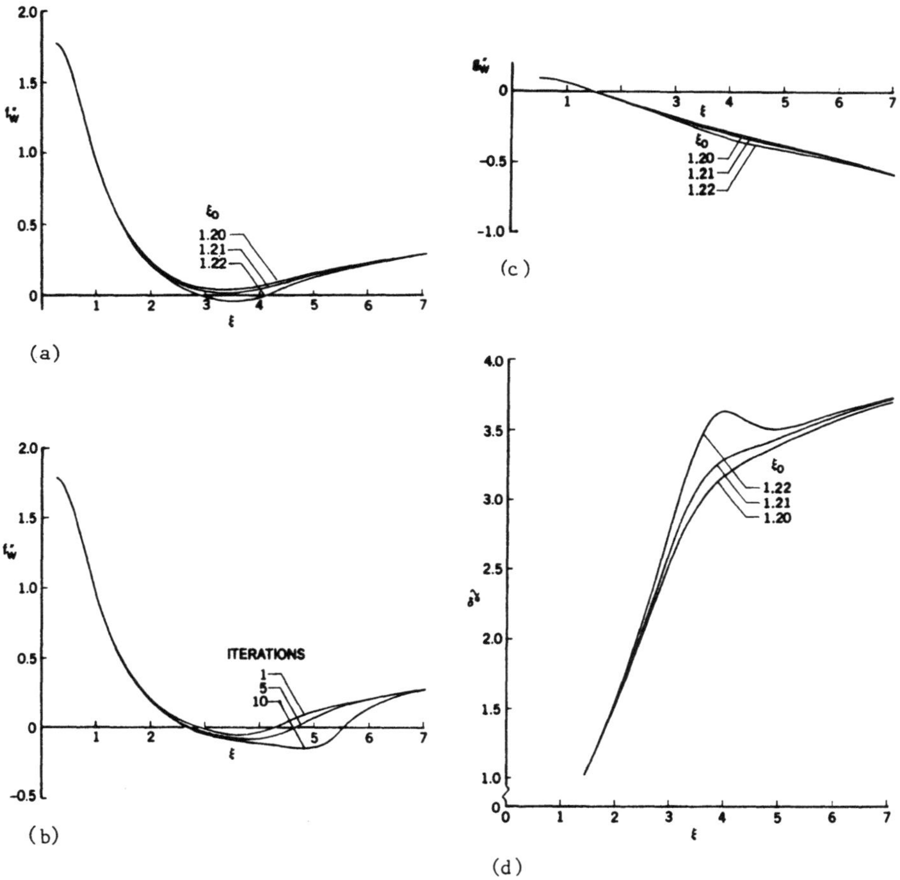

Figure 6. Results for $\lambda = 10°$ for various reduced angles of attack, ξ_0. Variation with ξ of: (a) streamwise wall shear parameter f_w'', (b) streamwise wall shear parameter f_w'' for several iterations for $\xi_0 = 1.230$, (c) spanwise wall shear parameter g_w'', (d) displacement thickness $\tilde{\delta}^*$.

plotted as f_w'' vs $\alpha - \alpha_c$. Here α_c corresponds to the angle of attack at which the noninteractive boundary layer first separates. The parameters A(0) and Γ, which are already scaled variables, are rescaled to match the results for $\lambda = 10°$ so that A(0) at $\Gamma = 0$ is equal to f_w'' at $\alpha - \alpha_c = 0$ and Γ at A(0) = 0 is equal to $\alpha - \alpha_c$ at $f_w'' = 0$.

As can be seen from Fig. 8, the trends of the three-dimensional flow results are similar to those for two-dimensional flows excluding multiple solutions. The solutions for $\lambda = 10°$, which are a good approximation for

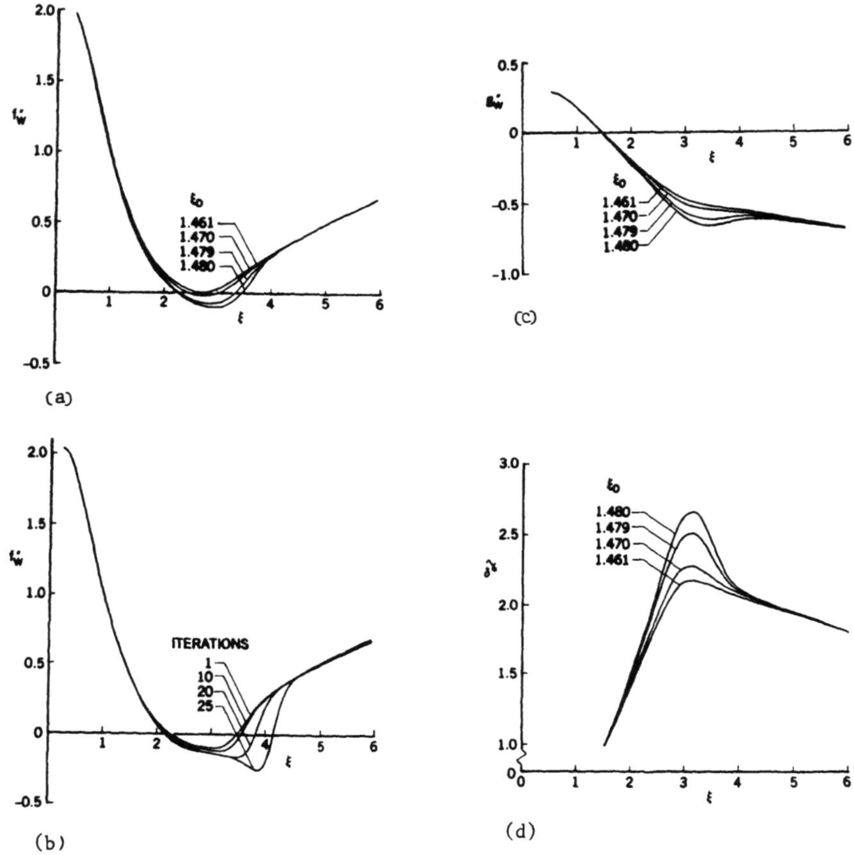

Figure 7. Results for $\lambda = 30°$ for various reduced angles of attack, ξ_o. Variation with ξ of: (a) streamwise wall shear parameter f''_w, (b) streamwise wall shear parameter f''_w for several iterations for $\xi_o = 1.481$, (c) spanwise wall shear parameter g''_w, (d) displacement thickness $\tilde{\delta}^*$.

two-dimensional flows are in good agreement with the upper branch of the rescaled triple-deck solutions.

Concluding Remarks

An interactive procedure for computing threedimensional flows with separation has been formulated with a quasi-three-dimensional approximation and tested with a model problem corresponding to a thin triaxial ellipse at

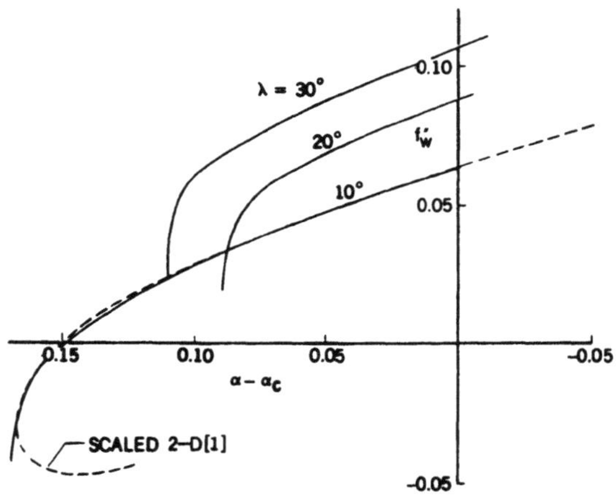

Figure 8. Variation of the interactively calculated wall shear parameter f_w'' with angle of attack at the location where the noninteractive boundary layer first separates; α_c corresponds to the angle where the noninteractive flow first separates.

incidence. The results indicate that, as in two-dimensional flows, laminar separation bubbles exist for only a limited range of angles of attack. The critical reduced angle of attack is a function of the sweep angle and varies from 1.22 in two-dimensional flows to 1.481 in three-dimensional flows in which the sweep angle variation is up to $\lambda = 30°$.

The study also indicates that the trends of the interactive flow solutions are in good agreement with those obtained by the triple-deck theory for reduced angles of attack up to its critical value. Whether the multiple flow solutions also exist for three-dimensional flows remains to be explored. It is difficult to assess this question by using Brown's recent study [7] for a quasi-three-dimensional flow on the plane of symmetry. Whereas the crossflow in her study was either all inflow or all outflow, it was of the through-flow type in the present study.

Acknowledgment This work was supported by Dr. J. D. Wilson of the Air Force Office of Scientific Research under contract F49620-84-C-0007.

References

1. Cebeci, T.; Stewartson, K.; Williams, P.G.: Separation and reattachment near the leading edge of a thin airfoil at incidence. AGARD CP 291, Paper 20, 1981.

2. Stewartson, K.; Smith, F.T.; Kaups, K.: Marginal separation. Studies in App. Math. 67 (1982) 45-61.

3. Stewartson, K.: Some recent studies in triple-deck theory. In Numerical and Physical Aspects of Aerodynamic Flows (T. Cebeci, Ed.), Springer-Verlag, 1982, 129- 144.

4. Cebeci, T.; Schimke, S.M.: The calculation of separation bubbles in interactive turbulent boundary layers. J. Fluid Mech. 131 (1983) 305.

5. Cebeci, T.; Clark, R.W.; Chang, K.C.; Halsey, N.D.; Lee, K.: Airfoils with separation of the resulting wakes. J. Fluid Mech. 163 (1986) 323-347.

6. Horton, H.P.: Redeveloping turbulent boundary layers behind yawed separation bubbles. The Aero. Quarterly 23, Pt. 3 (1972) 211-228.

7. Brown, S.N.: Marginal separation of a three-dimensional boundary layer on a line of symmetry. J. Fluid Mech. 158 (1985) 95-111.

8. Cebeci, T.; Kaups, K.; Khattab, A.A.: Studies of interactive procedures for three-dimensional flows. I. Separation and reattachment near the leading edge of a thin wing. Douglas Aircraft Co. Rept. MDC J3835, Sept. 1985.

9. Bradshaw, P.; Cebeci, T.; Whitelaw, J.H.: Engineering Calculation Methods for Turbulent Flows. Academic Press, London, 1981.

10. LeBalleur, J.C.: Couplage visquex-non visquex: Methode numerique et applications aux ecoulements bidimensionnels transsonique set supersoniques. Le Recherche Aerospatiale No. 1978-2 (1978) 65-76.

11. Carter, J.E.: A new boundary-layer inviscid interaction technique for separated flow. AIAA Paper 79-1450, 1979.

12. Veldman, A.E.P.: New quasi-simultaneous method to calculate interacting boundary layers. AIAA J. 10 (1981) 769-785.

13. Cebeci, T.; Stewartson, K.; Williams, P.G.: Separation and reattachment near the leading edge of a thin airfoil at incidence. AGARD CP 291 (1981) Paper 20.

14. Cebeci, T.: Separated flows and their representation by boundary-layer equations. Rept. ONR-CR 215-234-2, Mech. Eng. Dept., California State University, Long Beach, 1976.

15. Keller, H.B.: Accurate difference methods for two-point boundary-value problems, SIAM J. Numer. Analy. 11 (1974) 305-320.

16. Reyhner, T.A.; Flügge-Lotz, I.: The interaction of a shock wave with a laminar boundary layer. Int. J. Nonlinear Mech. 3 (1968) 173-199.

Separating Flow: Small-scale, Large-scale, and Nonlinear Unsteady Effects

F.T. SMITH

Department of Mathematics
University College London
Gower Street
London WC1E 6BT, UK

Abstract

The aim is to understand theoretically the structure of external separating flows and their tendency to promote rapid transition to turbulence, at increased Reynolds numbers. Large-scale separation involving a massive eddy behind a bluff body is considered first, with a view to comparing with recent Navier-Stokes based computations. Then the more commonly observed phenomenon of smaller-scale separation is studied, numerically and analytically. Finally, linear and nonlinear unsteady features possible in both types of separating flow are discussed, with particular emphasis on local nonlinear breakdown of the motion beyond separation and comparisons with experiments.

1. Introduction

This article is mostly about theoretical work in progress on small- and large-scale separating flows, and their nonlinear unsteady features, at high Reynolds numbers. It contains probably more questions than answers. These are motivated by two experimental observations, for example in aerodynamics and internal flows: the first is that boundary-layer separation so often occurs at medium-to-high Reynolds numbers, in practical configurations, and the second is that the separation quite commonly promotes boundary-layer transition to turbulence. We will concentrate on the aerodynamical context of external flow here.

The theory starts in section 2 by, perverse as it may be, moving away from reality to study the steady flow past a bluff body, where increasingly large-scale separation ensues at increased Reynolds numbers. This is a fundamental and tantalising problem and it acts as a stern test for theoretical ideas on the structure of large-scale eddies. In particular

F. T. Smith, S. N. Brown, (Eds.)
Boundary-Layer Separation
IUTAM Symposium London 1986

an attempt is made to connect up with Fornberg's recent computations of the Navier-Stokes equations for symmetric laminar flow past a circular cylinder.

A return to reality is intended in the work of sections 3, 4, with section 3 addressing smaller-scale steady separations of the local viscous-inviscid interactive kind which are often seen and measured in practice: see experimental comparisons in the reviews mentioned later. An attempt is made to describe the emergence of enhanced separation regions as the local disturbance, a hump shape or a corner for instance, is increased, and to link up with the larger-scale effects of section 2. Section 4 is concerned with linear and nonlinear unsteady effects, with a view to understanding more the increased instability due to separation and to comparing theory with the experimental observations, e.g. of Dovgal et al, Telionis and others recently. Qualitative comparisons seem fairly encouraging so far.

The Reynolds number Re used below is based on the airfoil chord (or cylinder diameter) and the free stream speed U_∞ , as are the nondimensional cartesian coordinates x, y , the associated velocity components u, v and the time t . The pressure p is also nondimensional, with respect to $\rho_\infty U_\infty^2$ where ρ_∞ is the freestream density, and for the most part we consider here two-dimensional, steady and unsteady, subsonic/ incompressible-fluid flow. Three-dimensional features are studied elsewhere: e.g. see references.

2. Large-scale separated flow

The understanding and determination of the large-scale separated laminar flow past a bluff body, be it a circular cylinder or a fixed airfoil, at high Reynolds numbers, seem to be hindered by three major factors. The first comes from the theoretical complexity of describing a massive closed eddy (or eddies) which, on the face of it, is predominantly inviscid but which nevertheless is influenced in shape and size by the viscous forces active in the shear layers around the eddy, and of matching the flow in and about that eddy with the flow closer to the body itself. Thus the whole motion is multi-structured and governed by viscous-inviscid interaction. The second main difficulty is the physical one, that at sufficiently high Reynolds numbers the large-scale separated laminar flow outlined above is seldom if ever observed in practice, because of the strong instabilities present (we refer to section 4), and hence we cannot necessarily appeal to physical intuition or experience to guide us through the theory then. The third hindrance to determining the motion is numerical, in the sense that direct computational treatments of the Navier-Stokes equations tend to flounder when confronted by the needs to resolve the extreme scales, both small and large, in the flow, to capture the unknown free-shear-layer location(s), and to maintain accuracy. Again, alternative computational treatments such as interacting-boundary-layer methods and related composite methods in principle are able to resolve the various scales and maintain accuracy, by their design, but as yet the satisfactory numerical capturing of the free shear-layers in large-scale separated flow remains a central difficulty there.

Progress towards understanding, and then determining, the separated bluff-body motion is rather tentative and conjectural, therefore. Below we summarize the recent developments in this fundamental problem, based on the work of Fornberg [1] and Smith [2], where more details can be found (see also Peregrine [3]). Fornberg's computations of the Navier-Stokes equations for symmetric flow past a circular cylinder, at increased values of Re , appear (so far) to point to an overall flow structure consisting mainly of a massive $O(Re) \times O(Re)$ eddy-scale motion and a smaller $O(1)$ body-scale motion, when Re is asymptotically large. All the various regions present in the flowfield match satisfactorily provided each of a sequence of problemsor tasks has an acceptable solution.

The first of these tasks is to find flow solutions of the inviscid massive-eddy form, a form proposed earlier by Sadovskii [4]. That yields a doubly-infinite family of inviscid solutions, as described below. The second step is expected to reduce the acceptable solutions to a singly infinite family, by requiring the viscous shear layers bordering the massive eddy (and governed by the boundary-layer equations) to satisfy a periodicity condition. The third task, which may render the entire flow solution unique, or at least reduce the number of acceptable large-Re solutions substantially, is to solve for the body-scale motion. This brings in the triple-deck criterion (see later) at the separation point and, among other things, the question of secondary separation and the influence of the backward $O(1)$ thick jet which the centre-line shear layer above sends into the rear end of the body-scale motion as a source of $O(1)$ vorticity there.

Here we concentrate on the first task, concerning the large-scale inviscid eddy. Its dimensions are $O(Re)$ in both x and y, so that its interior motion is as strong as the motion outside, yielding the governing equations

$$\nabla^2\overline{\psi} = -\omega \quad \text{(inside the eddy)} \;, \tag{2.1a}$$

$$\nabla^2\overline{\psi} = 0 \quad \text{(outside the eddy)} \tag{2.1b}$$

in scaled terms, with unknown scaled constant vorticity ω, subject to the boundary conditions

$$\overline{\psi} = 0,\ p \quad \text{continuous at} \quad Y = S(X) \;, \tag{2.1c}$$

$$S(0) = S'(0) = S(L) = S'(L) = 0 \;, \tag{2.1d}$$

and $\overline{\psi} = 0$ at $Y = 0$ for symmetry. Here $Y = S(X)$ gives the unknown eddy shape between the ends $X = 0,\ X = L$ in scaled variables; the unscaled eddy length is LRe. Also, the condition (2.1d) imposes tangential departure at the ends of the eddy, while (2.1c) allows the fluid speed [q_1, q_2 just outside and inside] to be discontinuous across the vortex sheet, or free shear layer, at $Y = S(X)$. The typical velocity and pressure variations throughout are $O(1)$. The solutions of (2.1a-d) depend on a single parameter, the end-pressure value

$$C_{II} = \frac{1}{2}\left[1 - q_1^2 + q_2^2\right] \tag{2.2}$$

from Bernoulli's theorem; alternatively one may use the vorticity parameter ωL, for instance, to characterize the flow. In any event there exists a double infinity of solutions. For convenience the unknown value of L can be normalized to 2. Solutions of (2.1a-d) appear to exist for the entire range of interest, $0 < C_{II} < 0.5$, according to the following. If C_{II} is small, corresponding to large $|\omega|L$ values, then the eddy is slender and so (2.1a) becomes $\partial^2\overline{\psi}/\partial Y^2 = -\omega$ to leading order, giving on integration a pressure gradient proportional to $\omega^2 SS'$, while (2.1b) yields linearized potential flow. Hence the pressure condition in

(2.1c) produces, to leading order, the integrated Benjamin-Ono equation

$$C_{II} - \frac{1}{8}\omega^2 S^2(X) = -\frac{1}{\pi}\int_0^L \frac{S'(\xi)d\xi}{(X-\xi)} \tag{2.3}$$

for $S(X)$, subject to (2.1d) still. Computational solutions of (2.3), (2.1d) may be obtained fairly readily [2], by a semi-inverse or a local updating method, and the results appear to be self-consistent with regard to grid resolution and accuracy. Moreover, the slender-eddy predictions from (2.3), (2.1d) happen to agree quite well in numerical terms with the earlier results of Sadovskii, even though the latter are concentrated towards the upper extreme $C_{II} = 0.5$ whereas the slender-eddy analysis is strictly for $C_{II} \ll 1$. This suggests basing a numerical procedure for the full problem (2.1a-d) on the approach used for the reduced problem (2.3), and such a procedure [2] is indeed found to work reasonably efficiently over the entire range of values of C_{II}. The results thus obtained can be summarized in terms of the dependence of ωL, and of the typical eddy width $S(1)/L$, on the pressure parameter C_{II}. For small C_{II} the results join to those from (2.3) (which have application later in this article), while as C_{II} increases the previous solutions [4] are retrieved (to within approximately graphical accuracy) and, at the upper extreme, as C_{II} tends to 0.5 the Pierrehumbert [5] solution with continuous velocity at the interface (from (2.2)) is approached. All these inviscid solutions are symmetric about the mid-station $X = L/2$. Nonsymmetric solutions appear to be unlikely [2].

The other two tasks mentioned previously, to impose viscous periodicity around the eddy, after which the dependence of the drag c_D on the eddy length can be determined from the far-wake features, and to solve for the body-scale motion, have still to be tackled. In the meantime an intriguing aspect arises from comparison of the massive-eddy solutions [2] and the Navier-Stokes computation [1], namely that the latter tends to point to the upper extreme value $C_{II} = 0.5$ as being the correct high-Reynolds-number limit. This suggestion comes first from comparing the ratios of the maximum eddy width to the eddy length in [1], [2], but along with this the Navier-Stokes computations show a relatively minor variation in the velocity near the dividing streamline at larger Re, again in keeping with a value $C_{II} = 0.5$, because of (2.2). The matter of the appropriate value of C_{II} is significant for the body-scale motion because there the effective farfield condition is

$$u \to (1 - 2C_{II})^{\frac{1}{2}} , \quad p \to C_{II} , \tag{2.4}$$

since (2.4) holds near the start of the large-scale eddy. Thus if $C_{II} = 0.5$ the velocities in the body-scale motion can be reduced by an order of magnitude, consistent with an asymptotically small drag coefficient c_D. The small velocities here are due to the setting up of a stagnation point in the massive-eddy solution, as $(X,Y) \to (0,0)$, when $C_{II} = 0.5$ and the conditions on S' in (2.1d) no longer apply; in fact $S \propto -X \ln X$ locally, as $X \to 0+$, and similarly as $X \to L-$, so that the massive eddy then has blunt ends. A check on the possibility that C_{II} is equal to 0.5 would come from the second task referred to earlier, concerning viscous periodicity, which should pin down the value of C_{II}.

Hopes seem justifably high at the moment that a self-consistent large-Re description will emerge from the above structure. It would be helpful to connect this with the analysis of cascade flows [6], incidentally, and to consider the possibilities of nonuniqueness [private communications from Prof. A. Acrivos, [7] and section 3 below] arising there.

3. Smaller-scale separations

An alternative strategy for increasing the understanding of large-scale separated laminar flow is to go down to smaller sized separations, of the viscous-inviscid interactive kind, which can be completely understood, and to build up from there. This has the added advantage that smaller-scale laminar separations tend to be more easily observable in reality anyway: see also section 4 below. A direct connection between the large- and the smaller-scale separations can also be set up in principle, as we shall see.

The foremost type of such local separations is the triple-deck type [26,8], occurring typically with length, velocity and pressure scales of the form

$$(x - x_o , y) = (Re^{-3/8}\bar{x} , Re^{-5/8} \bar{y}) , \tag{3.1}$$

$$(u,v,p) = (Re^{-1/8}U , Re^{-3/8}V , Re^{-1/4}P) \tag{3.2}$$

near a station $x = x_o$, in the lower-deck sublayer closest to the solid surface. The governing equations there are the boundary-layer equations

$$\frac{\partial U}{\partial \bar{x}} + \frac{\partial V}{\partial \bar{y}} = 0 \quad , \tag{3.3a}$$

$$U \frac{\partial U}{\partial \bar{x}} + V \frac{\partial U}{\partial \bar{y}} = - P'(\bar{x}) + \frac{\partial^2 U}{\partial \bar{y}^2} \tag{3.3b}$$

with

$$U = V = 0 \ , \quad \bar{y} = 0 \quad , \tag{3.3c}$$

$$U \sim \bar{y} + A(\bar{x}) + \alpha F(\bar{x}) \ , \ \bar{y} \to \infty \quad , \tag{3.3d}$$

$$P(\bar{x}) = \frac{1}{\pi} \int_{-\infty}^{\infty} \frac{A'(\xi) d\xi}{(\bar{x} - \xi)} \quad . \tag{3.3e}$$

Here (3.3c) is the no-slip condition at the surface, while (3.3d) comes from the unknown displacement $(\alpha - A)$ of the $O(Re^{-\frac{1}{2}})$ boundary layer and (3.3e) is the pressure-displacement law due to the subsonic potential flow in the upper deck outside the boundary layer. The given, local, disturbed surface shape is $\alpha F(\bar{x})$ in scaled terms, with $|F|$ being typically of $O(1)$ and α measuring the relative disturbance size. Both the scaled pressure $P(\bar{x})$ and the displacement function $-A(\bar{x})$ are unknown, so that any "separation" or flow reversal occurring is a regular process. The same goes for supersonic external flow, where (3.3e) is replaced by

$$P(\bar{x}) = - A'(\bar{x}) \ , \tag{3.3f}$$

and in both of the cases (3.3e,f) the local flow problem is elliptic.

The major question in these flows is what happens as α increases, especially in connection with the larger-scale separations discussed earlier. It is perhaps natural to consider α becoming large, although the author has noted before the possibility of peculiarities, e.g. lack of steady solutions or nonuniqueness, arising at finite values of α instead. There seems little doubt that, in addition to analytical attempts, more accurate numerical solutions of the interactive problems (3.3a-e or f) are needed. Towards that end, the author has developed the following finite-difference numerical scheme aimed mainly at by-passing some of the difficulties, of forward-backward marching, handling the ellipticity and/ or grid resolution, inherent in all existing numerical approaches. The idea is to consider the momentum equation (3.3b) in the form

$$\frac{\partial^2 U}{\partial \bar{y}^2} = P'(\bar{x}) + R(\bar{x}, \bar{y}) \quad , \tag{3.4a}$$

with the inertial contribution R assumed known at its latest guessed value. Then (3.4a) can be integrated in principle, with (3.3c), to give

$$U = \frac{1}{2}\bar{y}^2 \; P'(\bar{x}) + \int_0^{\bar{y}} \left\{ \int_\delta^{\bar{y}} R \; d\bar{y} \right\} d\bar{y} + b(\bar{x})\bar{y} \tag{3.4b}$$

where $b(\bar{x})$ is an unknown function of $\bar{x}$ and δ is a specified, suitably large, positive constant. The two outer constraints $U = \bar{y} + A + \alpha F$, $\partial U/\partial \bar{y} = 1$ are imposed at $\bar{y} = \delta$, modelling (3.3d), to yield the two equations

$$\tfrac{1}{2}\delta^2 P' + I + b\delta = \delta + A + \alpha F \quad , \tag{3.4c}$$

$$\delta P' + b = 1 \tag{3.4d}$$

for the three unknown functions P, b, A, where

$$I = \int_0^\delta \left\{ \int_\delta^{\bar{y}} R d\bar{y} \right\} d\bar{y} \; . \tag{3.4e}$$

The third equation necessary is (3.3e or f). For the supersonic case (3.3f), for instance, we obtain the differential equation

$$\tfrac{1}{2}\delta^2 P'' - P = (I' - \alpha F')(\bar{x}) \tag{3.4f}$$

for $P(\bar{x})$, to be solved subject to prescribed end values $P \to P(\pm\infty)$ as $|\bar{x}| \to \infty$. Once $P(\bar{x})$ is determined, $b(\bar{x})$ follows from (3.4d), so that U can be updated from (3.4b), V follows from (3.3a, c) and we may update the field $R(\bar{x}, \bar{y})$ from

$$R = U \frac{\partial U}{\partial \bar{x}} + V \frac{\partial U}{\partial \bar{y}} \quad , \tag{3.4g}$$

using relaxation. The term I in (3.4e) can then be updated, P is re-evaluated from (3.4f), and so on. An aim throughout is to reduce repeated numerical differentiation, however, and to accentuate the repeated integration present, and some analytical manipulations, e.g. in solving (3.4f) for $P(\bar{x})$ in "closed" form, assist this. This iterative method has also been applied to the subsonic/incompressible case (3.3e), for which (3.4f) is replaced by an integro-differential equation for $P(\bar{x})$ which is found to be solvable in closed form again; and the method should be readily extendable to unsteady and three-dimensional interactive motions, in the same spirit. So far it works apparently as well as the best existing methods (see e.g. in the reviews [26], [8]), that is, it produces reasonably accurate converged solutions with a fair amount of separated flow present, at increased values of α. Whether it can go further than previous schemes into the larger-α regime, or not, while retaining accuracy, remains to be seen, although in principle adequate grid resolution can be built into the method (3.4a-g) quite easily. Further study of this is under way.

Analytically, extended-Kirchhoff eddies with virtually uniform pressure have been proposed previously, for the large-α behaviour of (3.3a-e or f), but these appear to develop inconsistencies due to the promotion of a backward jet at the eddy closure. The main alternative, just as in section 2, is for the contained eddy motion to be as strong as the motion in the detached shear layer and outside, in which case the constant-vorticity argument again leads to a pressure gradient $\propto SS'$ where S is the eddy thickness. It can be argued also that $S = -A - \alpha F$, from use of (3.3d). Hence (3.3e) yields the forced integrated Benjamin-Ono equation

$$\bar{c} - \frac{1}{8}\bar{\omega}^2(A + \alpha F)^2 = \frac{1}{\pi} \int_{\bar{x}_0}^{\bar{x}_1} \frac{A'(\xi)d\xi}{(\bar{x} - \xi)} \tag{3.5a}$$

where the pressure constant $\bar{c}$, the constant vorticity $\bar{\omega}$ and the separation and reattachment positions $\bar{x}_0$, $\bar{x}_1$ are all unknown. Solutions of (3.5a) exist exhibiting tangential departure $(A = A' = 0)$ at $\bar{x} = \bar{x}_0 +$, $\bar{x} = \bar{x}_1 -$; for example, if we argue that the eddy size is much greater than the disturbance size then $\alpha F(\bar{x})$ can be neglected in (3.5a) and so (3.5a) reduces to the earlier solved form (2.3) with $S \simeq -A$. The novel feature here, however, compared with section 2, is that the boundary layer ahead of separation also plays a leading role. This boundary layer, like the detached shear layer, is driven by the pressure gradient that the eddy shape sets up. So if the displacement function $-A$ has the usual $(\bar{x} - \bar{x}_0)^{3/2}$ behaviour just beyond separation, the induced pressure gradient upstream $\propto(\bar{x}_0 - \bar{x})^{-\frac{1}{2}}$ becomes infinitely adverse and the boundary layer cannot survive up to separation. This rules out the massive-eddy limit above where $S \approx -A$ in particular, and it suggests that instead we should seek solutions of (3.5a) in which

$$-A \propto (\bar{x} - \bar{x}_o)^{5/2} \quad \text{as} \quad \bar{x} \to \bar{x}_o + \, , \tag{3.5b}$$

i.e. "smooth separation" is required at $\bar{x} = \bar{x}_o$. Solutions with this property are currently being sought.

It is worth noting here also that secondary separation is likely to arise, among other complications, and that the conjectured flow structure now has the Prandtl-Batchelor form. If this is true then presumably the Prandtl-Batchelor form continues to hold for larger disturbances, e.g. thicker airfoils, and so, given section 2, there is the possibility of nonuniqueness for separated flow past a bluff body and/or a cut-off/stall phenomenon involving a switch from the Prandtl-Batchelor to the massive-eddy structure of section 2 as, say, the body thickness is increased further. As in section 2, however, a number of steps addressing the inviscid and the viscous contributions to the flow structure have still to be followed through to check on the large-α proposal. The counterpart in supersonic flow, with (3.3f) holding, has also to be tied together. Further, certain of the flow features in the large-α proposal, such as a significant reversed sublayer and pressure fall in the eddy, have not been observed clearly (yet) in computational work for increased α , although again we could refer back to the comments at the beginning of section 2 to counter that doubt.

4. Unsteady effects (linear and nonlinear)

Both large- and smaller-scale separated motions are prone to instability and subsequent transition in practice, especially the large-scale forms. Here we discuss briefly their linear and nonlinear unsteady or instability properties, starting with the smaller-scale form and working upwards, as it were. Further details can be found in the referenced works.

The unsteady triple-deck problem is addressed first since it captures the structure both of separating flow (local or breakaway) (section 3) and of Tollmien-Schlichting (TS) instabilities [9] in the boundary layer approaching separation. Thus in scaled form, for subsonic flow, with (3.1), (3.2) and

$$t = Re^{-\frac{1}{4}}\bar{t} \quad , \tag{4.1}$$

we have (3.3a,c-e) again but with (3.3b) replaced by

$$\frac{\partial U}{\partial \bar{t}} + U\frac{\partial U}{\partial \bar{x}} + V\frac{\partial U}{\partial \bar{y}} = -\frac{\partial P}{\partial \bar{x}} + \frac{\partial^2 U}{\partial \bar{y}^2} \tag{4.2}$$

and $P(\bar{x}, \bar{t})$ etc. replacing $P(\bar{x})$ etc. now. The unsteady nonlinear system with (4.2) has been integrated numerically in recent work [10, 11], for nonlinear disturbances from Blasius flow ($U = Y$, $V = P = A = 0$), but if the starting steady flow is itself distorted, for example as in the separating motions considered in section 3, then the matter is more difficult, for linear or nonlinear unsteadiness. This is because the starting flow cannot be treated in the classical way as parallel: the separation length scale coincides with the TS length scale. In addition there are several instability mechanisms available for the separating flow, namely TS, inflexional Rayleigh, shear-layer Kelvin-Helmholtz (see below) and perhaps Görtler due to streamline curvature.

One question is whether the enhanced instability produced as a flow separates is due to the sudden onset of Rayleigh inviscid instability [12, and references therein] or to the accentuation of TS viscous-inviscid instability as the basic flow changes [13], or possibly to both. In the end nonlinear aspects almost certainly must decide the issue. Some insight into the question is provided by the properties of relatively-high-frequency TS disturbances, high frequencies corresponding to higher input frequency upstream or to movement downstream. If the typical frequency Ω is defined by $\partial/\partial\bar{t} = -i\Omega$ in the linear case or $\Omega = O(|\partial/\partial\bar{t}|)$ generally, then for high Ω values the small spatial growth rate for linear disturbances is found to be [13]

$$G = \frac{1}{\sqrt{2}} - \frac{3}{2}\frac{d\bar{A}}{d\bar{x}} \tag{4.3}$$

in scaled terms, where $-\bar{A}(\bar{x})$ is the basic-flow displacement. Thus separating flows, which produce positive displacement slopes $-d\bar{A}/d\bar{x} > 0$ such as those in section 3, force enhanced TS growth. The extra effect

$\propto -d\bar{A}/d\bar{x}$ in (4.3) is a nonparallel one. Note also that a recent study [14] has extended the two-dimensional result (4.3) to three-dimensional disturbances and to cross-flow effects. For three-dimensional disturbances it is found that as the basic flow becomes increasingly separated the direction of maximum disturbance growth gradually veers away from 0°, i.e. the freestream direction, towards an oblique angle

$$\theta = 64.7^o \tag{4.4}$$

from the freestream direction, with growth rates then exceeding those in (4.3). Further implications of this accentuated three-dimensionality have still to be followed through, and so below we stick to two-dimensional properties.

For breakaway separation, e.g. on a bluff body as in section 2, the basic displacement $-\bar{A} \sim \frac{2}{3} b\, \bar{x}^{3/2}$ downstream, as $\bar{x} \to \infty$, where $b \doteqdot 0.44$ [15-17]. Hence the growth rate G increases there as

$$G \sim \frac{3}{2} b\, \bar{x}^{\frac{1}{2}} . \tag{4.5}$$

This increasing TS growth leads to a new stage of fast growth at a distance $O(\Omega^{1/3})$ beyond separation. There the scaled wavenumber α satisfies [18]

$$\alpha^3\Delta - \alpha^2\Omega + \Omega^2 = 0 \tag{4.6}$$

where Δ is the normal distance of the detached shear layer from the flat surface; here $\Delta = (2b/3)\bar{x}^{3/2}$. At this stage therefore there is an interplay between TS instability (the second and third terms in 4.6)), Rayleigh instability (the first and second terms) and a form of Kelvin-Helmholtz instability (from the first and third terms). This yields the prediction

$$\bar{x}_{crit} = (\Omega/3b^2)^{1/3} \tag{4.7}$$

for the position, beyond separation, at which sudden increased instability occurs. Recent experiments [19] measuring the instability and transition of breakaway separating flow past a bluff body agree at least qualitatively, and perhaps quantitatively, with the prediction (4.7).

We turn now to nonlinear unsteady aspects. One option is to compute (4.2) directly, extending the earlier noted computational studies to separating basic flow. Another course however is to address relatively-high-frequency nonlinear disturbances. These provide insight into transition properties in attached boundary layers, pointing especially to the significance of an Euler stage [13,20]. For breakaway separating boundary layers a main non-linear interaction takes place in the region just beyond separation, where (4.6) - (4.7) hold if the disturbances remain small but for nonlinear disturbances we have, formally,

$$\frac{\partial S}{\partial \bar{t}} + \frac{\partial}{\partial \bar{x}}(\hat{U}S) = 0 \quad , \tag{4.8a}$$

$$\frac{\partial \hat{U}}{\partial \bar{t}} + \hat{U}\frac{\partial \hat{U}}{\partial \bar{x}} = -\frac{\partial P}{\partial \bar{x}} \quad , \tag{4.8b}$$

$$\frac{\partial}{\partial \bar{t}}(S + A) + (S + A)\frac{\partial}{\partial \bar{x}}(S + A) = -\frac{\partial P}{\partial \bar{x}} \quad , \tag{4.8c}$$

$$P(\bar{x}, \bar{t}) = \frac{1}{\pi} ⨍_{-\infty}^{\infty} \frac{\partial A}{\partial \xi}(\xi, \bar{t}) \frac{d\xi}{(\bar{x} - \xi)} \quad . \tag{4.8d}$$

Here $S(\bar{x}, \bar{t})$ is the eddy thickness, $\hat{U}(\bar{x}, \bar{t})$ is the eddy's streamwise velocity, zero initial vorticity is assumed inside the eddy, and all of $\hat{U}$, P, S, A are unknown functions of $\bar{x}, \bar{t}$. The nonlinear system (4.8a-d) can admit travelling-wave solutions if the original steady, breakaway-separating, flow is not disturbed too much. For sufficiently strong input disturbances however the solution of (4.8a-d) is expected to break down instead, at a finite time $\bar{t}_o$, focussed around a station $\bar{x} = \bar{x}_o$, in the form

$$[\hat{U},P,S,A] \sim [(\bar{t}_o - \bar{t})^{n-1}\hat{U}^* , (\bar{t}_o - \bar{t})^{2n-2}P^* , (\bar{t}_o - \bar{t})^{3n-2}S^*, (\bar{t}_o - \bar{t})^{3n-2}A^*], \tag{4.9a}$$

$$(\bar{x} - \bar{x}_o) \sim (\bar{t}_o - \bar{t})^{n}\bar{x}^* \quad , \tag{4.9b}$$

with $n > 0$ being unknown. The governing equations for the starred variables, as well as integral properties for all time, suggest that the appropriate value for n is $n = 1/2$, from joint work with Prof.S.N.Brown. Solutions can then be found analytically, for this special case as well as for nonzero initial vorticity and for the supersonic analogue where in effect (3.3f) replaces (4.8d). The nonlinear breakdown (4.9a,b) then points to an Euler stage next, and eventual transition, in the context of larger-scale separated flow (an alternative nonlinear mechanism, we note, pointing to transition, comes from marginal separations [25]). That context is described in a sense by the upper limit of the system (4.8a-d) in which (4.8a,b) still hold but instead of (4.8c,d) we have

$$P(\bar{x}, \bar{t}) = -\frac{1}{\pi} \int_{-\infty}^{\infty} \frac{\partial S}{\partial \xi}(\xi, \bar{t}) \frac{d\xi}{(\bar{x} - \xi)} \tag{4.10}$$

and $\bar{x}$ now represents distances on the $O(1)$ scale of the airfoil itself. The limit system with (4.10) again appears to blow up in the focussed form (4.9a,b). Thus for large-scale separation major unsteady-disturbance effects can occur just beyond the separation point, where the eddy is initially thin, with slope $O(\varepsilon)$ say, and then (4.10) applies, leading to the breakdown (4.9) and thereafter to an Euler stage of length scales $O(\varepsilon^{\frac{1}{2}})$ in both x, y. There vortex-sheet dynamics and possible roll-up come into operation. The overall process predicted seems to tie in with some experimental observations: see [19, 21, 22].

The same limit system, i.e. (4.8a,b) with (4.10), can be shown to apply (despite a change in boundary conditions) to the interactive stage [23] encountered after the breakdown [24] of the unsteady classical boundary layer on an impulsively started circular cylinder. So we expect that interactive stage also to terminate in the form (4.9a,b), and thence proceed into the Euler stage and the start of vortex shedding. We believe that the same collapse (4.9a,b) can describe the breakdown of the entire unsteady triple-deck motion of (4.2), in fact. This is despite the predominantly inviscid nature of some of the analysis above, the theoretical need to check on the viscous effects, and the need from practical considerations to tackle three-dimensional nonlinear features in view of properties such as (4.4).

References

1. Fornberg, B., 1985, J. Comp. Phys. 61, 297-320; also Fornberg's article in this volume.

2. Smith, F.T., 1985, J. Fluid Mech. 155, 175-191 (earlier,UTRC Rept. UT84-31) and 1986, J. Eng. Maths., 20, 271-292.

3. Peregrine, D.H., 1985, J. Fluid Mech. 157, 493-500.

4. Sadovskii, V.S., 1971, Prikl. Math. Mekh. 35. 773-779.

5. Pierrehumbert, R.T., 1980, J. Fluid Mech. 99, 129-144.

6. Smith, F.T., 1983. Utd. Tech. Res. Cent.. E. Hartford. CT. U.S.A., Rept. UT83-13; J. Mathl. Physl. Sci., 1985, 19, 1-80.

7. Milos, F.S. & Acrivos, A., 1986, Phys. Fluids 29 (5). 1353-9.

8. Smith, F.T., 1986, Ann. Rev. Fluid Mech. 18, 197-220.

9. Smith, F.T., 1979, Proc. Roy. Soc. A366, 91-109 and A368, 573-589.

10. Duck, P.W., 1985, J. Fluid Mech. 160, 465-498.

11. Smith, F.T., 1984, A.I.A.A. paper no. 84-1582; 1985, Utd. Tech. Res. Cent., E. Hartford, Ct, U.S.A., Rept. 85-36; J. Fluid Mech. (1986) 169, 353-377.

12. Smith, F.T. & Bodonyi, R.J., 1985, Aeron. J. of the Roy. Aeron. Soc., June-July.

13. Smith, F.T. & Burggraf, O.R., 1985, Proc. Roy. Soc. A399, 25-55.

14. Smith, F.T. & Stewart, P.A., 1986, Proc. Roy. Soc. A., in press.

15. Smith, F.T., 1977, Proc. Roy. Soc. A356, 433-463.

16. Korolev, G.L., 1980, Sci. J. TSAG 11 (2), 27-36.

17. Van Dommelen, L.L. & Shen, S.F., 1984, Proc. Symp. Num. Phys. Aspects Aerod. Flows, No. 2, Ed. T. Cebeci, Springer-Verlag (pub.).

18. Smith, F.T., 1985, Utd. Tech. Res. Center, E. Hartford, Ct, U.S.A., Rept. UT85-55; and Proc. Symp. Stab. Slowly-Varying And Nonparallel Flows 1985, Springer-Verlag (pub.).

19. Mezaris, T.B., Telionis, D.P., Barbi, C. & Jones, G.S., 1986, to appear in Proc. Roy. Soc. A, 1986-87.

20. Smith, F.T. & Stewart, P.A., 1986, Utd. Tech. Res. Cent., E. Hartford, Ct., U.S.A., Rept. UT86-26; in press, J. Fluid Mech., 1987.

21. Van Dyke, M., 1982, An Album of Fluid Motion. Parabolic Press (pub.).

22. Dovgal, A.V., Kozlov, V.V. & Simonov, O.A.,1987, article in this Volume.

23. Elliott, J.W., Cowley, S.J. & Smith, F.T., 1983, Geo. Astrophys. Fluid Dyn. 25, 77-138.

24. Van Dommelen, L.L., 1981, Ph.D. thesis, Univ. of Cornell, U.S.A.

25. Smith, F.T. & Elliott, J.W., 1985, Proc. Roy. Soc. A401, 1-27.

26. Messiter, A.F., 1983, Trans. A.S.M.E. J. Appl. Mech. 50, 1104-1113.

Session 6

Breakdown of Steady and Unsteady Interacting Boundary Layers

R.A.W.M. HENKES
Dept. of Applied Physics, Delft University of Technology, Delft, The Netherlands

A.E.P. VELDMAN
Dept. of Mathematics and Informatics, Delft University of Technology, Delft, and National Aerospace Laboratory NLR, Amsterdam, The Netherlands

Summary

It is known that the classical boundary-layer solution breaks down by the appearance of the Goldstein singularity in a steady solution or Van Dommelen's singularity in an unsteady solution. Interaction between the inviscid flow and the boundary layer removes the Goldstein singularity, until a new critical parameter is reached, corresponding to a marginal separation in the asymptotic triple-deck description. In earlier studies instabilities were encountered in interacting boundary-layer calculations of steady flow past an indented plate, which might be related to the breakdown of the marginal separation. The present study identifies them as numerical. Further, until now it was unknown whether the unsteady interacting boundary-layer approach would remove Van Dommelen's singularity in the classical boundary layer around the impulsively started cylinder. It is shown that its appearance is at least delayed. The calculations show the experimentally known individualization of a vortex, after which the solution grows without reaching a steady limit; a process which is likely to be related to dynamic stall.

1. Introduction

For large Reynolds numbers, the Navier-Stokes solution can be approximated by an asymptotic description. Critical combinations of parameters (e.g. angle of incidence, geometry size, Reynolds number) can exist where a certain asymptotic description fails to represent the physics and breaks down. This mathematical breakdown can be characterized by a singularity, the non-existence/non-uniqueness of the steady solution or the growth beyond all bounds at infinitely large time of the unsteady solution. In the worst case the asymptotic solution may diverge from the Navier-Stokes solution without a special warning. New asymptotic equations are required to continue the asymptotic description of the Navier-Stokes solution.
A breakdown might be related to the transition to another physical structure: the breaking-away of the boundary layer from the wall, the formation of a vortex or the dynamic stall process.

F. T. Smith, S. N. Brown, (Eds.)
Boundary-Layer Separation
IUTAM Symposium London 1986

The classical asymptotic description of Prandtl [1] omits the viscous terms everywhere, yielding the Euler equations, except in a thin layer along the wall, where the (classical) boundary-layer equations hold. Goldstein [2] proved that the steady classical boundary-layer solution, in which the pressure is prescribed, has a singularity at the point of zero wall shear stress, unless the pressure fulfils some special conditions. Van Dommelen [3] calculated the appearance of a singularity, not of the Goldstein type, in the unsteady classical boundary-layer solution within a finite time after the impulsive start of a cylinder. Sychev [4] and Smith [5, 6] removed the Goldstein singularity in the steady solution by the introduction of the triple-deck equations in the neighbourhood of the point of zero wall shear stress; the pressure is no longer prescribed, but results from interaction with the boundary layer. Sychev showed that the triple-deck description is consistent with the assumption that the Kirchoff free-streamline solution is the relevant Euler solution for the cylinder (Brodetski [7]); triple-deck equations describe finite Reynolds number effects in the neighbourhood of the position where the free stream-line leaves the wall (at 56°). This description, in which the boundary layer is continued as a free shear layer along the free streamline, agrees with Prandtl's suggestion that a large-scale-breaking-away of the boundary layer, i.e. the "separation" of the boundary layer, is found in the steady flow when the wall shear stress vanishes. Stewartson et al. [8] describe it as the transition from an "attached" boundary layer (viscous effects are only felt in an $O(R^{-\frac{1}{2}})$ layer close to the wall) to a "detached" boundary layer (viscous effects are also felt in layers further removed from the wall). It is emphasized that separation, attachment or detachment within this boundary-layer concept should not be confused with the stream-line separation point or reattachment point used in the real (Navier-Stokes) flow concept at a finite Reynolds number.

A singularity in the classical boundary layer does not necessarily imply its large-scale-breaking-away; finite Reynolds number viscous-inviscid interaction can prevent it, admitting a small recirculation region (bubble) in the steady boundary layer, but leaving it still attached. Considering triple-deck equations, Stewartson et al. [8] reserve the term "marginal separation" for this behaviour. Triple-deck equations were used by Dijkstra [9], to calculate a steady boundary layer with bubble along a wall with smooth step, and by Veldman & Dijkstra [10] to calculate the bubble along an indented plate. Stewartson et al. [8] used the same equations for a leading-edge bubble on a thin ellipse under angle of incidence. From their calculations it follows that a critical combination of parameters (thickness ratio of the ellipse, angle of incidence and Reynolds number) exists, where non-existence or non-uniqueness of the solution occurs. Smith [11] repeated the analysis with unsteady triple-deck equations; rapidly growing peaks in the displacement thickness are calculated, strongly indicating the appearance of a finite-time singularity. Smith suggested that this behaviour is related to the dynamic stall process, in which a small increase of the angle of incidence totally changes the streamline pattern, achieving the violently unsteady transition from an attached to a detached boundary layer.

One might want to go one step further and calculate attached boundary layers with bubbles that do not fit in the asymptotic triple-deck scalings. Veldman [12, 13] generalized the triple-deck idea to a composite formulation, valid in the whole boundary layer. Its unsteady, incompressible formulation, in boundary-layer scalings, reads

$$\frac{\partial u}{\partial x} + \frac{\partial v}{\partial y} = 0$$

$$\frac{\partial u}{\partial t} + u\frac{\partial u}{\partial x} + v\frac{\partial u}{\partial y} = -\frac{\partial p}{\partial x} + \frac{\partial^2 u}{\partial y^2} \tag{1}$$

with the initial- and boundary conditions

$$\begin{aligned} t=0 &: u,\ v \text{ components specified} \\ y=0 &: u = v = 0 \\ y \to \infty &: u(x,y,t) \to u_e(x,t) \end{aligned} \tag{2}$$

$$u_e(x,t) = u_{e,0}(x,t) + \frac{1}{\pi\sqrt{R}} \int_{x_b}^{x_e} \frac{\partial(u_e\delta^*)/\partial\xi}{x-\xi}\, d\xi$$

$x = x_b$: u-profile (without backflow) of the oncoming boundary layer specified.

The displacement thickness δ^* is defined as

$$\delta^*(x,t) = \int_0^\infty \left(1 - \frac{u(x,y,t)}{u_e(x,t)}\right) dy \quad . \tag{3}$$

The first part of the edge velocity, $u_{e,0}$, results from the Euler solution, assuming an attached vanishingly thin boundary layer. The second part, with the Cauchy integral, accounts for the viscous-inviscid interaction. The composite formulation (1)-(2) is referred to as the interacting boundary-layer equations. Its steady formulation was used by Carter & Wornom [14], Veldman [12, 13] and Edwards & Carter [15] to calculate the attached boundary layer with bubble along a plate with small indentation. Convergence problems are revealed for very large Reynolds numbers, which could be related to non-existence or non-uniqueness of the solution, similar to the breakdown of the steady triple-deck solution according to Stewartson et al.

Recently some aspects of the breakdown of the interacting boundary-layer solution for a trailing-edge configuration were treated by Rothmayer & Davis [16]; the interacting boundary-layer solution in this case reveals a breakdown similar to the breakdown of the triple-deck solution. Two new examples are given here. With an unsteady evolution, it is investigated whether the convergence problems of the steady interacting boundary-layer solution along the indented plate are related to a breakdown (finite-time singularity or unboundedness for $t \to \infty$ of the unsteady solution, implying the non-existence of the steady solution) or whether they have only a numerical cause. As a second example we consider a problem where the recirculation region is larger: investigated is the influence of interaction on Van Dommelen's finite-time singularity in the unsteady classical boundary-layer solution for the impulsively started cylinder. The numerical method used is described in Henkes [17].

2. Indented plate

The shape of the indented plate is given by (boundary-layer coordinates)

$$y_w = \alpha \sqrt{R} \operatorname{sech} 4(x-2.5) \qquad , \; x \geq 0 \; . \tag{4}$$

The Reynolds number is based on the oncoming velocity u_∞ and the characteristic plate length ℓ. Taken is $\alpha = -0.03$. The Euler solution, assuming an attached vanishingly thin boundary layer, is approximated by

$$u_{e,0} = 1 + \frac{1}{\pi\sqrt{R}} \int_0^\infty \frac{dy_w/d\xi}{x - \xi} \, d\xi \quad . \tag{5}$$

The steady interacting boundary-layer equations were solved on the interval $1 \leq x \leq 4$ to verify the results of Carter & Wornom [14] (R= 8 x 10^4) and Veldman [12, 13] (R= 3.6 x 10^5). Veldman could not obtain converged solutions for larger Reynolds numbers. Related problems were encountered by Veldman & Dijkstra [10], who treated the same geometry, but now within a triple-deck framework. However, recently Edwards & Carter [15] could calculate the case R= 6 x 10^5 with a first order scheme and with neglect of the convective term in x-direction in regions with backflow (FLARE approximation). In the present study it appeared possible to calculate the case R= 6 x 10^5 with a second order scheme, without neglect of any terms, by initializing the iterative process with the solution obtained in a more stable first-order scheme.

This strategy failed at R= 1.5 x 10^6, but another strategy was sucessful. A converged steady interacting solution for R= 1.5 x 10^6 can be obtained when an unsteady path is followed, starting with the steady solution at R= 6 x 10^5. As a rule three sweeps are made at every timelevel. The almost steady solution on the 61 x 41 spatial grid at t= 8.0, after 75 timesteps, is used to initiate a calculation with the steady equations for R= 1.5 x 10^6: another 58 sweeps are required to satisfy the criterion that the maximum norm of changes in $u_e\delta^*$ is below 10^{-4}.

Steady classical boundary-layer solutions with edge velocity $u_{e,0}$ end up in a Goldstein singularity for indentations $\alpha < -0.0085$. Herwig [18] suggests that the relevant Euler solution for indentations with such a Goldstein singularity is of the Kirchoff free-streamline type. The steady interacting boundary-layer solution, being an attached boundary-layer strategy, supports the suggested detached boundary-layer structure, at least for the indentation $\alpha= -0.03$. Increase of the Reynolds number leads to a smooth transition from a steady attached flow to a steady detached flow, i.e. without any clear type of breakdown. The detached flow characteristics in the steady interacting solution for increasing Reynolds number are summarized:

1. The recirculation region fills the indentation more and more (figure 1a).
2. The wall shear stress in the recirculation region tends to zero (figure 1b).
3. The pressure becomes constant in the recirculation region (pressure plateau), except close to reattachment (figure 1c).
4. Viscous effects are more and more concentrated in a free shear layer.

The core of the shear layer, defined as $\max_y \frac{\partial u}{\partial y}(x,y)$, approaches the shape of the suggested free streamline (figure 1d).

It is seen that the second contribution to the edge velocity in (2) does not vanish for very large Reynolds numbers, but corrects the wrong choice of the relevant Euler solution $u_{e,0}$, which assumed an attached boundary layer in the limit $R \to \infty$. The interacting boundary-layer equations are not a true asymptotic description for the indentation $\alpha = -0.03$, because an asymptotic potential layer arises for very large Reynolds numbers between the free streamline and the wall according to Herwig's detached flow model with finite Reynolds number effects, which is not captured in this composite formulation. The transition from the attached flow description to the detached flow description is smooth, implying that a range of Reynolds numbers exists where both descriptions are valid. Care must be taken with the interpretation of an attached flow solution for a very large Reynolds number; the interacting boundary-layer equations for the considered indentation do not seem to warn with a special breakdown behaviour, although the solution has possibly left the range of validity.

We have verified that the unsteady classical boundary-layer solution for the indentation $\alpha = -0.03$ reveals a finite-time singularity of the Van Dommelen type (details not presented here). The above results show that the interaction completely removes this singularity for the indentation considered; a steady state limit of the unsteady equations can be obtained. In analogy with the analysis of the triple-deck equations by Stewartson et al., it is expected that a critical combination of α, R-parameters exists, where a clear breakdown warning of the interacting boundary-layer solution is given: non-existence of the steady solution, with a finite-time singularity or the growth beyond all bounds for $t \to \infty$ in the unsteady solution. Instead of checking this for an indentation $\alpha < -0.03$, the unsteady interacting flow around the impulsively started cylinder is determined; its flow structure is believed to be comparable to the flow along the plate with indentation $|\alpha| = O(1)$.

3. Impulsively started cylinder

During recent years much interest has been given to the unsteady boundary-layer development around the impulsively started cylinder. Van Dommelen [3] calculated very accurately that the unsteady classical boundary-layer solution develops a singularity within a finite time, implying the non-existence of a steady classical solution. The formation of this singularity has been confirmed by Cowley [19] and Ingham [20].

The characteristic velocity U_0 and length ℓ are taken as two times the oncoming velocity and the cylinder radius respectively. The boundary condition u=0 at x=0 and x=π follows from the assumption that the flow is symmetric. $u_{e,0}$ is taken as sin(x), resulting from the Euler solution that assumes a symmetric potential flow with a vanishingly thin boundary layer along the wall. The first term in Blasius' [21] series expansion for small times after the impulsive start is used to initiate the calculation at t=0.01.

The unsteady classical boundary-layer calculation has been repeated. Our implicit treatment of the x-derivatives in the backflow region requires some sweeps through the computational region; the convergence is very fast, and as a result three sweeps are made at every timelevel. The displacement thickness in figure 2a (81 x 41 spatial grid, with a timestep 0.01 for $0.01 \leq t \leq 0.1$ and 0.05 for $t > 0.1$) agrees with Van Dommelen until shortly before the appearance of the singularity at t=3.0045.

Unsteady interacting boundary-layer solutions have been generated for R= 10^6, 10^5 and 10^4. One has to realize that only a simple interaction model is used, in which the effect of the finite cylinder curvature, for example, is neglected. Stream- and vorticity lines and the displacement thickness for R= 10^5 are depicted in the figures 2b and 3. At t=2.7 two peaks are formed in the displacement thickness, corresponding to the individualization of a vortex within the recirculation region. The gridsize and timestep used are too coarse to decide whether the growth of the displacement thickness for R= 10^5 is singular at a finite time, or only leads to an unbounded solution for $t \to \infty$. The interacting solution for R= 10^6 develops peaks in the displacement thickness at t=2.9, which also characterize the breakdown of Smith's unsteady triple-deck solution for a leading-edge bubble. The solution for R= 10^4 on the 81 x 41 grid can be obtained until t=2.9, where the sweep process diverged (on the 41 x 21 grid solutions can be obtained for larger times), probably because the boundary layer has become too thick for our interaction model to make sense.

Navier-Stokes solutions for R= 10^5 and 10^6 are unknown to the present authors. Comparison in figure 4 of the interacting boundary-layer solution with the Navier-Stokes solution of Dennis & Staniforth [22] for R= 10^4 (a more extended calculation was recently given by Ta Phuoc Loc & Bouard [23]) shows a good agreement of the wall shear stress at t=0.6 and t=2.0, but there is only qualitative agreement at t=3.0. The interacting solution gives an improvement compared to the classical solution. Experimental studies are only known up to $R=10^4$ as well. Qualitative agreement is found between the interacting solution for $R=10^5$ and the experiments of Bouard & Coutanceau [24] for $R=10^4$; in particular the individualization of a vortex, absent in the classical solution, is indeed captured in the interacting solution.

It is concluded that the interacting boundary-layer approach for the impulsively started cylinder does at least delay the appearance of the finite-time singularity. However the unsteady interacting solution is unbounded and a steady interacting solution does not exist. Van Dommelen's suggestion that the singularity in the unsteady classical solution corresponds to the individualization of a vortex in the experiment of Bouard & Coutanceau is strengthened by the interacting results for $R=10^5$. The results also support the idea (Van Dommelen [3], Smith [11]) that the singularity in the unsteady classical solution is the mathematical representation of an infinitely fast growing vortex in y-direction in the limit $R \to \infty$, indicating an abrupt transition from an unsteady attached to an unsteady detached flow. The interacting solution for a finite Reynolds number calculates the initial phase of the vortex growth (maybe ending in a singularity), achieving a violently unsteady transition to the unsteady detached flow structure (dynamic stall). When the vortex becomes too large, the attached, interacting boundary-layer approach loses its validity; new asymptotic structures have to be developed to describe the unsteady detached continuation of the flow until the steady detached Kirchoff free-streamline limit (if it exists) is reached.

4. Conclusion

For very small plate indentations ($-0.0085 \leq \alpha \leq 0$) the steady classical boundary-layer solution does not break down with a Goldstein singularity; the boundary layer remains attached, even in the limit $R \to \infty$. Interaction for the indentation $\alpha = -0.03$ removes the Goldstein singularity. An unsteady evolution prevents the numerical instabilities which were found in earlier calculations of steady solutions at very large Reynolds numbers. Increase of the Reynolds number leads to a smooth transition from a steady attached flow to a steady detached flow, although the interacting boundary-layer solution for very large Reynolds numbers is not expected to capture the potential core of the detached flow; there may be a breakdown without any warning.

Interaction does delay Van Dommelen's finite-time singularity in the unsteady classical solution for the impulsively started cylinder, and further analysis is required to decide whether its appearance is totally prevented. The unsteady solution is unbounded and hence a steady interacting solution does not exist. The individualization of a rapidly growing vortex in the interacting solution for $R= 10^5$, also found in experiments for $R=10^4$, supports the idea that the breakdown of the classical and interacting cylinder solution refers to a sudden transition from an unsteady attached to an unsteady detached flow.

The two examples of boundary-layer flow discussed in this paper both feature Goldstein's singularity (in case of steady flow) and Van Dommelen's singularity (in case of unsteady flow) when the classical boundary-layer equations are solved. The addition of interaction leads to a different response in the two examples. This different response of the interacting boundary-layer equations is in agreement with the asymptotic theory of Stewartson et al. and Smith, in which a critical combination of parameters appears, beyond which a steady solution ceases to exist.

References

1. Prandtl, L.: Über Flüssigkeitsbewegung bei sehr kleiner Reibung, Proc. 3rd Int. Math. Congr. Heidelberg (1904) 484-491.
2. Goldstein, S.: On laminar boundary-layer flow near a position of separation, Quart. J. Mech. Appl. Math. 1 (1948) 43-69.
3. Dommelen, L.L. van: Unsteady boundary layer separation, Ph.D. thesis, Cornell University, New York (1981).
4. Sychev, V.V.: Concerning laminar separation, Mekh. Zhid. i Gaza 3 (1972) 47-59 (in Russian), Fluid Dyn. 7, 407-417 (in English).
5. Smith, F.T.: The laminar separation of an incompressible fluid streaming past a smooth surface, Proc. Roy. Soc. London A.356 (1977) 443-463.
6. Smith, F.T.: Laminar flow of an incompressible flow past a bluff body: the separation, reattachment, eddy properties and drag, J. Fluid Mech. 92 (1979) 171-205.
7. Brodetsky, S.: Discontinuous fluid motion past circular and elliptic cylinders, Proc. Roy. Soc. London A.102 (1923) 542-553.
8. Stewartson, K., Smith, F.T. & Kaups, K.: Marginal separation, Stud. Appl. Math. 67 (1982) 45-61.
9. Dijkstra, D.: Separating, incompressible, laminar boundary-layer flow over a smooth step of small height, Proc. 6th Int. Conf. Num. Meth. Fluid Dyn., Lecture notes in physics 90, Springer-Verlag, New-York (1979) 169-176.
10. Veldman, A.E.P. & Dijkstra, D.: A fast method to solve incompressible boundary-layer interaction problems, Proc. 7th Int. Conf. Num. Meth. Fluid Dyn., Lecture notes in physics 141, Springer-Verlag, New-York (1980) 411-416.
11. Smith, F.T.: Concerning dynamic stall, Aer. Quart. 33 (1982) 331-351.
12. Veldman, A.E.P.: A numerical method for the calculation of laminar, incompressible boundary layers with strong viscous-inviscid interaction, NLR TR 79023 U (1979).
13. Veldman, A.E.P.: New, quasi-simultaneous method to calculate interacting boundary layers, AIAA J. 19 (1981) 79-85.

14. Carter, J.E. & Wornom, S.F.: Solutions for incompressible separated boundary layers including viscous-inviscid interaction, NASA SP 347 (1975) 125-150.
15. Edwards, D.E. & Carter, J.E.: A quasi-simultaneous finite difference approach for strongly interacting flow, Proc. 3rd symposium on numerical and physical aspects of aerodynamic flows, Long Beach, California, Januari 21-24 (1985).
16. Rothmayer, A.P. & Davis, R.T.: Massive separation and dynamic stall on a cusped trailing-edge airfoil, Proc. 3rd symposium on numerical and physical aspects of aerodynamic flows, Long Beach, Januari 21-24 (1985).
17. Henkes, R.A.W.M.: Computation of the separation of steady and unsteady, incompressible, laminar boundary layers, Report LR-483, Delft University of Technology, Department of Aerospace Engineering (1985).
18. Herwig, H.: Die Anwendung der Methode der angepassten asymptotischen Entwicklungen auf laminare, zweidimensionale Strömungen mit endlichen Ablösegebieten, Dissertation, Ruhr-University, Bochum (1981).
19. Cowley, S.J.: Computer extension and analytic continuation of Blasius' expansion for impulsive flow past a circular cylinder, J. Fluid Mech. 135 (1983) 389-405.
20. Ingham, D.B.: Unsteady separation, J. Comp. Physics 53 (1984) 90-99.
21. Blasius, H.: Grenzschichten in Flüssigkeiten mit kleiner Reibung, Z. Math. Phys. 56 (1908) 1-37.
22. Dennis, S.C.R. & Staniforth, A.N.: A numerical method for calculating the initial flow past a cylinder in a viscous fluid, Proc. 2nd Int. Conf. Num. Meth. Fluid Dyn., Lecture notes in physics 8, Springer-Verlag, New-York (1971) 343-349.
23. Ta Phuoc Loc & Bouard, R.: Numerical solution of the early stage of the unsteady viscous flow around a circular cylinder: a comparison with experimental visualization and measurements, J. Fluid Mech. 160 (1985) 93-117.
24. Bouard, R. & Coutanceau, M.: The early stage of development of the wake behind an impulsively started cylinder for $40 < Re < 10^4$, J. Fluid Mech. 101 (1980) 583-607.

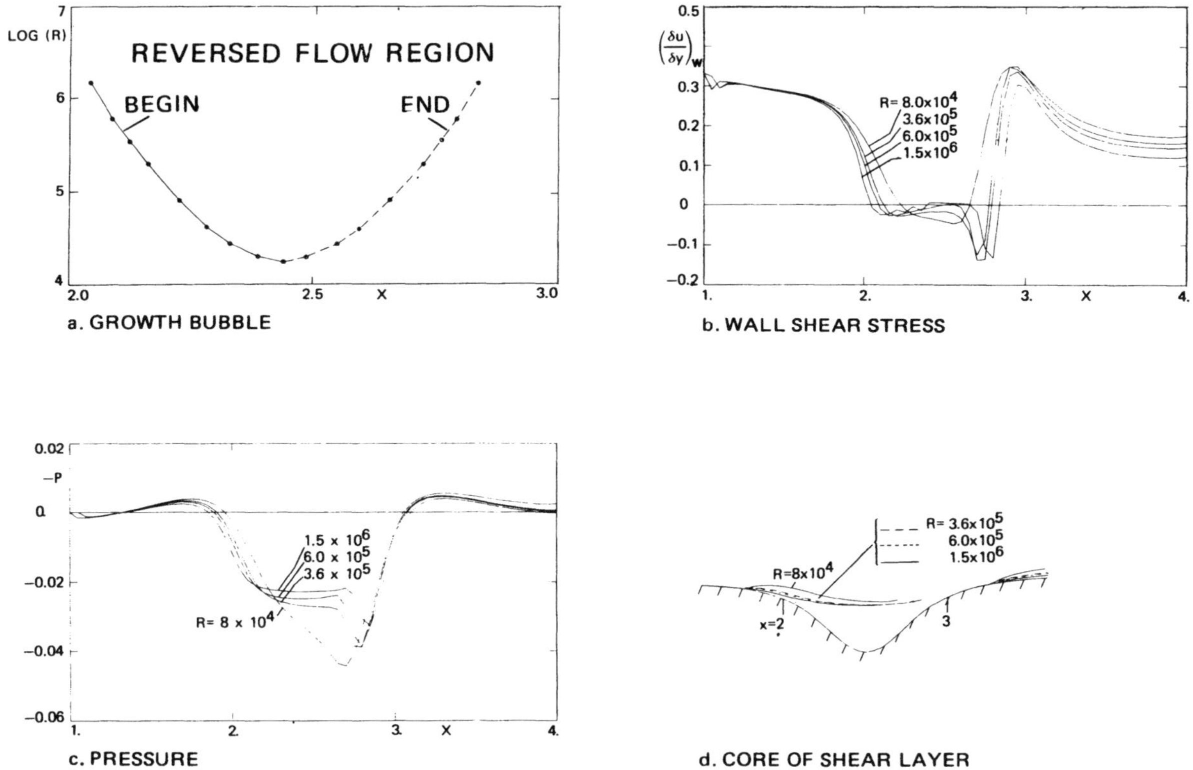

Fig. 1 Detached flow characteristics of steady interacting boundary-layer solution along indented plate

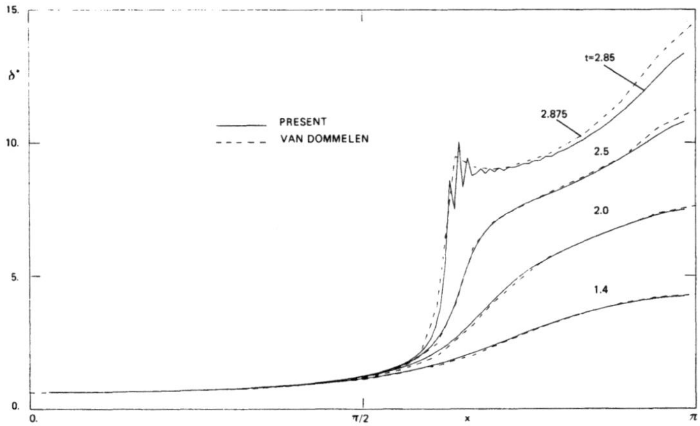

a. CLASSICAL BOUNDARY LAYER

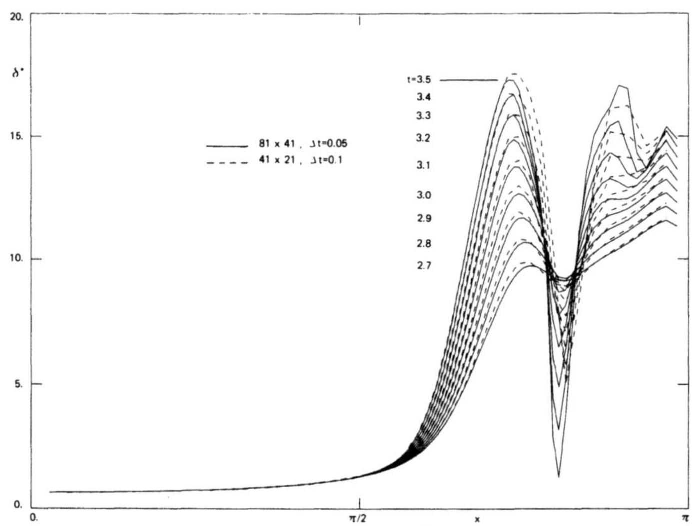

b. INTERACTING BOUNDARY LAYER FOR $R = 10^5$

Fig. 2 Displacement thickness for cylinder

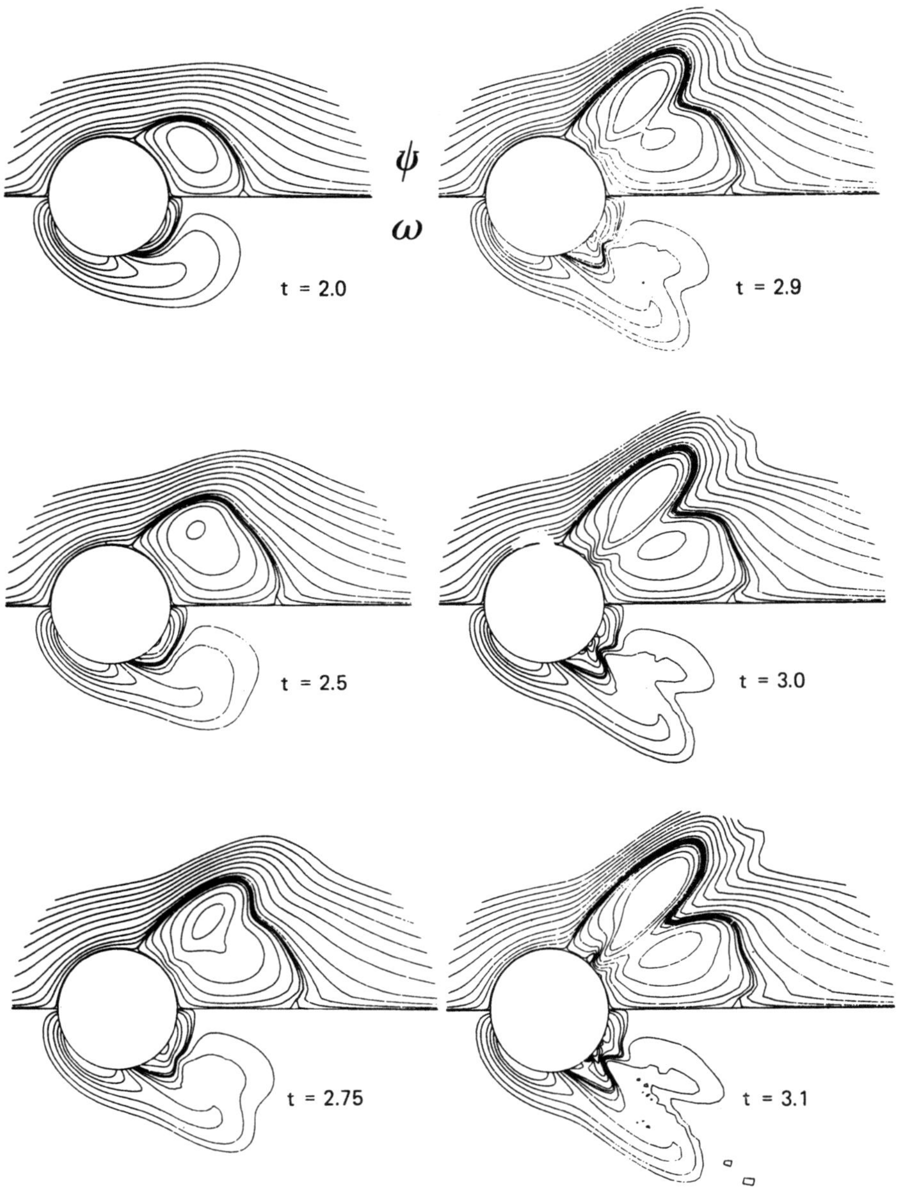

Fig. 3 Interacting boundary layer of cylinder for $R=10^5$; Stream- and vorticity lines. (radial coordinate of flow field is stretched with respect to cylinder radius)

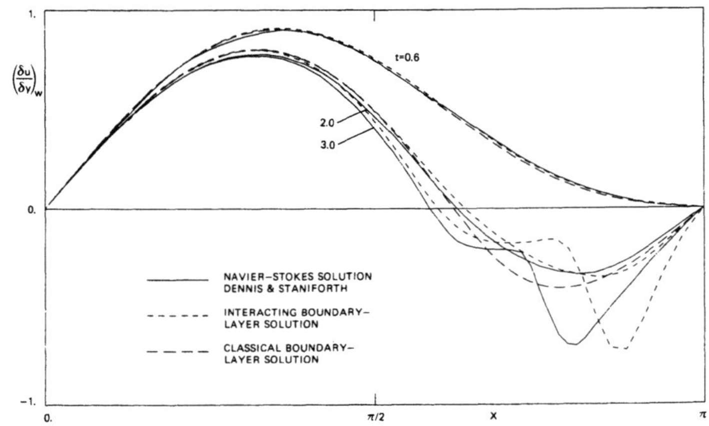

Fig. 4 Boundary-layer solution of cylinder for $R=10^4$ compared with Navier-Stokes solution

Asymptotic Separation from Slender Cones at Incidence

S. P. FIDDES, J. H. B. SMITH

Aerodynamics Department
Royal Aircraft Establishment
Farnborough, Hants, UK

Summary

An inviscid description of separated flow over elliptic cones at incidence in terms of vortex sheets and line vortices is developed within the framework of slender-body theory and with emphasis on solutions with lateral asymmetry. It is demonstrated that grossly asymmetric solutions with smooth separation exist for circular cones over a range of values of the incidence parameter, and properties of these solutions are displayed. For a typical incidence parameter and fixed symmetric separation position, it is shown that asymmetric solutions for elliptic cones only exist for cross-sectional thickness ratios greater than a critical value, at which they appear to bifurcate from the symmetric solution branch.

Introduction

Great advances have recently been made in our understanding of the interaction between the external inviscid flow and the separating boundary layer, as discussed in a recent review[1]. These advances are being applied fruitfully to technological problems in which the scale of the separation remains small and the flow is predominantly two-dimensional. If large-scale, essentially three-dimensional, separated flows, typically involving vigorous swirling motions, are to be treated similarly, models of the inviscid external flow are required in which rotation is represented. This paper presents some new developments in the use of one such model.

This is the vortex sheet model, applied in the framework of slender-body theory, with the infinitely rolled-up spiral core

F. T. Smith, S. N. Brown, (Eds.)
Boundary-Layer Separation
IUTAM Symposium London 1986
© Springer-Verlag Berlin, Heidelberg 1987

of the sheet replaced by a line vortex and a cut (or feeding sheet), as originally proposed by Mangler and Smith[2] for leading-edge separation from delta wings. The model was extended by Fiddes[3] to treat separation from smooth surfaces, providing the inviscid element of a calculation method for the separated flow over elliptic cones at incidence. His treatment explicitly represented the appropriate singular behaviour of the inviscid solution at the separation line[4], so that F.T. Smith's extension[5] of the triple-deck theory to three-dimensional flow could be used to describe laminar separation at high Reynolds numbers. The treatment of Ref 3 is confined to flows with lateral symmetry.

It subsequently became clear[6] that inviscid models of vortex flows over cones could shed light on the origins of the side force which is observed on slender pointed bodies with lateral symmetry at large angles of incidence and zero yaw. Accordingly, the treatment of the elliptic cone was modified to allow for lateral asymmetry in the flow and eventually an asymmetric solution for a circular cone with prescribed symmetric separation lines was found. The technique used to construct it has been described by Fiddes[7]. From this solution, a so-called 'second family' of solutions has been obtained by making small step-by-step changes in the parameters defining the solution. These parameters are illustrated in Fig 1: τ is the axis ratio of the elliptic cross-section of the cone, with $\tau = 1$ giving the circular cone and $\tau = 0$ the planar delta wing; θ_1 and θ_2 are the angular positions of the separation lines below the horizontal plane on the starboard and port sides; and a is the ratio of the angle of incidence, α, to the semi-angle, δ, of the horizontal section of the cone. First results from this second family of solutions for circular cones were presented by Fiddes and Smith[8]. Comparisons with observation showed agreement with a vapour-screen visualization of the vortex configuration, and with careful side-force measurements. Ref 7 showed agreement with pressure measurements. In all cases the observed separation line positions were used in the inviscid calculation. A review of the earlier work has been given by Smith[9].

Two new developments are now described. First, for a circular cone, we show smoothly separating asymmetric solutions of the second family. For these, the positions of the separation lines appear to be uniquely determined by the value of the incidence parameter. The sheets leave the surface without inducing adverse pressure gradients, so that the solutions are consistent with laminar boundary-layer separation in the limit of infinite Reynolds number. They also form the starting point for the prediction of separation position at finite Reynolds number[3,5]. Secondly, for elliptic cones oriented symmetrically to the onset flow and with separation lines chosen symmetrically, we show that the degree of asymmetry in second family solutions is reduced to zero as the thickness ratio, τ , is reduced. The second family appears to arise from a classifical bifurcation in the parameter τ . This result is consistent with the lower level of side force found in experiments on configurations with flattened noses.

Method

The method involves the use of slender-body theory for a vortex-sheet configuration in conical flow, the construction of a velocity potential by conformal mapping, superposition of singularities and the use of images, and the application of a generalized Newton-Raphson method to solve the set of non-linear equations which result. Accounts of these aspects are given in Refs 3 and 8, and will not be repeated here. However, if the results are to be understood, the discretization of the vortex sheets needs reiteration. Each sheet is constructed in a particular complex plane, obtained by first mapping the cross-flow plane exterior to the cross-section of the cone on to the upper half of the w-plane, with the separation line at the origin, $w = 0$, and then introducing a folding mapping:

$$\zeta^2 = sw \quad .$$

The sheet is constructed in the ζ-plane out of circular arc panels[2], arranged so as to provide continuity of slope and strength at their joins. Unless the base panel starting from

$\zeta = 0$ is actually flat, the sheet then has a square-root singularity in its curvature at the separation line in the physical plane, in accordance with Ref 4. The strength of this singularity is proportional to the curvature of the panel in the ζ-plane. The sheet boundary conditions are applied at the panel junctions.

The application of the theory requires that $\alpha \ll 1$ and $\delta \ll 1$; the results obtained for the geometry and strength of the vortex systems and for the variation in pressure round the cross-section of the cone are independent of Mach number. Discretization errors arise from replacing the infinitely rolled-up sheet by the model shown in Fig 1, from using a finite number of panels to represent the truncated sheet, and from terminating the Newton iteration at a non-zero tolerance. The last of these has no appreciable effect on any quantity of interest. Typical effects of the first two will be mentioned in context.

Smoothly separating second-family solutions for circular cones

In the asymmetric solutions presented so far[7-9], the separation lines have been typical of those observed, ie in the notation of Fig 1, θ_1 and θ_2 have been negative. The sheets then leave the surface abruptly, with large values of the coefficients $-A_1$ and A_2 of the curvature singularity at the base of the sheets. We should now like to find solutions with $A_1 = A_2 = 0$. From the description above, we see this is impossible, since it would entail the base panel lying in the surface of the cone. A fortiori , we cannot find solutions with $-A_1$ or A_2 negative, since the base panel would then lie within the cone. The best we can do is to demonstrate, for large enough values of the incidence parameter a , a reduction of $-A_1$ and A_2 through several decades, with no indication that further reductions are impossible. Conversely, we cannot say that for smaller values of a no such solutions exist: we can only report the failure of our search.

The form of the solutions of the second family resembles that sketched in Fig 1. We associate the more remote vortex with the starboard side and the suffix 1, the closer vortex with the port side and the suffix 2, though obviously the mirror image is equally possible. The general character of the dependence of the singularity coefficients A_1 and A_2 on separation position and incidence parameter is illustrated in Figs 2 and 3, both drawn for symmetric separation lines, $\theta_2 = \theta_1$. Fig 2 shows the variation of $-A_1$ and A_2 with a , for $\theta_2 = \theta_1 = 0.35 \doteq 20°$. The values of $-A_1$ and A_2 are not very different, despite marked asymmetry in the vortices. However, the difference between them is consistently of the same sign, and of about the same magnitude over the range of a . Using more panels on the same length of sheet has little effect on the difference between $-A_1$ and A_2 but using a shorter sheet increases the difference. The small values of $-A_1$ at the bottom of the range of values of a suggest that we are close to smooth separation. Fig 3 shows the variation of the same quantities with the position of symmetric separation for a fixed value, $a = 4.8$, of the incidence parameter. Again we see a consistent small difference between the coefficients for a given separation position. Since it emerges that A_1 depends primarily on θ_1 and A_2 primarily on θ_2 , we can read across the plot to deduce that the difference in separation position to achieve equal values of singularity coefficients is also small and fairly constant.

Unfortunately, the nearly linear decrease of $-A_1$ and A_2 with θ_1 and θ_2 does not persist to even lower values. Instead, larger increases in separation angle are required to produce similar reductions in singularity strength. However, we have found a pair of values of θ_1 and θ_2 for which both $-A_1$ and A_2 are less than 10^{-4} for values of a between 4.6 and 10, and these are taken to represent the positions of smooth separation, θ_{1s} and θ_{2s} . These values are shown in Fig 4. Again we see a consistent but very small difference between the port and starboard positions - just over 1° at most. Extending the length of the sheets by about half a turn about the line vortices reduced this difference by about a quarter.

More surprisingly, perhaps, the position of smooth separation varies little - less than 9° - over this large range of incidence parameter values.

Attempts to find a smooth solution for $a = 4.0$ and 4.4 failed. As θ_1 and θ_2 were increased, the Newton iteration eventually failed to converge, despite the use of very small steps. This failure to converge was associated with increasingly rapid variations in the solution as θ_2 changed. The smallest value of A_2 obtainable exceeded 0.17 at $a = 4$ and 0.08 at $a = 4.4$. The steepening of the variations of θ_{1S} as a falls, shown in Fig 4, suggests that solutions are running out. In contrast, both smooth symmetric solutions and non-smooth asymmetric solutions have been found down to $a = 3$.

Features of smooth solutions of the second family are shown in Figs 5 to 8. The sheet shapes and core positions are drawn in Fig 5 for three widely-spaced values of a. They clearly show a family resemblance. The almost symmetric separation positions are indicated. Arrows (A) point to the position of the attachment line on the leeward side, and we see that its position is much more sensitive to a. Fig 6 shows the variation of the circulation of the vortices (referred to Us) over the same range. Separate values are shown for the circulation of the line vortex and for the whole system. The greater strength of the port system agrees with the impression created by its larger diameter in Fig 5. The port system also has a larger proportion of its total circulation in the line vortex.

The strength of the sheets drawn in Fig 5 is shown in Fig 7. The arc length σ (referred to s) is measured from the separation line, so that if $\Gamma > 0$, $d\Gamma/d\sigma < 0$. Each sheet has an initial portion of almost constant strength, which generally extends from the separation line to the start of the rolled-up core. The exception is the starboard sheet for $a = 10$, which is perhaps affected by the proximity of the port core. These initial levels are larger on the port side. Beyond the initial portion, the strength falls and then fluctuates in phase with the local inclination of the sheet, as in the results for wing flows[10].

Fig 8 shows the pressure variation round the cone for four values of a , including those of Figs 5 and 7. The curves are displaced vertically for clarity and each shows the reduction in pressure coefficient, based on $\frac{1}{2}\rho U^2$, from its value at the windward generator. Note that the position of the windward attachment line is close to the windward generator in each case, so the difference in pressure between them is not apparent in the plot. The single suction peak, which lies under the port vortex, is conspicuous at each value of a , as in the earlier solutions[7]. The starboard vortex is apparently too remote to induce a separate peak. Also conspicuous are the pressure plateaus lying to leeward of the separation lines, S . The plateau level is higher on the port side, associated with the stronger sheet on that side, and contributing significantly to the side force. We point out that the separation is indeed smooth, with no adverse pressure gradient, showing the consistency of the modelling. The arrows along the pressure distributions show the directions of flow - from attachment to separation - and we note that the pressure gradients are favourable to boundary layer development except from the suction peak to the port separation line. Here the real flow would separate, producing a secondary vortex. A further vortex sheet could be incorporated in the model, but this might only lead to the appearance of a further adverse gradient, as found by Hoeijmakers[11] in a wing flow.

Finally in Fig 9 we show some evidence of the sensitivity of the overall flow to the major approximation in the model. For a case fairly close to smooth separation, two solutions are shown, one with an extra half-turn of each sheet about the cone. The differences are not large. Evidence from wing flow calculations[12] suggests that further increases in sheet length would reduce these differences.

Effects of thickness ratio

We turn now to a more practical aspect of the study - the reduction of side force attainable by shape changes. The family of shapes considered is that of cones of elliptic cross-section, defined by the values of the thickness ratio, τ , of the cross-section, as in Fig 1. We concentrate on a single value of the incidence parameter, $a = 4$, at which the second family solution for the circular cone, $\tau = 1$, is markedly asymmetric. The variation is most conveniently regarded as being with thickness, for fixed planform and angle of incidence. We also concentrate on a single, symmetric separation position, $\theta_1 = \theta_2 = 0$, chosen for compatibility with separation from the flat delta wing given by $\tau = 0$.

Calculated vortex configurations are shown in Fig 10 for $\tau = 1$, 0.8, 0.6 and 0.52. Comparison of Fig 10a with the configurations of Fig 9 suggests that the changes in incidence and separation position have not changed the character of the solution. However, as τ is reduced the asymmetry becomes less marked and has apparently disappeared at $\tau = 0.52$. To bring out the nature of the behaviour involved, Fig 11 has been prepared. This shows the heights of the vortices (referred to s) above the axis of the cone for $0.43 \leqslant \tau \leqslant 1$, and the coefficient of side force, C_Y , (based on planform area).

Looking first at the behaviour of the side force, we see it decreasing monotonically as τ decreases, eventually dropping rapidly to zero at $\tau = \tau_b$, where $0.47 < \tau_b < 0.48$. As far as can be seen, the derivative, $\partial C_Y/\partial\tau$, is infinite at $\tau = \tau_b$. For $\tau \leqslant \tau_b$, the side force is zero, and the other curves show that only a single vortex height is possible. For $\tau > \tau_b$, three vortex heights are possible. Along the central curve, only part of which is shown, the solutions are symmetrical. From this symmetrical solution, the two vortex heights of the second family of solutions branch off at $\tau = \tau_b$. The sign of the determinant of the Jacobian used in the Newton iteration changes sign in the interval (0.47, 0.48), confirming that a classifical bifurcation occurs. The appearance of a symmetric solution for $\tau = 0.52$ in Fig 10 does not indicate the absence of asymmetric solutions for this value of τ. The reduction of side force on slender pointed bodies with reduction in thickness ratio is familiar to aerodynamicists, and a flattened nose, corresponding to $\tau < 1$, is a feature of the highly-manoeuvrable F20 'Tigershark' aircraft. The effect has been attributed to the reduction in side area over which such a force may act, but the present study reveals a more subtle underlying interaction between shape and flow.

References

1. Smith, F.T.: Steady and unsteady boundary-layer separation. Ann. Rev. Fluid Mech. 18 (1986) 197-220.

2. Mangler, K.W.; Smith, J.H.B.: A theory of the flow past a slender delta wing with leading-edge separation. Proc. R. Soc. Lond. A251 (1959) 200-217.

3. Fiddes, S.P.: A theory of the separated flow past a slender elliptic cone at incidence. Paper 30 in Computation of viscous-inviscid interactions, AGARD CP291 (1980).

4. Smith, J.H.B.: Behaviour of a vortex sheet separating from a smooth surface. RAE Technical Report 77058 (1977).

5. Smith, F.T.: Three-dimensional viscous and inviscid separation of a vortex sheet from a smooth non-slender body. RAE Technical Report 78095 (1978).

6. Dyer, D.E.; Fiddes, S.P.; Smith, J.H.B.: Asymmetric vortex formation from cones at incidence - a simple inviscid model. Aeron. Quart. 33 (1982) 293-312.

7. Fiddes, S.P.: Separated flow about cones at incidence - theory and experiment. ICASE/Langley Workshop on vortex dominated flows (1985), Springer, to be published.

8. Fiddes, S.P.; Smith, J.H.B.: Calculations of asymmetric separated flow past circular cones at large angles of incidence. Paper 14 in Missile Aerodynamics, AGARD CP336 (1982).

9. Smith, J.H.B.: Theoretical modelling of three-dimensional vortex flows in aerodynamics. Aeron. J. (1984) 101-116.

10. Smith, J.H.B.: Improved calculations of leading-edge separation from slender, thin delta wings. Proc. R. Soc. Lond. A306 (1968) 67-90.

11. Hoeijmakers, H.W.M.: Methods for numerical simulation of leading-edge vortex flow. ICASE/Langley Workshop on vortex dominated flows (1985), Springer, to be published.

12. Smith, J.H.B.: Achievements and problems in modelling highly-swept flow separations. In Numerical methods in aeronautical fluid dynamics, ed. P.L. Roe, Academic Press (1982).

****

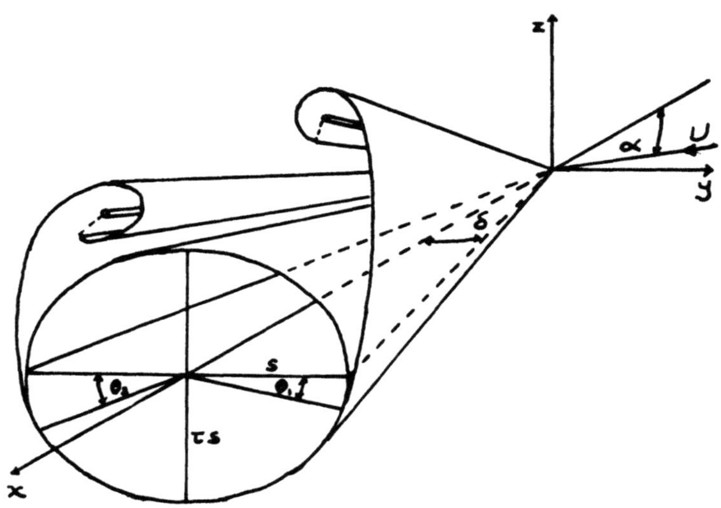

Fig.1 Configuration and notation

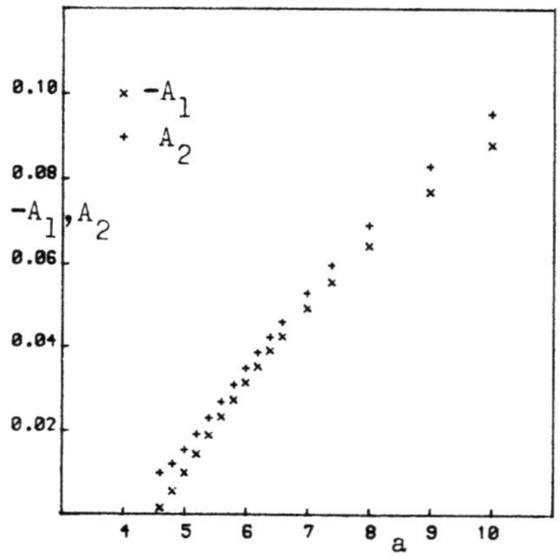

Fig.2 Singularity strength for $\theta_1 = \theta_2 = 0.35$

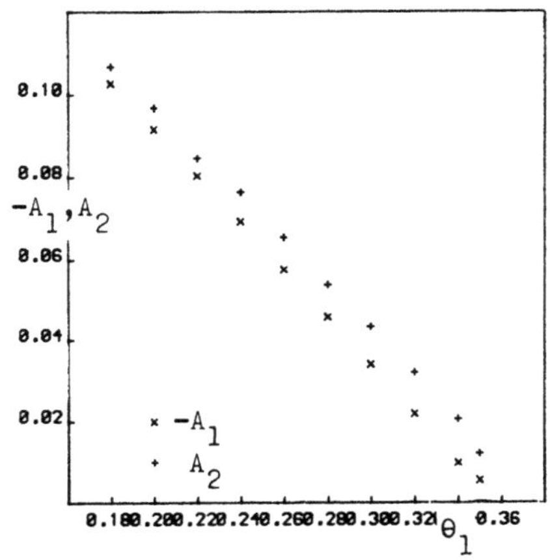

Fig.3 Singularity strength for $a = 4.8$ and $\theta_2 = \theta_1$

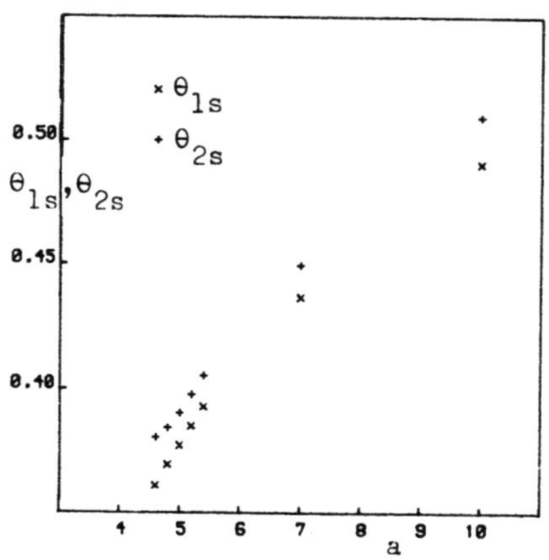

Fig.4 Smooth separation positions

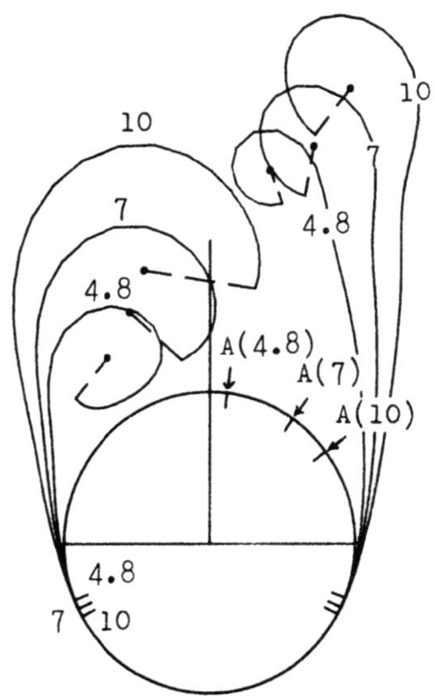

Fig.5 Smooth solutions for $a = 4.8, 7, 10$

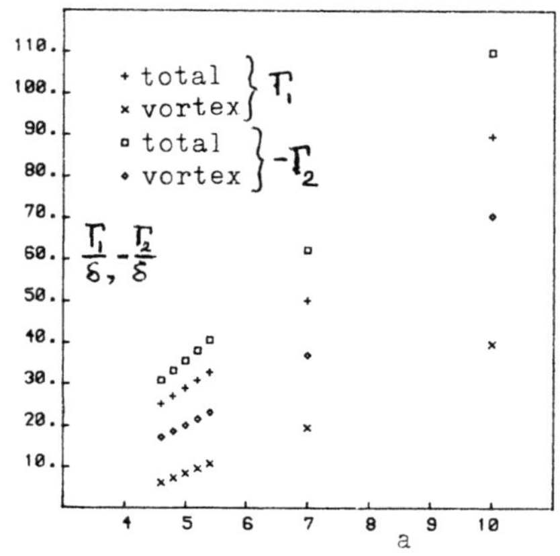

Fig.6 Circulation in smooth solutions

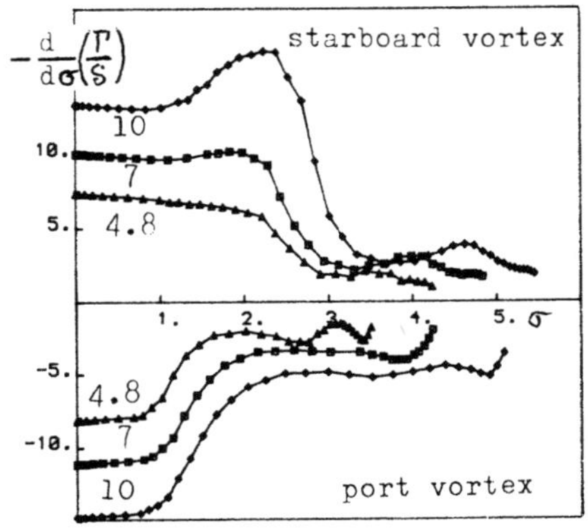

Fig.7 Sheet strength in smooth solutions

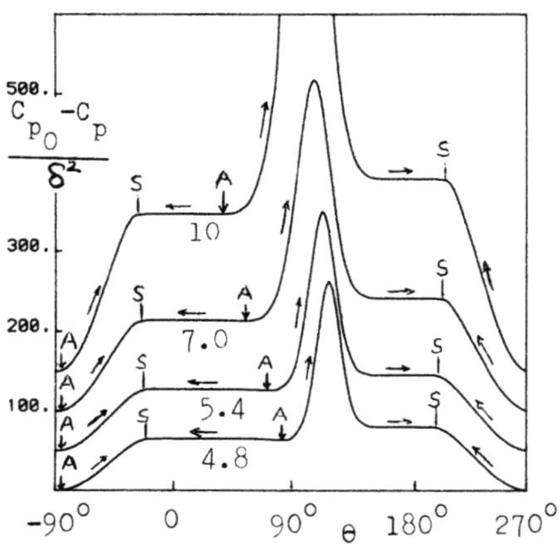

Fig.8 Surface pressure variation for smooth solutions

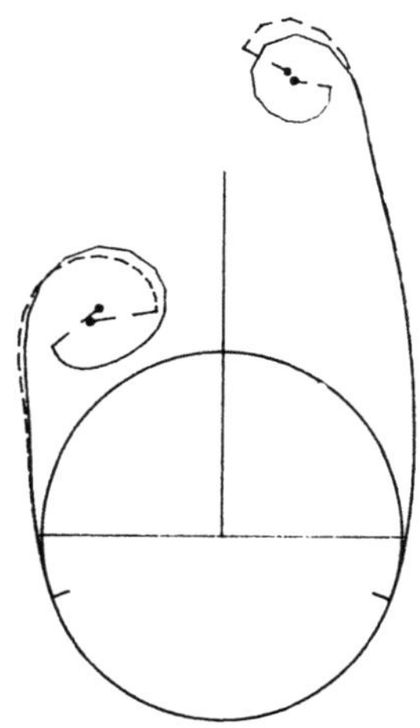

Fig.9 Effect of sheet extent: $a = 4$, $\theta_1 = \theta_2 = 0.35$

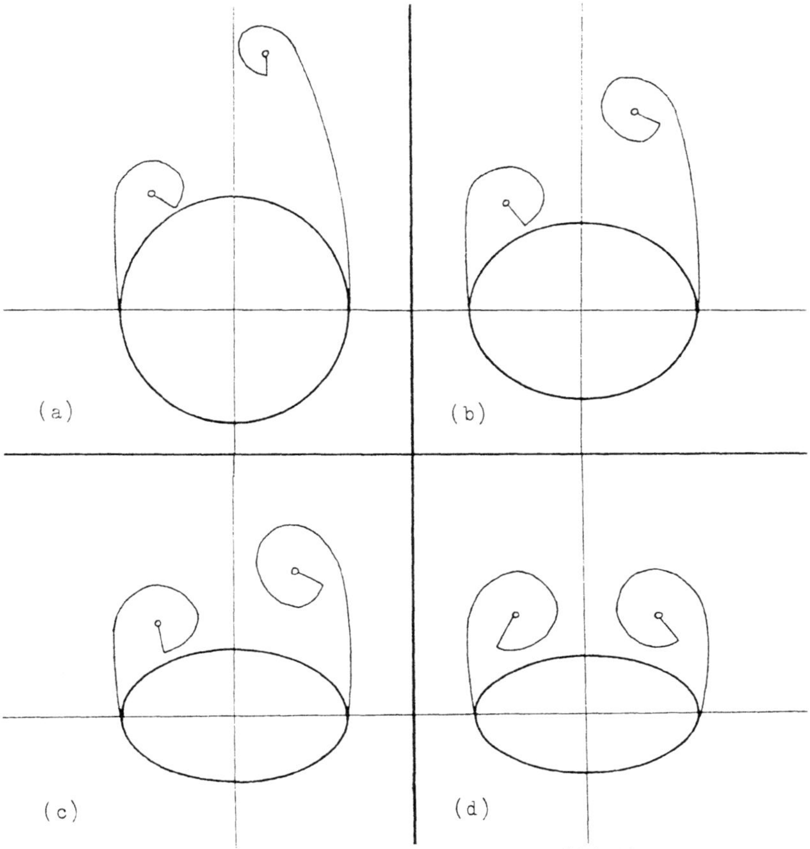

Fig.10 Vortex configurations on elliptic cones: $a = 4$, $\theta_1 = \theta_2 = 0$; $\tau = 1.0$(a), 0.8(b), 0.6(c), 0.52(d)

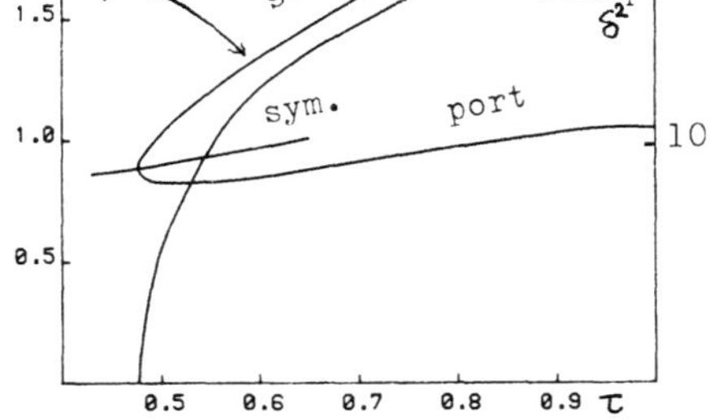

Fig.11 Bifurcation of solutions with thickness: $a = 4$, $\theta_1 = \theta_2 = 0$

Poster Session

Experimental Investigation of Topological Structures in Three-dimensional Separated Flow

H. Bippes
Deutsche Forschungs- und Versuchsanstalt
für Luft- und Raumfahrt e.V.
Institute for Experimental Fluid Mechanics
Bunsenstr. 10, D-3400 Göttingen, FR Germany

The simplest structures of three-dimensional separated flows are those where the vortices induced by the separating free shear layers are aligned with the oncoming flow. Such steady separation patterns are known to appear on aircraft wings and slender bodies at small angles of incidence and they are already modelled for numerical treatment. The present experiments are devoted to a second class of three-dimensional separated flows where a system of vortices develops with vortex filaments not everywhere aligned to the oncoming flow so that unsteady flow areas originate.

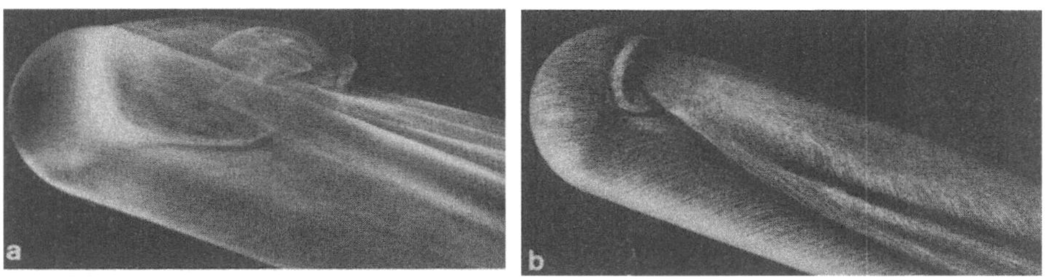

Fig. 1: Typical separation structure dealt with in the present experiments
a) Spatial structure visualized by dye in low speed water flow
b) Related skin-friction pattern visualized in wind tunnel flow

Fig. 2: Separation pattern on the upper side of a rectangular wing at incidence. Angle of incidence $\alpha = 22.5°$, $Re_c = 2.4 \times 10^6$

F. T. Smith, S. N. Brown, (Eds.)
Boundary-Layer Separation
IUTAM Symposium London 1986

A simple example of this type of separation is visualized on a hemisphere cylinder at incidence. Fig. 1a shows the spatial flow field and Fig. 1b the related skin-friction pattern. Besides the vortices along the flanks of the hemisphere cylinder a second pair of vortices emanates from spiral nodes in the skin-friction field. The similarity of the skin-friction fields developing on a hemisphere cylinder (Fig. 1b) and on a rectangular wing (Fig. 2) indicates that the discussed class of separated flows is not restricted to a special model geometry but is of more general significance. Unsteady effects superimposed on the apparently complex spatial structure may have prevented any attempts of modelling such flows for a long time. However, velocity field measurements averaged over a short period [1] and oil-flow visualizations [2] show a steady flow field.

An attempt has been made to detect and classify possible topological structures of the class of three-dimensional separated flows defined above (see also [3]). By varying systematically the angle of incidence, Reynolds number and Mach number a man-

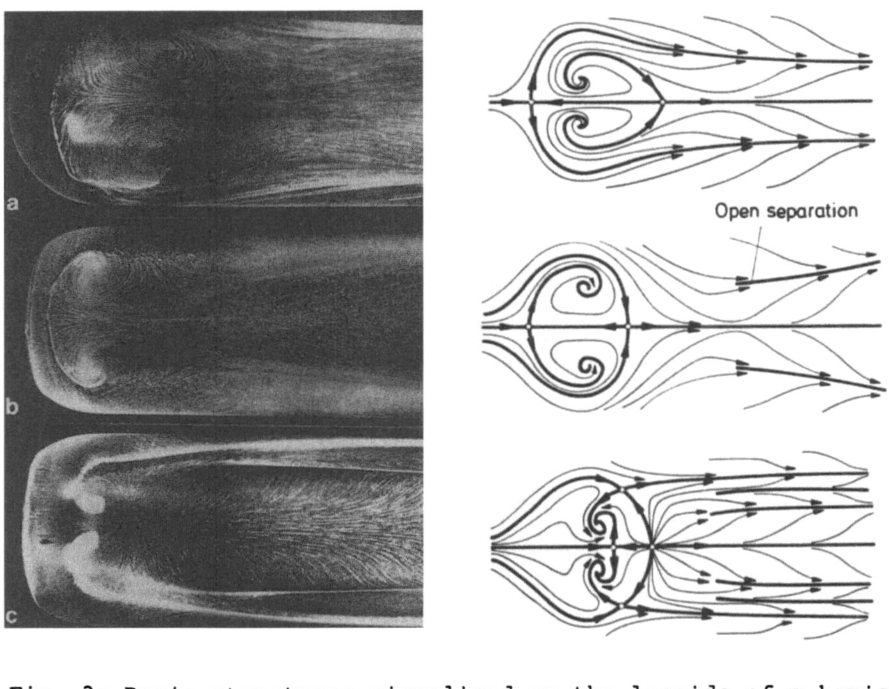

Fig. 3: Basic structures visualized on the leeside of a hemisphere cylinder; a) $\alpha = 10°$, $Re = 2.1 \times 10^6$, $M = 0.8$; b) $\alpha = 15°$, $Re = 1.9 \times 10^6$, $M = 0.7$; c) $\alpha = 33.5°$, $Re = 1.7 \times 10^6$, $M = 0.6$

ifold of skin-friction patterns was established on the hemisphere cylinder. The analysis revealed that even the most complicated configurations can be deduced from a simple basic structure by subsequent splitting of saddles into saddle-node-saddle combinations as illustrated in Fig. 3.

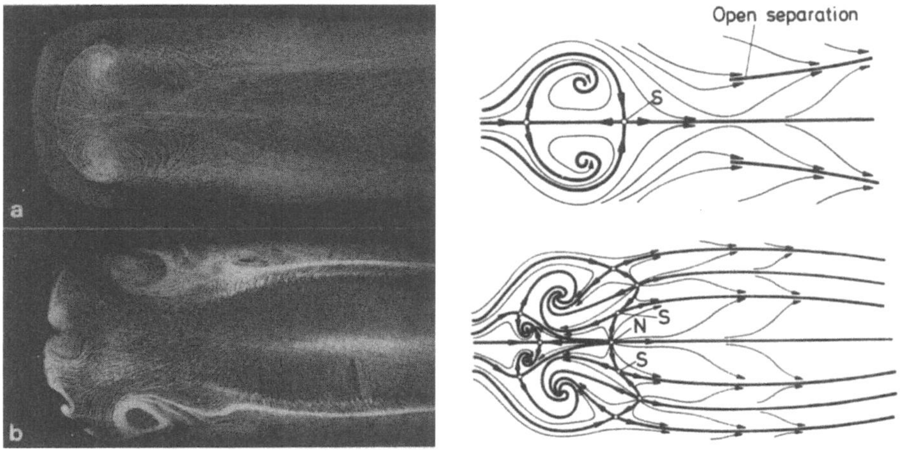

Fig. 4: Development of complicated separation patterns from a basic structure; a) $\alpha = 10°$, Re = 1.9×10^6, M = 0.7; b) $\alpha = 33.5°$, Re = 2.2×10^6, M = 0,95

On the hemisphere cylinder three different basic structures are identified. The first one, Fig. 4a has been observed only in compressible flow if separation is induced by a shock. The second one develops in stable laminar flow (i.e. if the flow persists in laminar state after separation) as well as in compressible flow not only on the hemisphere cylinder but also on the rectangular wing. It is therefore considered as the most stable one. The third basic structure could only be observed on the hemisphere cylinder in compressible turbulent flow.

References

1. H. Bippes: Experimental Study of Vortical Features in Three-Dimensional Separated Flows. 9th US-FRG DEA-Meeting, White Oak, MD., pp. 97-117, 1984.
2. H. Bippes, M. Turk: Oil Flow Patterns of Separated Flow on a Hemisphere Cylinder at Incidence; DFVLR-FB 84-20, 1984.
3. H. Hornung, A.E. Perry: Some Aspects of Three-Dimensional Separation, Part I: Streamsurface Bifurcations. ZFW 8, Heft 2, pp. 77-87, 1984.

Experimental Investigation of Three-dimensional Separation on Ellipsoid Bodies at Incidence

B. CHANETZ

OFFICE NATIONAL D'ETUDES ET DE RECHERCHES AEROSPATIALES
BP 72. F - 92322 Châtillon Cedex, France

The present research concerns the phenomenon of incompressible three-dimensional separation on two types of obstacle :

- the first model is an Ellipsoid Cylinder of Revolution (E.C.R) which was tested in the O.N.E.R.A. F2 subsonic atmospheric wind tunnel (see Fig. 1).
- the second model is an Oblate Ellipsoid Cylinder (O.E.C) which was used by using 3D LDV system in the O.N.E.R.A S2Ch wind tunnel (see Fig. 2).

Fig. 1 – Model installation in the O.N.E.R.A. F2 wind tunnel.

Fig. 2 – Model installation in the O.N.E.R.A. S2Ch wind tunnel.

The ECR model is made of a half ellipsoid of revolution (semi axis : $a = 0.8$ m, $b = c = 0.2$ m) followed by a cylindrical part.

Evolution of separation on the E.C.R. model with various angle of attack α and Reynolds number has been carefully investigated by close examination of the surface flow properties [1]. Thus, Fig. 3 shows the surface flow pattern observed for $\alpha = 40°$ and an upstream velocity equal to 90 m/s. The skin friction line pattern reveals the existence, as main features, of a primary separation line S1, of a secondary separation line S2 and, between S1 and S2 of an attachment line A. The drawing of Fig 4 schematizes this complex flow, in a plane perperdicular to the model axis, and helps in interpreting the photograph of Fig. 3.

F. T. Smith, S. N. Brown, (Eds.)
Boundary-Layer Separation
IUTAM Symposium London 1986

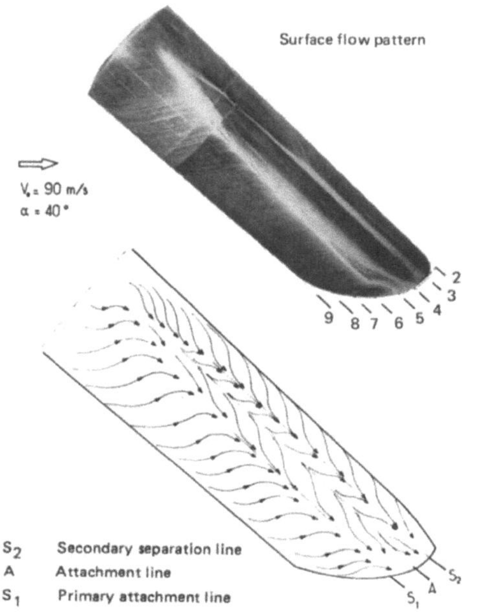

Fig. 3 – Separation phenomenon at high incidence.

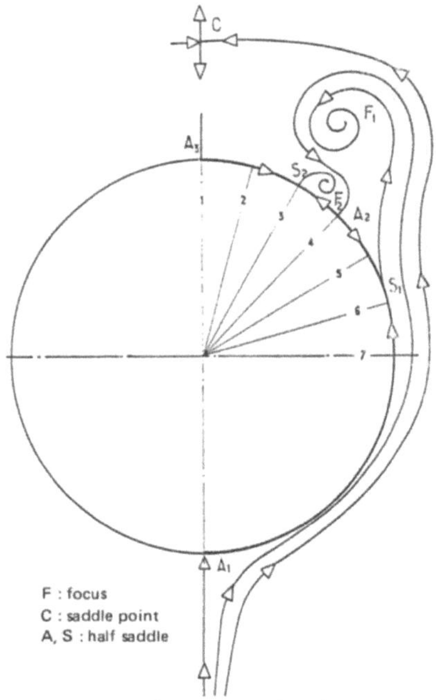

Fig. 4 – Sketch of the flow in a perpendicular plane X = 0.8 m.

The O.E.C. model is made of an oblate half cylinder (semi axis : a = 0.8 m, b = 0.4 m, c = 0.2 m) with a cylindrical extension terminated by a flat base. The detailed flow field explorations were carried out for the following conditions : $\alpha = 30°$, $V_o = 50$ m/s. These conditions lead to the development of a well separated flow on the leeward side of the model. The phenomenon gives rise to two main vortices located symmetrically with respect to the vertical symmetry plane. The flow field in the region of vortex formation was probed in fifteen planes normal to the model axis by using a 3D LDV system. Fig. 5a shows vector plot of the mean velocity component contained in the plane normal to the model axis located at X = 0.96 m (the most upstream measurement plane). The contour lines of the normalized turbulent kinetic energy $k/V_o^2 = 0.5\,(\overline{u^2} + \overline{v^2} + \overline{w^2})/V_o^2$ for the same plane are represented in Fig. 5b. Fig. 6 shows the same results in the most downstream plane (X = 1.2 m). There, the vortex is well organized and well detached from the wall as revealed by the cross flow structure (Fig. 6a). At the same time, as shown in Fig. 6b, the turbulent energy level has increased.

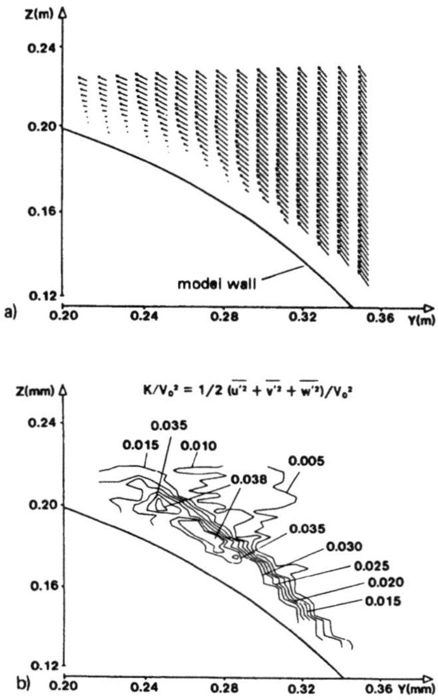

Fig. 5 – Vortex in the plane
X = 0.96 m.

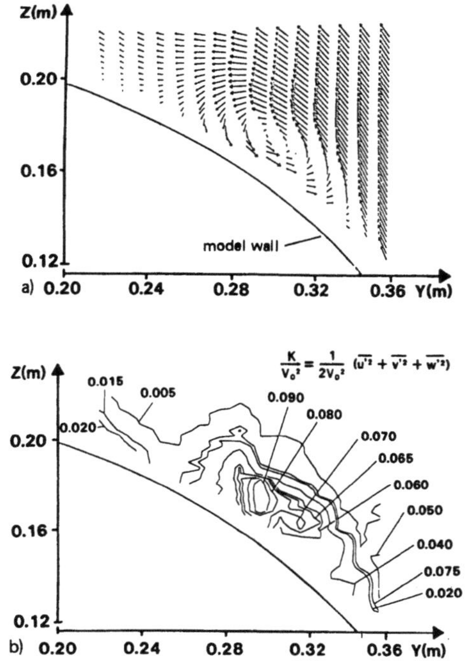

Fig. 6 – Vortex in the plane
X = 1.2 m.

The experimental investigation performed on simple shaped bodies provides fundamental information on turbulent separation in three-dimensional flows. The data obtained contribute to a better physical understanding of the flow structure and are used to validate theoretical models [2].

REFERENCES

1. Chanetz B., Contribution à l'étude du décollement tridimensionnel en écoulement turbulent incompressible. Thèse de Doctorat, Université de Lyon 1, septembre 1986.

2. Barberis D., 3-D boundary layer computation on arbitrary obstacles with direct and inverse methods. Aerospatial Research, 1986-3, pp. 169.195.

Three-dimensional Separated Flow Structures

U. Dallmann, W. Kordulla, H. Vollmers

DFVLR-Institute for Theoretical Fluid Mechanics,
Bunsenstr. 10, D-3400 Göttingen, FRG

In the following we emphasize some selected problems associated with the description and analysis of three-dimensional separated flow phenomena. An example is given with the help of the numerical simulation of the transonic laminar flow past a hemisphere cylinder. Further informations are given in [1,2], and details have to be published elsewhere.

Surface oil-flow visualizations certainly provide necessary but not sufficient information about the flow structure. The non-uniqueness of the relation between skin friction lines and the three-dimensional streamline pattern within the outer flow field requires extensive flow field measurements or the numerical simulation of the flow topology. But in three-dimensional separated flows principle numerical problems arise due to streamline sensitivity on initial conditions. Therefore bifurcating elementary topological structures and their streamsurface representations have been investigated on the basis of dynamical equations locally equivalent to the equations of motion. Using these elements we conjecture so called (instantaneous) singular streamsurfaces (fig. 1) for representing the flows given by Bippes' oil-flow visualizations (see this volume).

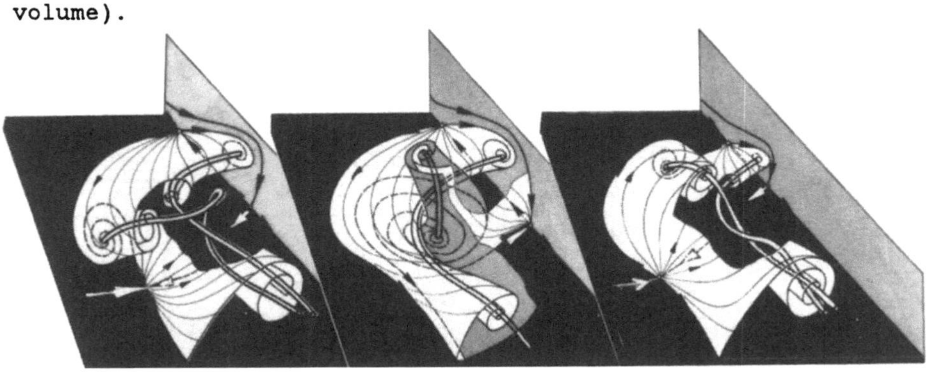

Fig. 1: Conjectured singular streamsurfaces corresponding to fig. 1ab, 3a, 3c of Bippes' oil-flow visualizations (this volume).

F. T. Smith, S. N. Brown, (Eds.)
Boundary-Layer Separation
IUTAM Symposium London 1986

Two different flow fields are called topologically identical if they show qualitatively the same distribution and connections of singular (critical) points and singular streamsurfaces. Hence, comparison of experimental and numerical flow data should be accompanied by comparison of the flow topologies. Severe numerical difficulties may be encountered [1] in the simulation of viscous three-dimensional separating flows. It is very difficult to obtain a perfectly two-dimensional high Reynolds-number flow with separation in experiments, too. The slightest asymmetry in the numerical procedure may cause three-dimensional vorticity separation and may trigger unsteadiness. A numerical simulation which was described in detail in [2] examplifies such a sensitivity to form three-dimensional separating flow around an axisymmetric configuration at zero angle of attack. At a Mach number of 0.9, and a Reynolds number based on the radius of the sphere of 212500 a quasi-2D separation bubble was expected to form around the body for zero angle of attack, and was supposed to deform somehow for 19°. But a very complex flow results (fig. 2). Locally swirling flow is not always associated with separation of boundary layer vorticity away from the body only. It appears that streamlines spiral into as well as out of foci indicating vortical flow attachment as a detail of the separation phenomenon. (See also fig. 1 and Bippes' contribution).

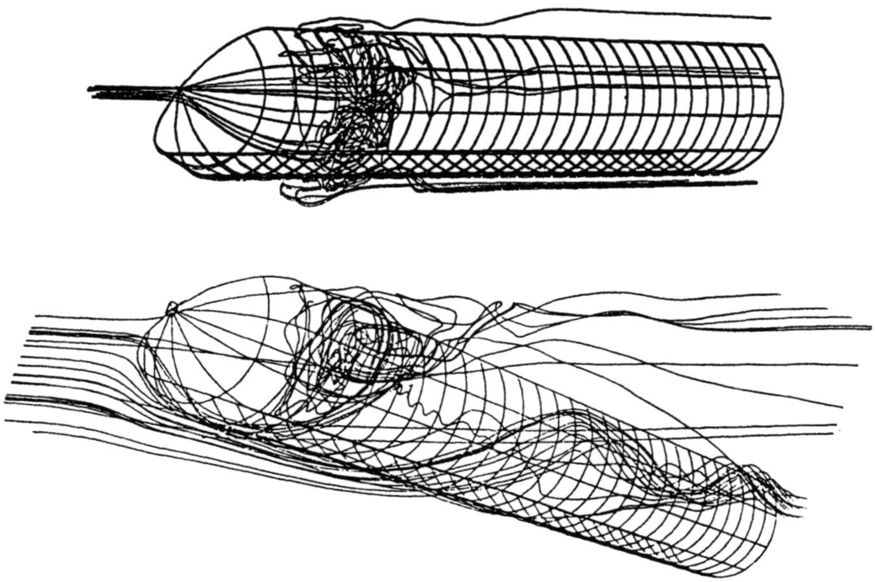

Fig. 2: Navier-Stokes simulation of flow around hemisphere-cylinder , Ma = 0.9, Re = 212500 and $\alpha = 0°$ (above), 19° (below).

One has to choose appropriate points close to the critical points within the instantaneous flow field to integrate the corresponding instantaneous streamlines (fig. 2) in order to compare the numerical data with experiment via flow topology. Otherwise the complexity of those swirling motions cannot be resolved at all because of sensitivity to initial conditions. Following [2] we have been interested in the topological changes between axisymmetric and fully three-dimensional flow separation. Only a finite number of different elementary topologies bifurcates from the axisymmetric recirculating separation bubble. In fig. 3 we only indicate the associated regions of vortical (swirling, spiralling) motions. Complex separating flows (fig. 2) can be resolved into such well-defined details that can be compared with experiments (see [2]). The vorticity field created within the boundary layer or within curved shocks often induces additional gradual vortex formation associated simply with local streamsurface folds. So called open separations form in a similar way but details have to be discussed elsewhere.

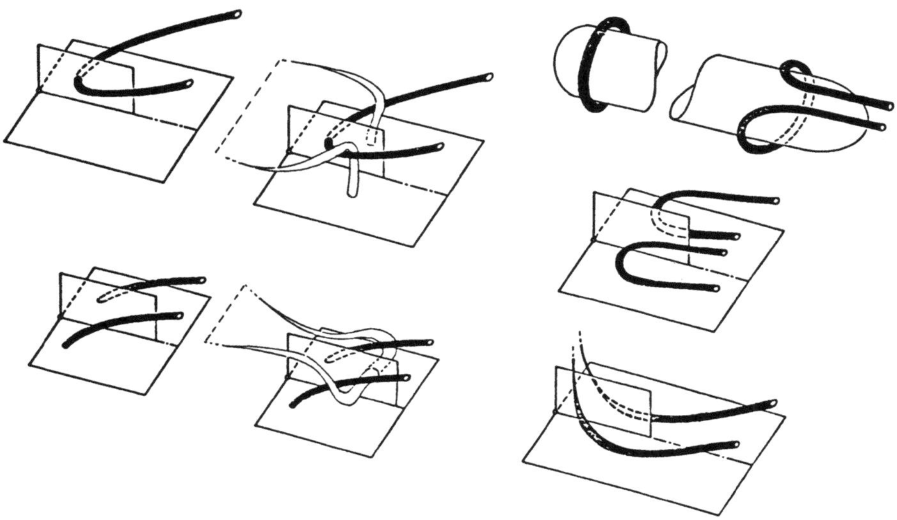

Fig. 3: Principle topological changes of axisymmetric separation bubble leading to streamwise vortical flow. (Schematic)

1. Dallmann, U.: Three-dimensional vortex separation phenomena. A challenge to numerical methods. Proc. Fifth GAMM Conf. on Numerical Methods in Fluid Mech., Vieweg (1983).
2. Kordulla, W.; Vollmers, H.; Dallmann, U.: Simulation of 3-D transonic flow with separation past a hemisphere-cylinder configuration. AGARD-CP 412 (1986).

Asymptotic Theory of High Reynolds Number Turbulent Flows

E. DERIAT
OFFICE NATIONAL D'ETUDES ET DE RECHERCHES AEROSPATIALES
BP 72. F - 92322 Châtillon Cedex, France
J-P. GUIRAUD
University Paris VI, Mécanique Théorique
Tour 66, Place Jussieu
75230 PARIS CEDEX 05

Summary

The communication reports joint work by both authors which will be published at length elsewhere [1, 2]. It concerns asymptotic studies of turbulent boundary layer flow. One part of the work [1] deals with model equations while the other part [2] deals with turbulence in boundary layers independently of any model and discusses what meaning may be given to turbulence in boundary layer when the thin layer structure is accounted for, asymptotically.

Asymptotics According to Model Equations

We consider two-dimensional steady, incompressible, turbulent flow around a streamlined body. A standard k-ε model is used throughout, except in the wall layer where we turn to the mixing length one. Three parameters are involved : a Reynolds number $R = U^*_C L^*_C/\nu$ and two turbulent intensity parameters $T = k^*_C/U^{2*}_C$ and $T^{BL} = k^{*BL}_C/U^{2*}_C$; they are defined with a characteristic macroscopic velocity U^*_C, a characteristic length L^*_C for the body, two levels of turbulent kinetic energy k^{*BL}_C inside the boundary layer and k^*_C outside it, and the kinematic viscosity ν. The basic assumption is that R^{-1}, T and T^{BL} are all small and $T \leq T^{BL}$. Matched asymptotic expansions are carried over in three regions : the main one which is, to leading order, inviscid, the defect part of the boundary layer which has a relative thickness of the order $(T^{BL})^{\frac{1}{2}}$, and the wall layer with a relative thickness of the order $R^{-1}(T^{BL})^{-\frac{1}{2}}$. The equations for the inviscid region consist in the Euler ones, decoupled from two ordinary differential equations along each streamline which rule the turbulent kinetic energy and the turbulent isotropic dissipation. This may be considered as giving asymptotic support to all the works based on the inviscid equations. The defect nature of the velocity within the main part of the boundary layer is shown to arise from the necessity to keep the production of turbulence within limits compatible with the level of turbulent intensities whatever it is. The turbulent term in the momentum

F. T. Smith, S. N. Brown, (Eds.)
Boundary-Layer Separation
IUTAM Symposium London 1986

equation and the production one involve contributions from the wall curvature and from the exterior velocity gradient which are usually

Nonuniqueness of Triple-deck Solutions for Axisymmetric Supersonic Flow with Separation

PH. GITTLER and A. KLUWICK

Institut für Strömungslehre und Wärmeübertragung
Technische Universität Wien, Austria

The triple-deck equations (e.g.[1]) for axisymmetric supersonic laminar boundary layers,[2]

$$\frac{\partial u}{\partial x} + \frac{\partial v}{\partial y} = 0 \quad , \quad u\frac{\partial u}{\partial x} + v\frac{\partial u}{\partial y} = -\frac{dp}{dx} + \frac{\partial^2 u}{\partial y^2}$$

$$u=v=0 \text{ for } y=F(x),\ u=y \text{ for } x\to-\infty,\ u=y+A(x) \text{ for } y\to\infty$$

$$p(x)=-A'(x) + \frac{1}{a}\int_{-\infty}^{x} W\left(\frac{x-\xi}{a}\right)A'(\xi)\,d\xi \ , \quad W(z)=\int_0^\infty \frac{\exp(-\lambda z)}{K_1^2(\lambda)+\pi^2 I_1^2(\lambda)}\frac{d\lambda}{\lambda} \ ,$$

$$\text{where } F(x) = \begin{cases} 0 & \text{for } x<-\rho \\ \alpha\left(\frac{x^2}{4\rho} + \frac{x}{2} + \frac{\rho}{4}\right) & \text{for } -\rho\leqq x\leqq\rho \\ \alpha x & \text{for } x>\rho \end{cases}$$

describes a smoothed cylinder-cone configuration, have been solved for $a=1.$, $\rho=1.$ and moderately large values of the scaled flare angle α. Here a denotes the scaled radius of the cylinder.

If α is positive and sufficiently large the boundary layer separates upstream of the corner and the formation of a pressure plateau similar to the two-dimensional case is observed, Fig.1.

In contrast to the case of planar flow, however, separation can occur also if $\alpha<0$ owing to the axisymmetric effect of overexpansion and recompression. The separation point then is located downstream of the corner and, most remarkable, it is found that the solutions of the asymptotic interaction equations are not unique and consist of three branches if $-11.28\geqq\alpha\geqq-12.42$, Fig.2. If $\alpha>-11.42$ one of the solutions describes attached flow while all possible flow patterns include regions of separation if $\alpha\leqq-11.42$, Figures 2 and 3. Unfortunately, the numerical scheme did not yield converged solutions with long separation bubbles (branch (c)) for values of α less than -11.7 but the results

F. T. Smith, S. N. Brown, (Eds.)
Boundary-Layer Separation
IUTAM Symposium London 1986

flow. We have investigated the consequences of this way of attacking the problem of boundary layer turbulence at various levels. The first one deals with the decomposition of the field into means and fluctuations. The second deals with the process of simplifying the averaged equations by taking into account the thin layer assumption. The third level concerns the derivation of higher order averaged equations and their asymptotic structure. In order to do that we have had to discuss the process of getting equations for the fluctuating quantities and their simplification under the assumption of thin layer. We have been faced with a delicate issue : could we neglect, in some averaged equations, terms coming from neglected terms in the equation for fluctuating quantities from which the averaged equation is issued through averaging ? Despite the seemingly inconsistent character of the answer we have shown that one cannot and we have discussed what terms could be neglected and what needed to be retained. Finally the last level is the one of closure. Asymptotics is evidently not sufficient for dealing with closure and it provides rather limited help. Nevertheless we have discussed this issue. Our view is that closure should be understood as the search for function relations between averages of turbulent quantities and the mean velocity profile as a function of the sole transverse coordinate. This functional should be a non local one and we have discussed its approximation taking into account of the double deck structure of the boundary layer. We express the view that the most important step in future research will be the one about modeling the Reynolds shear stress in the wall layer by using a combination of asymptotic analysis, analytical calculations on models, numerical simulation and ingenuity.

References

1. E. DERIAT : Analyse asymptotique du modèle k-ε pour la couche limite turbulente. To appear in : La Recherche Aérospatiale 1987 (French and English editions).

2. E. DERIAT and J-P. GUIRAUD : On the asymptotic description of turbulent boundary layers. To appear in : Journal de Mécanique Théorique et Appliquée. Special issue 1986 (English edition).

are in reasonable good agreement with an asymptotic structure for $|\alpha| \to \infty$, [2]. Since α is a scaled variable containing the free stream Machnumber and the Reynoldsnumber these results indicate the occurence of a hysteresis phenomenon if α is varied continuously from $\alpha > -11.28$ to $\alpha < -12.42$ and back to $\alpha > -11.28$.

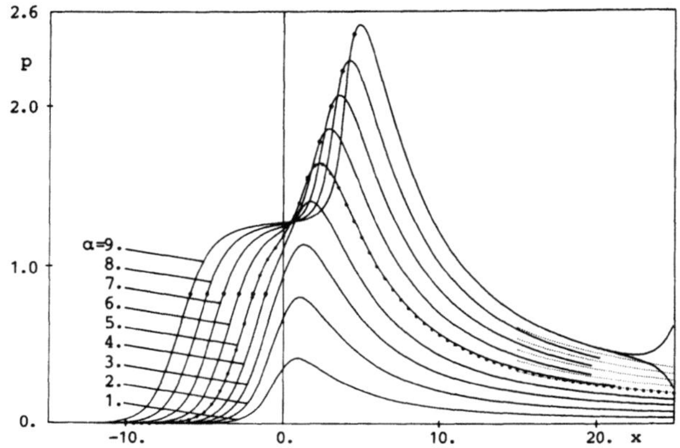

Fig.1. Pressure distribution for various positive cone angles
—— $\alpha=1.\to4.$ } spectral method, —— $\alpha=5.\to9.$ finite difference method (cf.[2])
··· $\alpha=5.$

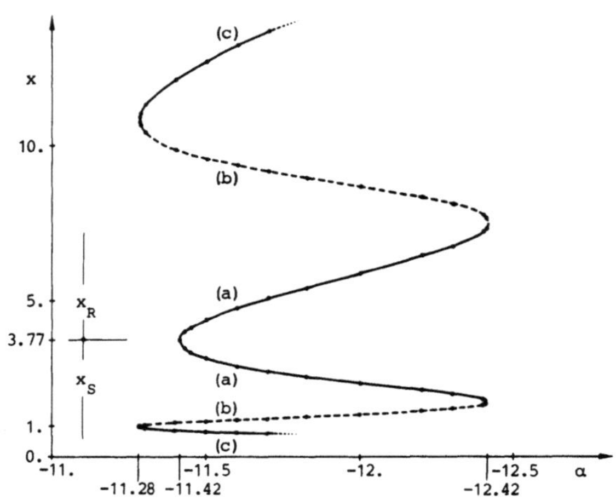

Fig.2. Position of the separation point x_S and the reattachment point x_R for various negative cone angles α

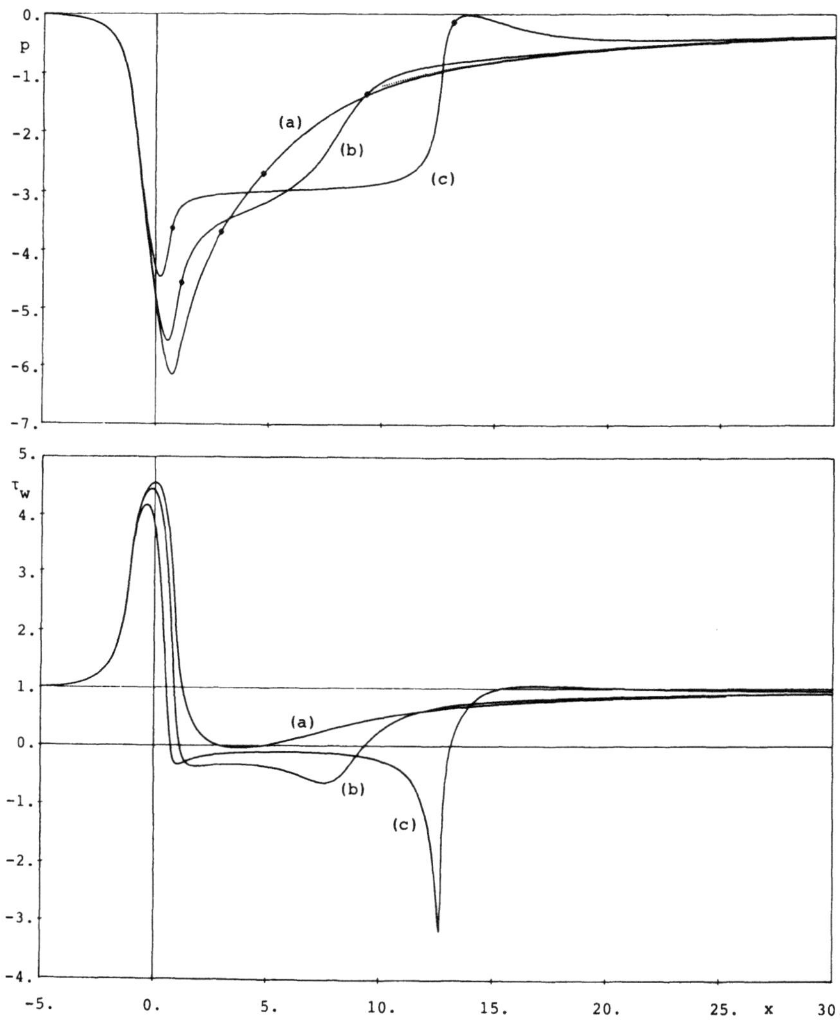

Fig.3. Pressure and wall-shear-stress distribution for $\alpha=-11.6$ showing the nonuniqueness of the solutions corresponding to the three different branches depicted in Fig.2.
◆ separation and reattachment points

References

1. Smith, F.T.: On the High Reynoldsnumber Theory of Laminar Flows, IMA J. Appl. Math. 28 (1982), 207-281.
2. Gittler, Ph. and Kluwick, A.: Triple-Deck Solutions for Supersonic Flows past Flared Cylinders, J. Fluid Mech. (1987), in press.

The Dynamic Interaction Between Dune Shape, Flow Field and Sediment Motion

Harald Mikkelsen

Institute of Geology, Aarhus University, Denmark

The field experiment.

Location of the field experiment. The almost two-dimensional dune lies across the prevailing wind direction.

Onshore winds erode sand from the windward side of a dune and deposit it on the lee side. The deposition occurs where flow divergence leads to a fall in the surface stress. This divergence can also produce separation. Such deposits alter the surface roughness and the dune shape, but these, in turn, affect the flow and stress fields.

In order to study this complex interaction it is necessary to have a very clear picture of the stress distribution across the dune under different wind conditions.

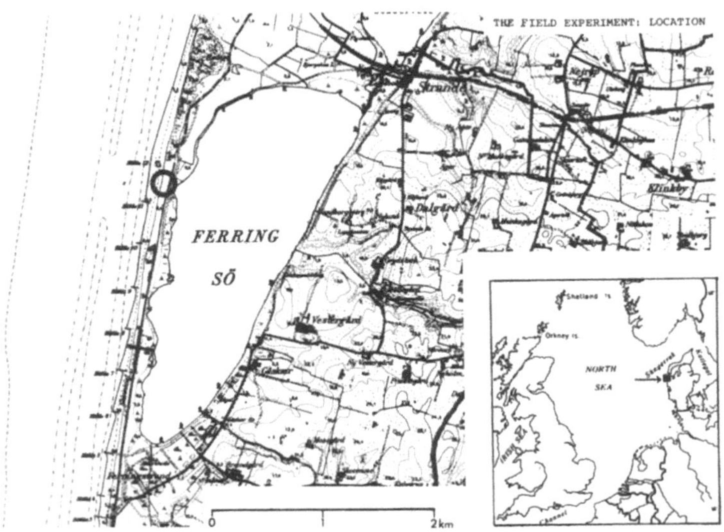

F. T. Smith, S. N. Brown, (Eds.)
Boundary-Layer Separation
IUTAM Symposium London 1986

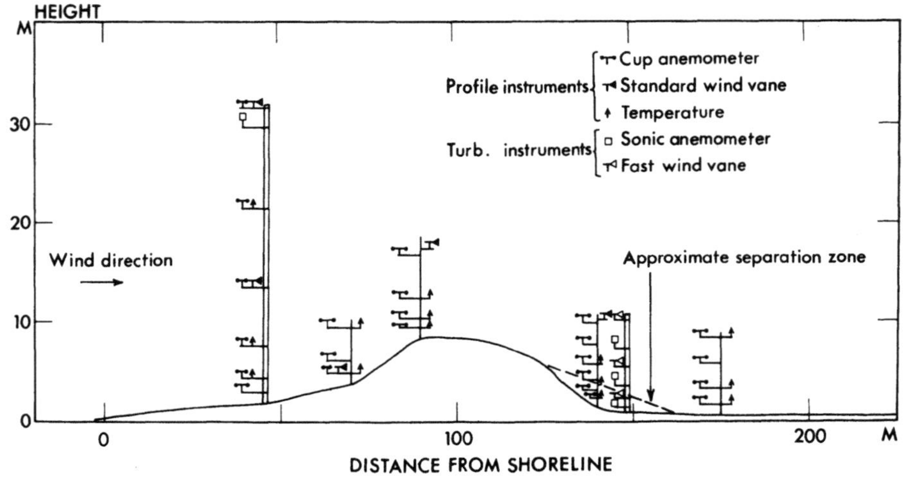

The field experiment: Cross-section of the dune, showing the instrumentation for measuring the wind profile and turbulence parameters.

Modelling the flow.

a) A two-dimensional second order closure model was used for calculating the mean flow and turbulence parameters. The model was adapted from one developed by the Risø National Laboratory for studies of flow over large hills (Zeman & Jensen, 1986).

b) Preliminary results suggest that on the upwind and central part of the dune mean flow can be adequately described in terms of present theories (e.g. Jackson and Hunt, 1975).

c) Using a forward intergration technique the model can predict the stress distribution on these parts of the dune, too, but it breaks down on the steepest part of the lee slope where separation occurs. Furthermore, the model shows that the development of separation on the lee side is very sensitive to upwind surface roughness.

Future work.

The preliminary results have raised the following three problems:

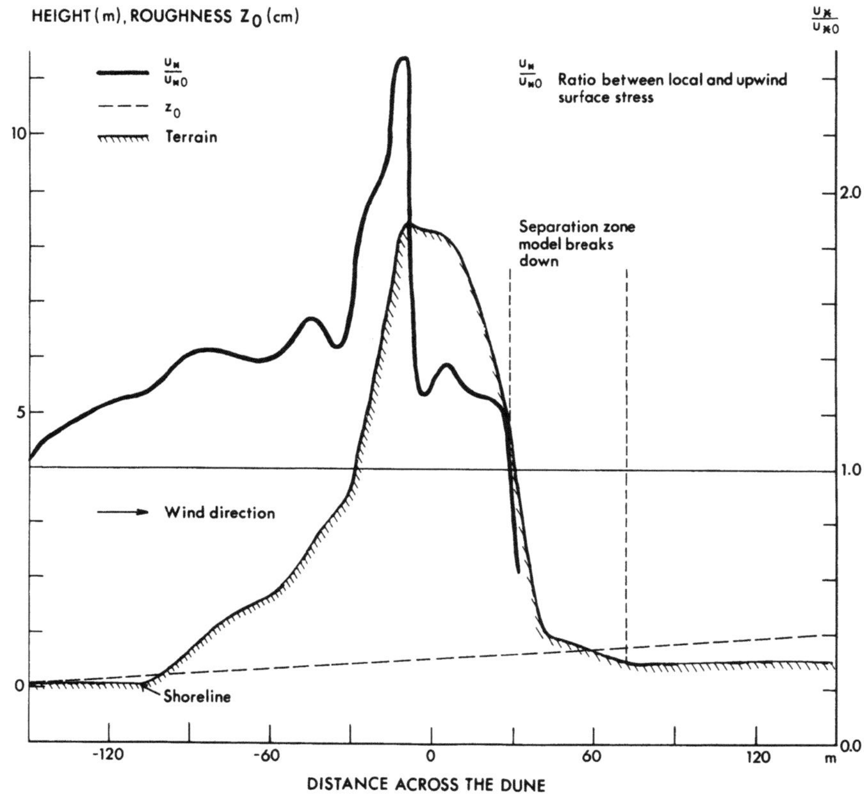

1) In what way is the separation related to dune morphology and surface roughness?

2) The dynamics of the lee slope depend not only on the mean flow, but also on the turbulent variations. In particular, there is a need for a model of separation turbulence.

3) The sediment transport extracts momentum from the flow. How does this affect the structure of the separation zone?

This research is carried out at the Institute of Geology, University of Aarhus, Denmark and at the Department of Wind Energy, Risø National Laboratory. The project is funded by the Danish Technical Research Council.

The photograph of the steepest part of the lee slope was taken after a force 10 westerly storm.

The light coloured areas are sand deposited below the shoulder of the dune. (The dune crest is out of sight)

References.

Jackson, P.S. and Hunt, J.C.R., 1975: Turbulent wind flow over a low hill. Quart. Journ. Roy. Met. Soc. 101, pp. 929-955.

Zeman, O. & Jensen, N.O., 1986: Modification of turbulence statistics in flow over hills. To appear in Quart. Journ. Roy. Met. Soc.

Structure of confined Wakes Behind a Circular Cylinder

N.K. Mitra, M. Fiebig, J. Franz
Institut für Thermo- und Fluiddynamik
Ruhr-Universität Bochum FRG

ABSTRACT

The structure of threedimensional wakes behind a circular cylinder partially enclosed in a rectangular channel have been investigated for Re (based on the average incoming velocity and the cylinder diameter) less than 800 from numerical solutions of incompressible Navier Stokes equations. Figure 1a shows particle traces of a typical wake structure On the symmetry plane the wake resembles the separation bubble of a twodimensional flow. Between the bottom wall and the symmetry plane a vortex tube in the shape of a tilted funnel with its apex on the bottom wall appears. Figure 1b shows the vortex tube as well as the forward separation point and the horse shoe vortex on the bottom plate. Figure 2a shows the limiting streamlines on the cylinder surface, 2b magnifies the spiral singular point and 2c presents the scheme of the different art of sigular points that appear. Figures 3a, b and 4a, b show limiting streamlines on the bottom plate and the scheme of singular points respectively. Figure 3a shows periodic (asymmetrical) flow. Figure 3 is symmetric, only one spiral point has been drawn.

Acknowledgement: This work has been supported by the Deutsche Forschungsgemeinschaft.

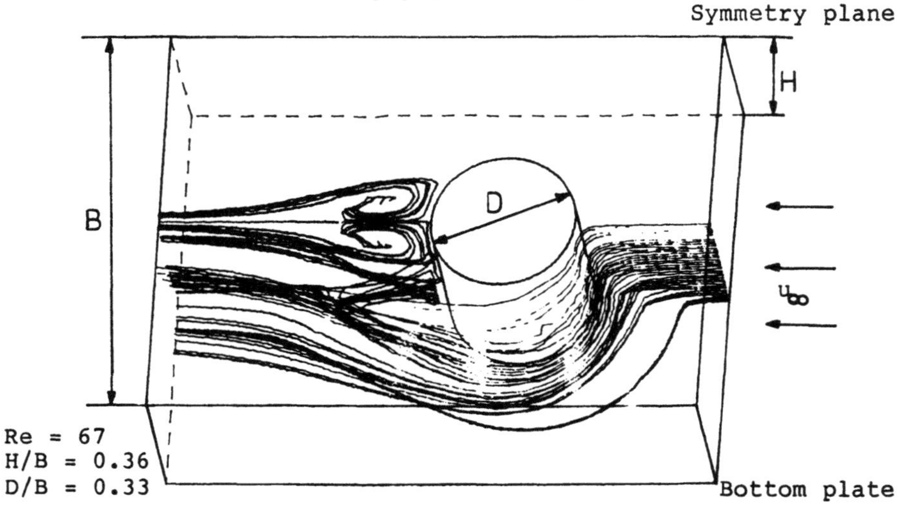

Fig. 1a Particle traces show 2-D like structure on symmetry plane and vortex tube between the bottom plate and symmetry plane

F. T. Smith, S. N. Brown, (Eds.)
Boundary-Layer Separation
IUTAM Symposium London 1986

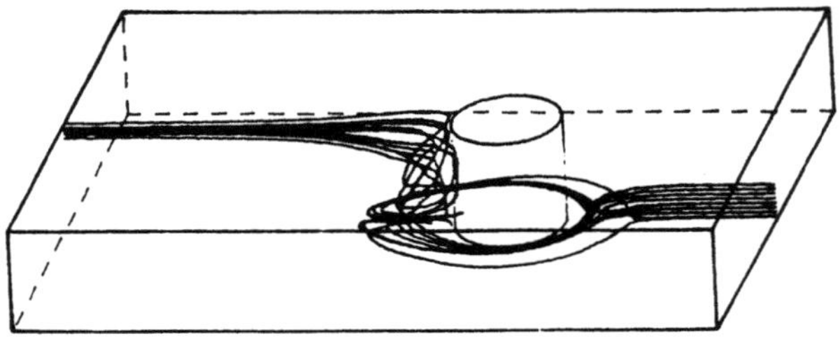

Fig. 1b Vortex tube in the wake and the horseshoe vortex on the bottom plate for the case of fig. 1a.

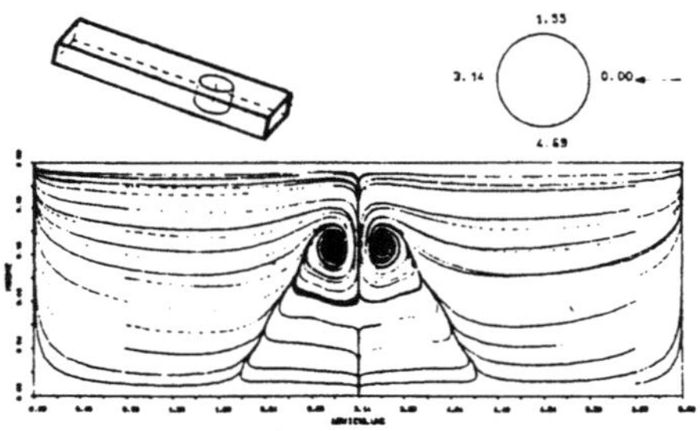

Fig. 2a Limiting streamlines on the cylinder surface

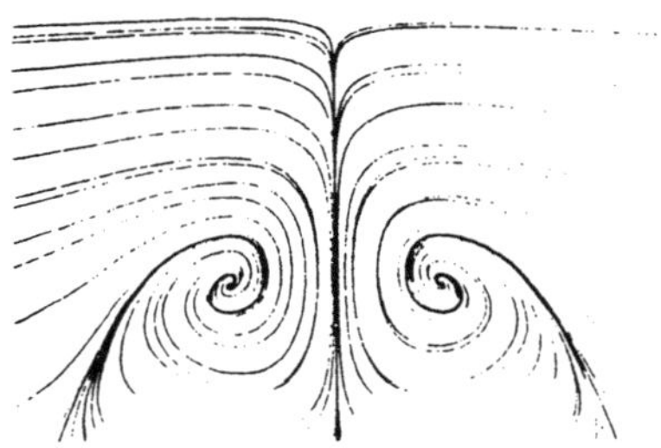

Fig. 2b Details of the singular spiral point of fig. 2a

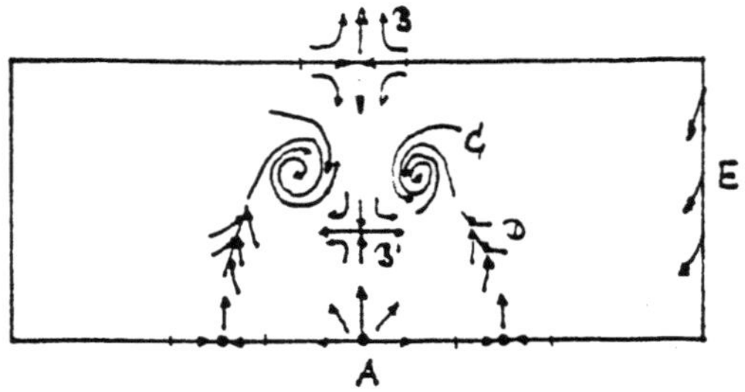

Fig. 2c Scheme of the observed singular points for fig. 2a
A: Node, Ü: Saddle, C: Spiral, D: Asymptotic to separation, E: Asymptitic to reattachment.

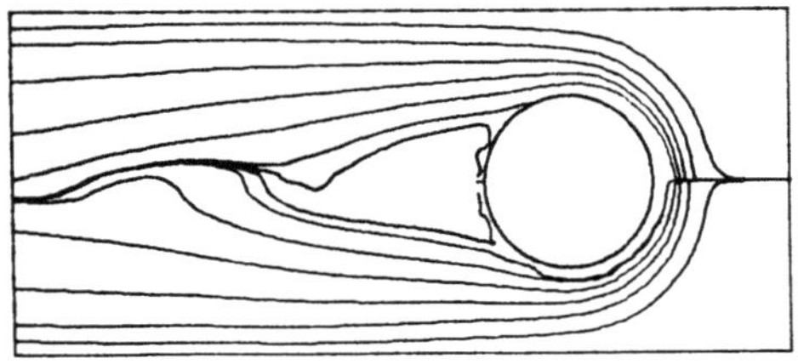

Fig. 3a Limiting streamlines on the bottomplate
H/B = 0.5, D/B = 0.5, Re = 440

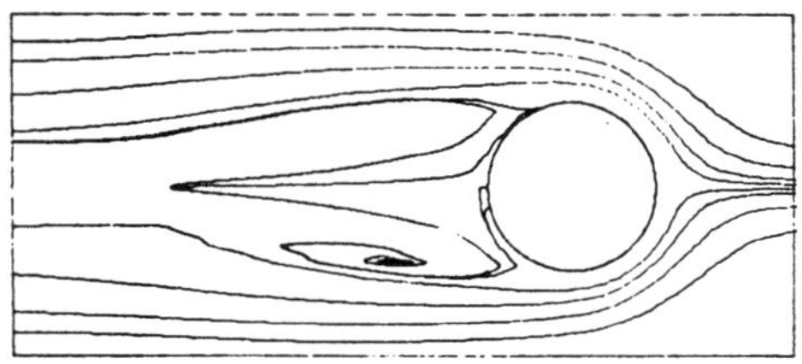

Fig. 3b Limiting streamlines on the bottomplate
H/B = 0.22, D/B = 0.5, Re = 330

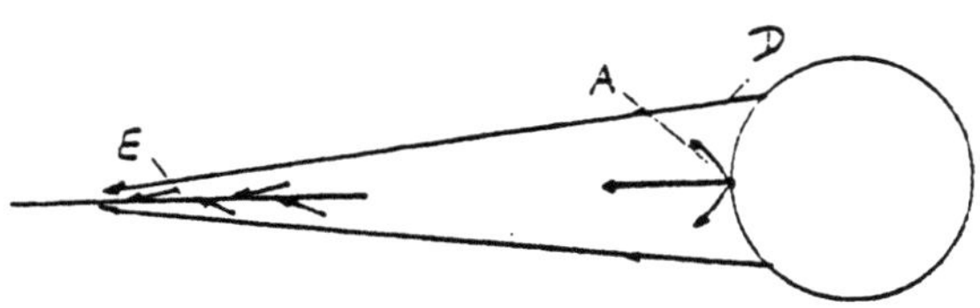

a) H/B = 0.5, B/D = 2, Re = 440

Fig. 4a Scheme of observed singular points for fig. 3a (for legend see fig. 2c)

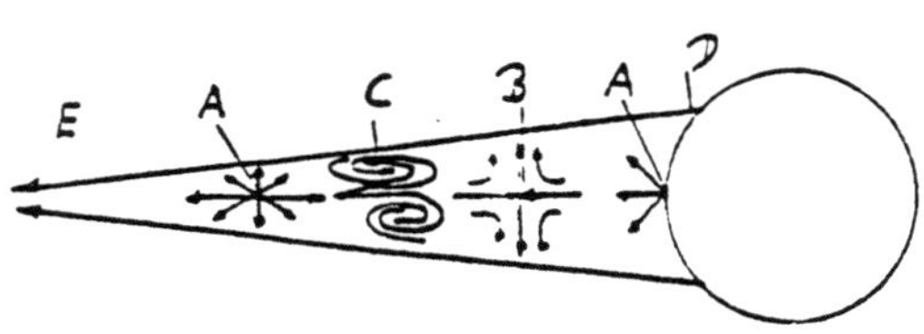

c) H/B = 0.22, B/D = 2, Re = 330

Fig. 4b Scheme of observed singular points for fig. 3b (for legend see fig. 2c)

Low-Rossby-number Flow Past a Circular Cylinder in a Rotating Frame

MICHAEL A. PAGE

Department of Mathematics, Monash University,
Clayton, Victoria, Australia 3168.

Introduction

The flow past a circular cylinder in a rapidly rotating frame is considered when the Rossby number Ro is $O(E^{\frac{1}{2}})$, where E is the Ekman number. Attention concentrates on the $E^{\frac{1}{4}}$ layer on the surface of the cylinder, which has a similar rôle to the classical high-Reynolds-number boundary layer in nonrotating fluid. The boundary layer in this case has thickness of $O(\delta)$, where $\delta = d(\frac{1}{2}E^{\frac{1}{2}})^{\frac{1}{2}}$ and d is the scaled depth of the fluid, and a second parameter $\lambda = Ro/2E^{\frac{1}{2}}$ measures the ratio of inertial effects to Ekman friction. The extra degree of freedom provided by the second parameter enables the development of flow separation in a steady flow to be examined, by increasing λ from zero upwards through finite values.

Governing equations

Using the usual definitions for the $E^{\frac{1}{4}}$ layer (for details see Page [1,2]), the velocities $(\bar{u}, \bar{v})$ in that layer satisfy the equations

$$\lambda(\bar{u}\frac{\partial \bar{u}}{\partial \bar{x}} + \bar{v}\frac{\partial \bar{u}}{\partial \bar{y}}) = \lambda \bar{u}_e \frac{d\bar{u}_e}{d\bar{x}} + \bar{u}_e - \bar{u} + \frac{\partial^2 \bar{u}}{\partial \bar{y}^2} \qquad \text{and} \qquad \frac{\partial \bar{u}}{\partial \bar{x}} + \frac{\partial \bar{v}}{\partial \bar{y}} = 0,$$

where $\bar{x}$ is measured from the forward stagnation point and $\bar{y}$ is the coordinate normal to the cylinder, scaled by the layer thickness δ. The tangential velocity $\bar{u}$ is zero on the cylinder and tends to the external value $\bar{u}_e = 2\sin\bar{x}$ as $\bar{y} \to \infty$, determined by the potential flow outside the $E^{\frac{1}{4}}$ layer.

Parameter regimes

The $E^{\frac{1}{4}}$-layer flow for $\lambda = 0$ is determined by the linear Stewartson-layer solution, $\bar{u} = \bar{u}_e[1 - \exp(-\bar{y})]$, which has constant displacement thickness (Walker & Stewartson [3]). Once $\lambda > 0$, inertial effects in the boundary layer lead to the displacement thickness decreasing on the upstream side of the cylinder and increasing on the downstream side. The skin friction τ remains positive all around the cylinder, but tends to decrease on the lee side with $d\tau/d\bar{x}$ becoming smaller at the rear stagnation point ($\bar{x} = \pi$) as λ

F. T. Smith, S. N. Brown, (Eds.)
Boundary-Layer Separation
IUTAM Symposium London 1986

is increased. The flow remains regular for $\lambda \leq \frac{1}{4}$ but beyond this value a singularity develops at the rear stagnation point (Walker & Stewartson [3]). There is no evidence that the flow separates beyond $\lambda = \frac{1}{4}$; in fact, the skin friction τ varies smoothly through that value and remains positive everywhere with $\bar{x} < \pi$ up until $\lambda = \frac{1}{2}$. Details of the flow for $\frac{1}{4} < \lambda \leq \frac{1}{2}$ are elucidated in Page [2] and Page & Cowley [4], where the structure of the singularity at $\bar{x} = \pi$ is examined. In particular, it is found the flow near that point splits into three sub-layers, the innermost of which is viscous, while the other two sub-layers are essentially inviscid. In addition, the $E^{\frac{1}{4}}$ layer imports fluid for all $\bar{x} < \pi$, and then forms a wake behind the cylinder. Once $\lambda > \frac{1}{2}$ the $E^{\frac{1}{4}}$ layer separates from the cylinder upstream of the rear stagnation point and the outer potential flow solution which corresponds to $\bar{u}_e = 2\sin\bar{x}$ is no longer appropriate. Page [1] describes how this separated flow can be calculated using a free-streamline theory similar to that proposed for nonrotating fluids by Smith [5]. An example of the resulting flow when $\lambda = 2$ is illustrated in figure 1; the flow within the bubble is stagnant and its shape is determined by a Bernoulli condition across the free streamline.

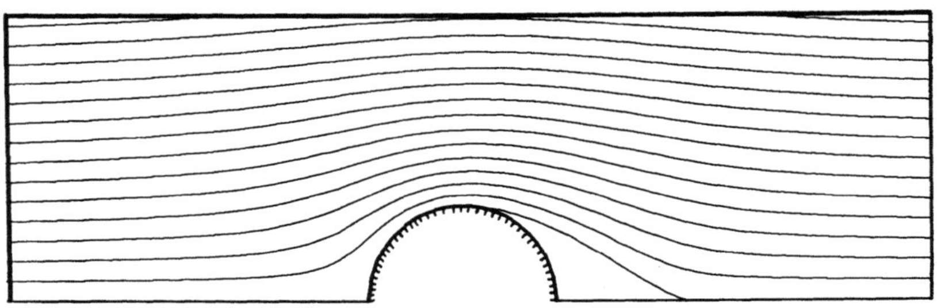

Figure 1. Separated outer flow past a cylinder for $\lambda = 2$; for details see Page [1].

References

1. Page, M.A.: Separation and free-streamline flows in a rotating fluid at low Rossby number. To appear in *J. Fluid Mech.* (1986).

2. Page, M.A.: On the low-Rossby-number flow of a rotating fluid past a circular cylinder. *J. Fluid Mech.* **156** (1985) 205-221.

3. Walker, J.D.A. & Stewartson, K.: Flow past a circular cylinder in a rotating frame. *Z. angew. Math. Phys.* **23** (1972) 745-752.

4. Page, M.A. & Cowley, S.J.: On the flow near the rear stagnation point of a circular cylinder. Submitted to *J. Fluid Mech.* (1987).

5. Smith, F.T.: Laminar flow of an incompressible fluid past a bluff body: the separation, reattachment, eddy properties and drag. *J. Fluid Mech.* **92** (1979) 171-205.

Closing Remarks

JAMES LIGHTHILL

University College London

Gower Street
London WC1E 6BT, UK

I should like if I may to congratulate Frank Smith on three achievements: first of all for his 1984 success in convincing the International Union of Theoretical and Applied Mechanics that it would be extremely timely to organise a major IUTAM Symposium on Boundary Layer Separation; next for his 1985 success, as Chairman of the International Scientific Committee appointed by IUTAM, in helping to put together a programme for the Symposium that would triumphantly demonstrate that timeliness; and finally for his 1986 success in organising, with the very strong support and extremely active participation of Susan Brown and the devoted assistance of many other colleagues, the exceptionally fine IUTAM Symposium that we have just heard. It has impressed us by its remarkable range, with at least six major themes pursued in depth both experimentally and theoretically; and also by its brilliant illustration of the value of some of the most modern techniques, experimental and theoretical, in the context of those themes. These modern techniques included advanced Laser Doppler Anemometry and advanced Computational Fluid Dynamics, alongside some of the most recent developments in Interactive Boundary Layer Theory and in Triple Deck Theory.

Out of the six major themes treated in the Symposium we may note first of all the attempts to describe complete separated flows around bluff bodies. The remarkable power of present-day developments in Computational Fluid Dynamics founded on the Navier-Stokes equations was demonstrated triumphantly in this context: Professor Fornberg showed how effectively a complete steady-flow field could be determined on a supercomputer at Reynolds numbers around 10^3 by using a parabolically-shaped grid appropriate to the actual distribution of vorticity; while

F. T. Smith, S. N. Brown, (Eds.)
Boundary-Layer Separation
IUTAM Symposium London 1986

Professor Ghia showed that computation based on the Navier-Stokes equations for unsteady motion is also extremely powerful in the vorticity-equation form. His calculation proceeded from an irrotational flow as the initial condition and analysed early developments of the wake in good agreement with experiments by Bouard and Coutanceau; while his results on asymptotic behaviour for large time were close to Professor Fornberg's remarkable steady-flow conclusions. In the meantime a wide variety of complete steady flows was impressively illuminated also from the different viewpoint of advanced Triple Deck Theory by Professor Ryzhov and by Frank Smith himself, while Dr Rothmayer demonstrated the effectiveness of Interactive Boundary Layer Theory in a certain range of cases. Professor Dennis concentrated on a moving-boundary problem, analysing flows around rotating cylinders with emphasis once again on their development in time and on a comparison with experiment.

A second major theme of the Symposium has been Shock Wave Boundary Layer Interaction. Yet again the power of modern Computational Fluid Dynamics techniques was exemplified; this time by Dr Girodroux-Lavigne's success in devising a program effective for those transonic flows that may exhibit oscillations of shock waves on a stationary aerofoil in a steady stream. Similarly, Professor Bogdonoff's experimental techniques were shown to be extremely powerful both for two-dimensional and for three-dimensional flows; and, once more, demonstrated the emergence of unsteady flow behaviour. Professor Burggraf showed us how Triple Deck descriptions can be obtained today even for complicated three-dimensional interactions, while Professor Neiland's lecture illustrated the power of Interactive Boundary Layer Theory in hypersonic-flow cases involving strong interactions.

Several brilliant descriptions of sophisticated experimental researches dominated the discussion of our third major theme: the characteristics of separation bubbles including their instability and unsteadiness and the presence of "large-scale structures" within them. Professor Kiya demonstrated the complexity of the unsteady behaviour which is characteristic of vorticity concentrations in separation bubbles, while Professors Dovgal and Kozlov focussed attention on important aspects of the instability of flows exhibiting such bubbles. Some useful light was thrown on these phenomena from the standpoint of triple-deck theory in the papers by Frank Smith and by Dr Duck.

A fourth major theme was the interpretation of three-dimensional separated flows with the comprehensive aid of intricate surface-streamline patterns and their subtle topology. This is a subject of a type which in an International Symposium is exceptionally well suited to treatment in the poster sessions, many of which exhibited and discussed large numbers of experimentally derived surface-streamline patterns of outstanding interest. In this field as in others, however, Computational Fluid Dynamics proved also to be a rich source of similar material as Professor Kuwahara's lecture demonstrated. A variety of surface-streamline patterns were interpreted convincingly by Professor Burggraf and Dr Cebeci from the standpoint of triple-deck theory and by Dr Edwards from the standpoint of Interactive Boundary Layer theory.

The Symposium's last two major themes were concerned with analytical problems of a particular kind. Programs for the numerical analysis of turbulent shear layers pre- and post-separation were valuably reviewed by Dr Simpson with some illuminating experimental comparisons, while Dr Lock described certain refined versions of such programs. The power of this type of analysis had earlier been demonstrated in the course of the sophisticated Shock Wave Boundary Layer interaction studies by Dr Girodroux-Lavigne already referred to.

Finally, the creative use of certain other rather simple ways of modelling rather complicated flows was exemplified in an interesting paper by Professor Drazin; and the Symposium ended with a brilliant account by Jeremy Smith of a further development of the classical Smith-Mangler technique. That highly successful approach to the modelling of flows with rolled-up vortex sheets is most convincingly extended in its latest version to allow for asymmetric vortex shedding; with, yet again, a most satisfactory comparison with experiment.

It is a special pleasure for me to thank all the speakers, all the discussants and all the organisers (but above all Frank Smith and Susan Brown) for making this IUTAM Symposium held at UCL such a memorable survey of mid-1980's knowledge and ideas on boundary-layer separation. I am proud to have been associated in two capacities, as President of IUTAM and as Provost of University College London, with a Symposium of which I confidently predict that the Proceedings will long be found indispensable to those working in this extremely important field.

MIX
Papier aus verantwortungsvollen Quellen
Paper from responsible sources
FSC® C105338

If you have any concerns about our products,
you can contact us on
ProductSafety@springernature.com

In case Publisher is established outside the EU,
the EU authorized representative is:
Springer Nature Customer Service Center GmbH
Europaplatz 3, 69115 Heidelberg, Germany

Printed by Libri Plureos GmbH
in Hamburg, Germany